MORE THAN JUST DIRT

A Fundamental Manual and Reference for
SOIL CONCRETE ASPHALT

Thomas Cope

innovativeink PUBLISHING
A Division of Kendall Hunt

Cover images © Shutterstock.com

innovativeink
PUBLISHING
A Division of Kendall Hunt

www.innovativeinkpublishing.com
Send all inquiries to:
4050 Westmark Drive
Dubuque, IA 52004-1840

Print ISBN: 979-8-3851-2116-8
Ebook ISBN: 979-8-3851-2117-5

Published in the United States of America

Dedication

To my wonderful wife of many years
and my family, who has supported me
during all the years that I put in many hours
of work learning the trade and information
I share in this book.

CONTENTS

PREFACE

This book has been written to provide general working knowledge of construction materials used in the construction industry, specifically Soils, Concrete, and Asphalt. It has been written with entry-level construction management and civil engineering technician students in mind, but hopefully, it will provide a good reference throughout their careers. I have not found a book that covers these three materials under one cover and addresses the concerns needed by the construction professional. Most books written are much more technical, focusing on the Geotechnical Engineer.

The material in this book differs from other books on this subject in several ways; it covers fundamental aspects of the materials, discusses practical issues with the materials, and is intended to serve the student as a reference for their career in the construction industry. The fundamentals of the materials will provide the student with how the materials are used in the construction industry without getting into great depths that the Geotechnical and Materials Engineers generally require. Construction professionals need a good understanding of the materials, leaving the details, design, and recommendations to the Engineering professionals. The student will be shown practical uses of the materials. These uses will include how, when, and why certain materials are used and how the adjustment of the materials influences the outcome. This book will also show the use of different equipment and how it affects the materials. The appendices will contain reference material that the Construction management professional will find useful throughout their career concerning the materials discussed in this book.

1

1.1 Introduction

This chapter will introduce the student to classroom and field safety. The next subject addressed will be symbols and units. The units will discuss how they apply in the industry and how different units are used by both the United States and the metric systems. Measurements will show how the units are derived. An overview of a few of the governing agencies will be discussed.

1.2 Safety

The safety section in this book is not intended to be a catch-all discussion. First, one must understand that safety is a large field and is always changing due to new inventions and approaches to construction. Students and professionals in the construction field should understand that every individual is responsible for safety. Everyone should look around the site, classroom/laboratory and identify potential safety risks. Once the risk is identified, proceed to the area, being cognizant of the risks while performing the tasks needed. If the individual perceives the risk to be too great to be able to perform their work safely, one should not enter the unsafe area, and their supervisor should be informed.

1.2.1 Classroom/Laboratory

A classroom/laboratory is a relatively safe environment. There are still safety risks that everyone should be aware of, such as heavy items, items that contain several pieces that may come apart, items that are hot or cold, water that may create slipping hazards, and more. When working in a construction materials lab, the individual should wear shoes that cover the toes and are resistant to slipping. There are other times when eye and hand protection is required.

1.2.2 Field

Fieldwork involves many more hazards, including construction vehicles/equipment, construction tools, tripping hazards, electrical hazards, hazards from above, fall hazards, and many more, depending on the individual's duties in the field. The Occupational Safety and Health Agency (OSHA) has jurisdiction over safety in the field and workplace. Individuals on construction sites are generally required to wear the following minimum safety gear: hard hats, safety vests, gloves, eye protection, and safety footwear.

Individuals should be trained in the equipment they are being asked to use. The training should include a brief discussion on the safety equipment needed. For example, a safety harness and tether would be required if an individual is required to work in areas where a falling hazard may occur. The supervisor should ensure the individual is properly fitted and understands the use of the safety equipment.

While walking through a site containing construction vehicles and equipment, you must be aware of all equipment movement around you. Before crossing a traffic area or the path of some construction equipment, I suggest you make eye contact with the driver of the equipment and signal to that driver that you are going to cross their path. The driver will normally acknowledge your intent and let you know whether to cross or wait.

1.3 Symbols

A list of symbols is provided in Appendix B. An attempt has been made to select those symbols typically listed in standard textbooks and used in the field.

1.4 Units

Two units are dominantly used throughout the world.

- International System of Units (SI).
  - This system is a revised version of the metric system.
- United States Customary System (USCS) units.
  - This system may also be called the Inch-Pound units (I-P units).
  - As the name implies, this system is generally used in the United States.

This book primarily uses USCS units. The following section will discuss how these units can be converted to SI units.

1.5 Unit Conversions

Commonly used conversion factors are listed in Appendix C. Appendix C lists numerous unit conversions, not just USCS to SI but also conversions within its own system. For example, 4 quarts equals one gallon. To convert the units, set the conversion as a multiplication problem, including the units. The units you presently have should be on the left side of the equal sign. Multiply that number by the conversion factor shown in Appendix C set up as a fraction, using the units in the equation. The unit you want to change should be the denominator (the bottom portion of the fraction). The unit you want your final response should be the numerator (the top portion of the fraction). Perform the multiplication canceling units to provide the answer with the correct units. This allows flexibility in the conversion process by converting either direction and additional conversions, such as from seconds to minutes or hours.

1.5.1 Examples

1. You have 128.6 cubic feet of material and want to know how many cubic meters.
   $128.6\ ft^3 \times 1\ m^3 / 35.31466\ ft^3 = 3.6\ m^3$

2. You have 678.27 grams of soil. How many ounces is this?
   678.27 gm x (1 oz/28.349523 gm) = 23.93 oz

3. You are going 35.4 fps. How fast is this in mph? In this example, convert feet to miles and seconds to hours.
   35.4 ft/sec x (1 mi/5280 ft) x (60 sec/min) x (60 min/hr) = 24.1 mi/hr

1.6 Decimal Places

Decimal places refer to the number of digits to the right of the decimal place. This is also a means to identify the precision of the number. The more decimal places shown, the more precise the answer or measurement. This statement is true if the answer shows no more decimal places than the least decimal places in the calculation. The answer can be no more accurate than the least-accurate number in the data. Rounding should be done only on the final calculation results. It should not be done on interim results. The correct number of decimals places is known as the significance figure.

Note that the correct decimal places are shown in the three equations above. In equation #1, the cubic feet were measured in one decimal place. Even though the conversion factor had five decimal places, the correct answer had one decimal place. Equation #2 measured the soils to two decimal places; therefore, the answer is also two decimal places since the conversion factor contained more than two decimal places. Equation #3 provides the feet per second with one decimal place. The conversion factor for feet to miles, seconds to minutes, and minutes to hours appears to have no decimal place. Those particular conversion factors equate to a whole digit. This can be represented with no decimal point or a decimal point followed by any number of zeros. Whole numbers do not have a bearing on the significant figure.

A common mistake in the laboratory or field is not recording the number of decimal places, which may affect the results.

1.6.1 Example

The scale in the concrete batch plant reads 2400.00 # of aggregate, but the employee records 2400 #. When the final batch weight is calculated, it is 10 cy, the correct volume for the 10-cy concrete truck to transport the load to the site. However, if the correct decimal places were applied, the final batch weight was determined to be 10.15 cy. The concrete truck is now over its allotted volume.

The weight is the same whether it is recorded as 2400.00 # or 2400 #. However, the decimal rule is used for the answer, which now impacts the outcome of the results.

When recording results in the laboratory or the field, it is important to show the results with the decimal place based on the precision provided by the equipment used.

1.7 Measurements

Measurements can be determined with various equipment. The equipment can be as simple as a tape measure or as sophisticated as precise measuring surveying equipment to measure length. In this book, measurements will be provided in USCS units. Table 1 below will show the most common measurements used in the industry, a brief description, and the units that may be used.

TABLE 1.1 Measurements

MEASUREMENTS	DESCRIPTION	UNITS
Length	Measure of distance.	Inches (in) Feet (ft) Yards (yd)
Area	Measure of a region's size on a surface. Length and width.	Square inches (in^2) Square feet (ft^2) Square yards (yd^2)
Volume	The measure of three-dimensional space. Length, width, and height	Cubic inches (in^3) Cubic feet (ft^3) Cubic yards (yd^3) Fluid Ounces (oz) Quarts (qt) Gallons (gal)
Density	Measure of mass per unit of volume.	Pounds per cubic foot (pcf) Pounds per cubic yard (pcy)
Unit Weight	Measure of the mass per unit volume of a material. The unit weight of water is 62.43 pounds-force per cubic foot at 39°F	Pounds per cubic foot (pcf)
Weight (Mass)	Weight is a measure of the force acting on the object due to gravity. It is measured by a scale. Mass is an intrinsic property of matter, whereas weight is a *force* that results from the action of gravity on matter: it measures how strongly the force of gravity pulls on that matter. However, in most practical everyday situations, the word "weight" is used when "mass" is strictly meant.	Weight = Pound-force (lbf) Mass = Pounds (lb or #) Ounces (oz) Tons (ton)
Pressure	Measure of force distributed over an area.	Pounds per square inch (psi)

1.8 Agencies

The main purpose of testing construction materials is to ensure minimum standards are met with the product provided and the workmanship. To ensure consistency in the results, several agencies were created. These agencies may be governmental or organizational in nature. This book will discuss the more common agencies that will be dealt with in the field and laboratory.

- Occupational Safety and Health Administration (OSHA)
    - OSHA is a governmental agency under the U.S. Department of Labor that ensures workers' safe and healthful working conditions by setting and enforcing standards and providing training, outreach, education, and assistance.
    - This agency may provide periodic reviews of sites but is generally only involved if an accident has occurred or someone notifies the agency concerning safety on a site.
    - This agency governs the minimum safety gear required for the type of work performed.

- American Society for Testing and Materials (ASTM)
    - ASTM is a nonprofit organization that develops and publishes technical standards, covering the procedures for testing and classification of materials of every sort.
    - The intent of ASTM is to assure tests are performed accurately and with reapable consistent results.
    - ASTM is constantly updating its test procedures. The labeling for an ASTM test is the designation number followed by a hyphenated number (D698 – 99). The hyphenated number represents the latest year the test procedure was modified. This book will leave off the latest year in the test designation.

- American Association of State Highway and Transportation Officials (AASHTO)
    - AASHTO is a nonprofit, non-partisan association representing highway and transportation departments in the 50 states, the District of Columbia, and Puerto Rico. It represents all transportation modes, including air, highways, public transportation, active transportation, rail, and water. Its primary goal is to foster the development, operation, and maintenance of an integrated national transportation system.
    - AASHTO works to educate the public and key decision-makers about the critical role that transportation plays in securing a good quality of life and a sound economy for our nation.
    - AASHTO serves as a liaison between state departments of transportation and the Federal government.
    - AASHTO is an international leader in setting technical standards for all phases of highway system development. Standards are issued for designing and constructing highways, bridges, materials, and other technical areas.

- American Concrete Institute (ACI)
  - ACI is a non-profit technical society and standard developing organization.
  - ACI provides standards, technical resources, education, training, and certification in the concrete industry.
  - ACI's mission is: "ACI develops and disseminates consensus-based knowledge on concrete and its uses."

- The Asphalt Institute (AI)
  - The Asphalt Institute is the international trade association of petroleum asphalt producers, manufacturers, and affiliated businesses.
  - Through education, engineering, technical development, environmental stewardship, and marketing leadership, the Asphalt Institute promotes the safe use, benefits, and quality performance of petroleum asphalts in a unified voice for its membership.

- Department of Transportation (DOT)
  - DOT is a governmental entity directed by each state. The agency is known as the Colorado Department of Transportation (CDOT) in Colorado.
  - DOT has goals similar to AASHTO's goals, but only at the local state level.

- Local Jurisdictions
  - This could include the County, City, improvement districts, special tax districts, and any other entity that dictates the construction industry's codes, rules, and regulations.
  - Most local jurisdictions will follow the lead of the above-mentioned agencies as well as the building codes that the jurisdiction has adopted.

Problems – Chapter 1

1. Who is responsible for safety on a job site? Choose all that apply.
   a. The employer
   b. OSHA
   c. The individual
   d. All of the above

2. What should you do if you are asked to work in an area you feel is unsafe to work in? Choose all that apply.
   a. Contact your insurance agent
   b. Contact your supervisor
   c. Phone home
   d. Do not enter the area

3. List at least three safety concerns you should look for in a laboratory.

4. List at least five safety concerns you should look for on a field job site.

5. Match the corresponding letter before the symbol with the description.

____ 1) Degree of saturation	a) e
____ 2) Water content	b) PI
____ 3) Void ratio of soil	c) V_s
____ 4) Liquid limit	d) γ_d
____ 5) Dry weight of soil	e) LI
____ 6) Volume of soil	f) S
____ 7) Dry unit weight of soil	g) W_s
____ 8) Plasticity index	h) LL
	i) ω

6. What does SI stand for?

7. What does USCS stand for?

8. What unit system is mostly used in the United States?

9. Using the correct decimal place, convert the following:
   a. 79 in into ft.
   b. 12.9 yd^3 into ft^3.
   c. 14°C into °F.
   d. 21,269.7 gm/m^3 into pcf.

10. What does ASTM stand for?

11. What does ACI stand for?

2

2.1 Soils

This book will not get into the geologic properties found on Earth. However, it is important to understand that Earth's surface is made up of rock and water. Soil is formed by the weathering of the rock. Weathering can occur due to changes in mineral composition (chemical weathering), changes in temperature (mechanical weathering), or a combination of both. Over time, the soil is transported and deposited into the stratification of the surface soils, also known as the soil profile. In general, there are three distinct zones in the natural soil profile.

A horizon:	The upper soil layer with humus or organics (topsoil).
B horizon:	The layer comprises mostly fine soil particles transported from the A horizon.
C horizon:	Contains the less weathered soils extending into the bedrock.

Soils are tested because they may have the most important impact on structures of any material in the construction industry. Yet soils are the least predictable of all construction materials because it is not homogeneous nor man-made. As we learn more about soils, we will see it is highly variable. Each variation may impact the structure differently: settlement, heave, or little movement. Soils are impacted by the amount of water, force applied to the soils, containment of the soils, and other physical effects. For example, similar clay samples will react very differently if the clay soils are wet versus dried out. Soil and or rock is what supports the foundation of all structures. The structure may be supported directly by the soil, or it may be supported by piers that will be bearing on the soil or rock and impacted by the soil along the length of the piers.

Subsurface soil investigations are generally required to be performed by a Geotechnical Engineer to identify the soil's characteristics. These characteristics include identification of the in-situ (in place) soils: classification, moisture content, density, settling/swell potential, and additional testing depending on the soils encountered and the proposed use of the structure. Once the characteristics are determined, the Geotechnical Engineer will make recommendations on the type of foundation (spread footings, concrete piers, etc.), provide the Structural Engineer with anticipated loads the soils can support, and discuss treatments of the soils for the anticipated use of the construction. It is important to note most structures will be moved by the soil. The Geotechnical and Structural Engineer tries to minimize the movement and reduce the potential for differential movement from one location on the structure to another. This book will not get into the process or how to determine the recommendations provided in a subsurface soil investigation. However, the information you learn from this book should provide you with sufficient knowledge to understand much of the subsurface soil investigation report.

The most effective way to minimize the effects of the soil on structures is to attempt to control the soil with development and or construction standards. The most common standards include moisture control, density requirements, and soil type. Appendix D is a sample site and grading specifications. We will get into much more detail about the potential problems that may occur with different soil, moisture content, and density later in this chapter. However, the two greatest effects of the structure are settlement and expansion.

2.2 Properties

The physical properties of soil are divided into three phases:

1. Solids
   a. This component is the soil particles.
2. Liquids
   a. The liquid portion is usually water that is contained in the void spaces between the soil particles.
3. Air/Gas
   a. The third phase is gas, such as air, that is also contained in the void spaces between the soil particles. Soil that is saturated (liquid fills all the void spaces in the soil particles) contains no open gas voids.

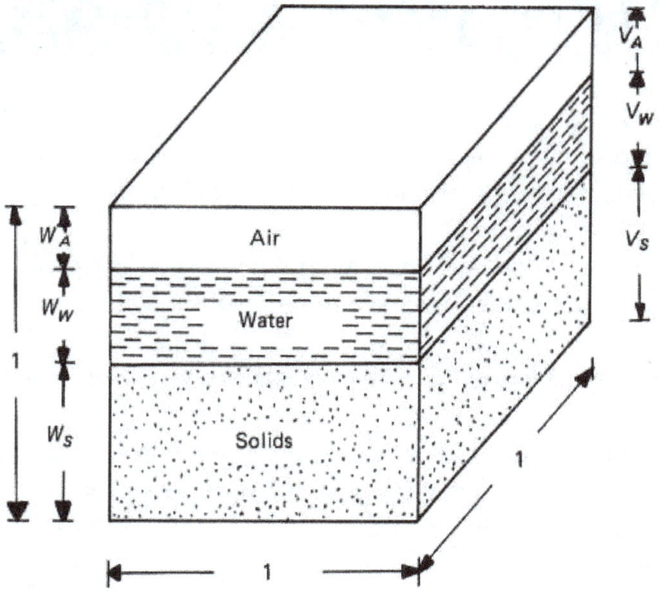

FIGURE 2.1 Elementary unit block
Adapted from reference 3.

Figure 2.1 shows the three phases of the elementary unit volume. The figure also shows the weight relationship on the left and the volume on the right. The total volume of the element is the volume of each phase.

$$\text{Total Volume (V)} = \text{Solids Volume } (V_S) + \text{Water Volume } (V_W) + \text{Air Volume } (V_A)$$

It is common to discuss voids in soils. Voids refer to the portion that is not solids. Therefore, voids consist of water and air.

$$\text{Void Volume } (V_V) = \text{Water Volume } (V_W) + \text{Air Volume } (V_A)$$

The total weight can be shown as the sum of the weight of the solid plus the weight of the water. The weight of the air is considered negligible.

$$\text{Total Weight (W)} = \text{Solid Weight } (W_S) + \text{Water Weight } (W_W)$$

2.3 Moisture Content

Moisture content is also known as water content. The test to determine water content is probably the most common and simplest soil test. This test can be performed on specimens of disturbed soil (soils not in their original state, fill) or undisturbed soil (soils in their original natural state, in situ).

Water is contained in the soil's natural state and infill material. The percentage of water filling the void of the sample is known as the degree of saturation of the soil.

$$\text{Degree of Saturation (S)} = \text{Water Volume } (V_W) \,/\, \text{Void Volume } (V_V)$$

If the degree of saturation equals zero, the sample contains no water in the soil. This rarely happens in the field due to an attraction of electrons between the soil particles and the water. This can happen in the laboratory by driving the water out of the soil with heat. The weight of this sample is equal to just the weight of the soil. You will see this is an important idea when determining the water content in the soil sample. When the voids in the soil are partially filled with water, the soil is said to be partially saturated. The soil is said to be fully saturated when the voids are fully filled.

Water content is determined by drying the soil until all the water is no longer found in the soil sample. The soil is initially weighted to determine the weight of the soil and water and then re-weighted once the water is cooked out of the soil. Water content is shown as a percentage calculated as the weight of the water divided by the weight of the soil.

$$\text{Water Content } (\omega) = \text{Water Weight } (W_w) \,/\, \text{Soil Weight } (W_s)$$

The weight of the water is calculated by subtracting the weight of the soil from the original weight of the soil sample.

$$\text{Water Weight } (W_w) = \text{Weight of sample (W)} - \text{Soil Weight } (W_s)$$

Inserting the water weight formula into the water content equation, we have the following:

$$\omega = (W - W_s) \,/\, W_s$$

2.3.1 Test Procedure

The ASTM designation for this test is ASTM D 2216. The ASTM procedure will have additional information than what is provided in this book. The lab procedures below summarize the ASTM procedures. Data Sheet 1 can be used while performing this test. The laboratory test procedure is as follows:

1. Using a scale, obtain the weight of a clean and dry specimen container W_c. Usually, the specimen container is made of metal. Both the container number and the weight are recorded on the datasheet.
2. Place the wet soil within the container. Be careful not to overfill the container, which will cause the sample to fall out, providing a false dry soil weight. Table 2.1 provides guidelines on the minimum amount of soil that should be used based on the largest particle dimension.
3. Using a scale, obtain the weight of the container and wet soil W_{wc}. Place the container and wet soil in an oven and then dry the soil overnight (12 to 16 hours) at a temperature of 230°F ± 10°F.
4. Remove the container and dry soil from the oven and use the scale to determine their weight in W_{dc}.
5. The water content can be calculated with the following equation:

$$\omega(\%) = 100M_w/M_s = 100\,(W_{wc} - W_{dc})/(W_{dc} - W_c)$$

TABLE 2.1 Minimum Soil Sample Weight for Water Content Testing
Adapted from reference 4.

Largest particle dimension (mm)	Sieve size corresponding to particle dimension	Recommended minimum mass of wet soil specimen for water content to be reported to the ±0.1% (g)	Recommended minimum mass of wet soil specimen for water content to be reported to the ±1% (g)
2.00	No 10	20	--
4.75	No. 4	100	--
9.5	3/8 in.	500	50
19.0	3/4 in.	2,500	250
37.5	1-1/2 in.	10,000	1,000
75.0	3 in.	50,000	5,000

Note: Table based on ASTM D 2216-19 (2019).

Some common laboratory errors in determining water content are as follows:

1. Inaccurate or poorly calibrated scale,
2. Loss of soil between first and second weighing,
3. Loss of moisture from the sample before first weighing,
4. Addition of moisture to the sample after drying and before the second weighing,
5. Incorrect oven temperature,

6. Sample size too small,
7. Incorrect container weight,
8. Specimen removed from the oven before reaching a constant oven-dry weight,
9. Organics cooked off during the drying procedure.

2.3.2 EXAMPLE

We obtained a soil sample from a test pit near the existing flagpole. The sample consists of silty sand passing a No. 4 sieve. We need to determine its water content. See data sheet 1 below. For this example, the calculations are shown on the datasheet in red. Normally, the calculations would not be shown. Data Sheet 1, Appendix J, was used in the lab to record the weights.

WATER CONTENT **Data Sheet 1**

Project: Moisture Test **Project No.:** AEC 2530

Location: Pit 20' S and 50' W of Flag Pole

Soil Description: Silty Sand

Tested By: Tom C **Date of Test:** 6/1/2023

Sample/Boring No.	Sample #1				
Sample depth	2'				
Soil description	Silty Sand				
Container No.	A				
Wt. of container (W_c)	6.6 g				
Wt. of container + wet soil (W_{wc})	45.2 g				
Wt. of container + dry soil (W_{dc})	43.8 g				
Wt. of dry soil (W_s) $W_s = W_{dc} - W_c$	37.2 g				
Wt. of water (W_w) $W_w = W_{wc} - W_{dc}$	1.4 g				
Water content % (ω) $\omega = (W_w/W_s) \times 100$	3.8				

2.3.3 *Moisture effects on soil*

Water has a major impact on different types of soil. To better understand the effects of water on soils, it is beneficial to have a general understanding of their molecular makeup. All soils are made up of elements that have positive and negative charges. Depending on the soil particles, they may be attracted to the elements found in water. Water binding to the soil minerals' solid surface is called absorption. Soils can be classified as cohesive or cohesionless soils.

2.3.3.1 Cohesive Soils

Cohesion is an inherent physical property of soil that causes individual grains to stick to each other when subjected to wetting and separate when drying. The force for keeping the fine-grained soils together is due mostly to the electric charge within the minerals and water known as ionic bonds, van der Waals bonds, and hydrogen bonds. The gravitational attraction also plays a role in this attraction.

The presence of clay minerals gives soil deposits their cohesive properties. The amount of cohesion depends on the grain size distribution of the soil and the amount and type of clay minerals present. Generally, most soils with the grains passing the No. 200 sieve will exhibit cohesion.

Cohesive soils can be visually identified by rubbing the soil between your fingers and noting its stickiness. If the soil is dry, it may not be able to stick together; however, when wet, it will become apparent.

Cohesive soils are generally formed in layers of flat plates; see Figure 2.2.a. The shape of the minerals that form may not be flat; however, when the particles bond to each other, they act as flat plates. Water tends to bond between the plates due to the electrical charge. Sometimes, the bonding of the plates is so great that the only way to introduce water into the soil is through mixing and breaking up the plate structure, allowing the charged particles to accept the water molecules between the plates. The water molecules cause a separation between the plates, causing the soil to exhibit an increased volume change, see Figure 2.2.b. Enough water introduced between the plates can cause a significant heave in the soil, damaging the structure supported by the soil. This action is known as swelling or heaving soil.

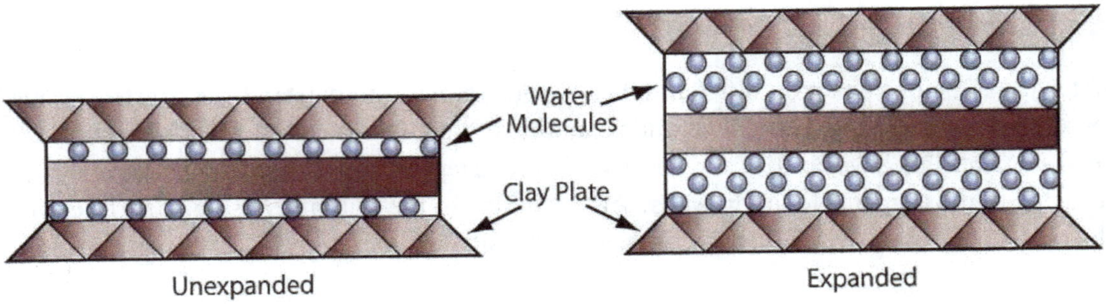

Expanding clay plates; view is magnified from a microscopic level. Modified from Hart (1974).

FIGURE 2.2.A Clay particle plates **FIGURE 2.2.B** Clay particle plates with additional water
Adapted from reference 20.

A swell-consolidation test is a common laboratory test performed during the subsurface soil investigation. The swell potential and swelling pressure are measured by placing a sample confined in a metal ring allowing only movement in a vertical direction. The sample is loaded with weight that simulates the overburden pressure. Water is applied to the sample, allowing the sample to move in a vertical direction and measuring the percent swell (increase volume change) or percent consolidation (decrease volume change). If the sample swells, the increase percentage from when the water is added to when the sample no longer increases in length is known as the swell potential based on a given initial load. Additional weight is added to the sample till the original length of the sample is obtained; this is known as the swell pressure. See Figure 2.3. The results of each loading are plotted on the chart, with the y-axis depicting the percent swell and the x-axis being a log scale depicting the applied pressure.

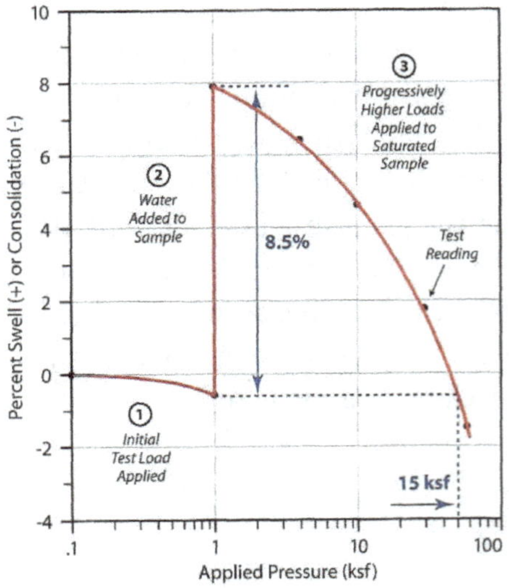

Graph of results for a swell-consolidation test. In this example, the swell potential is 8.5% and the swell pressure is 15,000 pounds per square foot (15 ksf).

FIGURE 2.3 Swell-Consolidation curve
Adapted from reference 20.

2.3.3.2 Cohesionless Soils

Soils in which the individual grains are not held together by any forces when dry are called cohesionless. Cohesionless soils are generally encountered as single-grain structures. The most important characteristic of a single-grained structure is the arrangement of the individual grains or packing. A soil's state of denseness or looseness depends on three factors: the grain shape, the grain size distribution, and the grain relative positions.

Grain shapes can range from angular to very round. The grain size distribution affects the packing of granular soils. Soil is well-graded if an assortment of small particles fills the voids formed by the larger particles, thus allowing for denser packing. Soils are considered poorly graded if the voids are not occupied by grains, creating a looser sample.

Cohesionless soils are generally sand and gravel. They can be identified by the gritty feel when rubbed between your fingers. Silt is generally noted as a cohesionless soil even though the particle size passes a #200 sieve. However, silt feels like flour when wet and does not stick to itself.

The swell-consolidation tests can also be performed on cohesionless soil. It is more common to see a decrease in the soil volume or consolidation of the sample. This is due to the water allowing the soil particles to move more freely with the lubrication from the water. The curve looks similar to Figure 2.3, except that the next reading after adding water with no load added is downward instead of upward, as shown above. This downward movement indicates a soil sample consolidated in the laboratory and may have the same results in the field when water and a load are added. It is important that the lab technician identifies consolidation due to the shape of the soil particles sticking to each other and then quickly releasing as opposed to a slower consolidation of the soil. If the soil quickly releases in the laboratory, this can be identified by a crunching sound and instantaneous consolidation; there most likely is a void in the soil created during the preparation of the sample, and it generally does not indicate this will occur in the field. Those test results should either be noted or not included in the soil report. This phenomenon may occur in the field, but it will generally happen during the initial construction, and the soil particles will be adjusted at that time with no long-term effect.

2.4 Identification

A soil identification (classification) system arranges different soils with similar properties into groups and subgroups based on their application. Classification systems provide a common language to express the general characteristics of soils briefly. Most soil classification systems for engineering purposes are based on simple index properties such as particle size distribution and plasticity.

Soils can be classified in their purest form as granular (gravel and sand), silt, or clay. However, most soils found in the field are a mixture of two or more. The different classification systems we will discuss take this into account.

This book will focus on four different engineering soil classification systems as follows:

1. Unified Soil Classification System (USC). This system was developed by the U.S. Army Corp of Engineers. The USC is the most widely used classification system in the United States and abroad.

2. AASHTO Classification System (AASHTO). This system was created by the American Association of State Highway and Transportation Officials. It is the classification of soils most widely used for State and Federal transportation projects. Local jurisdictions also commonly use this system when referring to State and Federal project specifications.
3. USDA Classification System (USDA). This system is the least used soil classification system when working on engineering projects. However, the U.S. Environmental Protection Agency (EPA) does refer to this classification system in the design of on-site wastewater disposal systems (septic systems) and stormwater management permitting.
4. OSHA Classification System (OSHA). This system was developed by OSHA to ensure the excavation is properly sloped or benched for the purpose of on-site safety during construction. OSHA provides instructions for proper sloping, benching, shoring, and bracing for excavated soil as identified by the OSHA classification method. Additional information can be found at the OSHA's website (www.OSHA.gov).

Table 2.2 shows the grain size limits for the three commonly used classification systems. Note how the identification of the four soil types is similar but varies slightly from each classification system.

TABLE 2.2 Grain Size Limits for Soil Identification

	Grain Size - mm and (in.)		
Soil	USC[a]	AASHTO[b]	USDA[c]
Gravel	76.2 (3) to 4.75 (0.2)	76.2 (3) to 2 (0.08)	> 2 (0.08)
Sand	4.75 (0.2) to 0.075 (0.003)	2 (0.08) to 0.075 (0.003)	2 (0.08) to 0.05 (0.002)
Silt	Fines <0.075 (0.003)	0.075 (0.003) to 0.002 (8×10^{-5})	0.05 (0.002) to 0.002 (8×10^{-5})
Clay	Fines <0.075 (0.003)	<0.002 (8×10^{-5})	<0.002 (8×10^{-5})

[a]Unified Soil Classification System (U.S. Army Corp of Engineers)
[b]American Association of State Highway and Transportation Officials
[c]U.S. Department of Agriculture

Adapted from reference 3.

2.4.1 Particle Size Distribution

The distribution of the particle size range is a very important characteristic of a given material. The distribution is generally depicted graphically on a curve called the gradation or grain size distribution curve. The grain diameter (log scale) is plotted versus the dry weight, expressed as a percent. Figure 2.4 is a typical grain size distribution curve.

Two laboratory test methods are used to determine the grain size distribution of soils. The Mechanical (Sieve) Analysis test is performed on soils with particle sizes larger than (0.075 mm). The Hydrometer Analysis is performed on soils with particle sizes smaller than (0.075 mm).

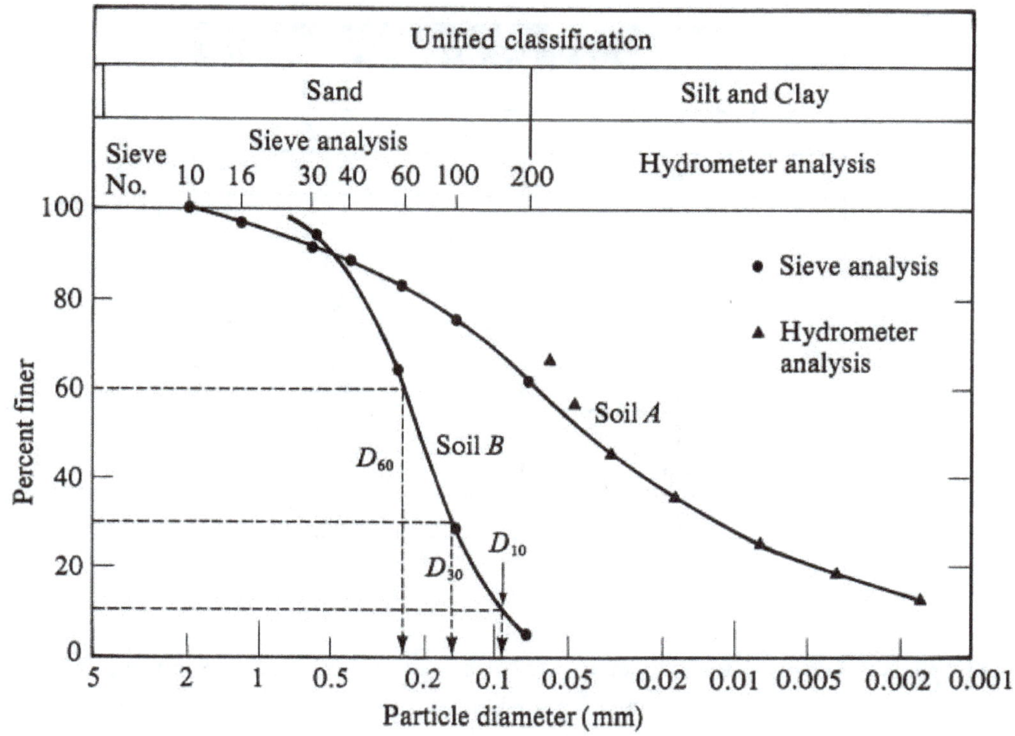

FIGURE 2.4 Particle Size Distribution Curve
Adapted from reference 1.

2.4.1.1 Sieve Analysis (ASTM D 6913)

A sieve is laboratory equipment that consists of a metal ring with a screen (square woven wire mesh) attached to the bottom of the ring. See Fig 2.5. The opening in the screen is designed to allow soil of a particular size to pass through the openings. It is important that the screen opening represents the correct size of the opening desired. The screen opening is designated either in inches or the size of the number of openings per square inch. For example, a 1" sieve has 1" by 1" wide openings.

FIGURE 2.5 Typical Soil Sieve

A No. 100 sieve has 100 openings per square inch. Table No. 2.3 lists the common sieve sizes and their associated size of opening.

TABLE 2.3 Soil Sieve Opening Size

SIEVE NO.	OPENING SIZE (mm)
4	4.75
8	2.36
10	2.00
16	1.18
20	0.85
30	0.60
40	0.425
50	0.30
60	0.25
100	0.15
200	0.075

2.4.1.1.1 Test Procedure

The soil sample is well mixed and properly split (discussed in Section 2.7, Sampling) to provide a representative test sample of the original sample obtained. Three samples should be obtained from the original well-mixed sample. One sample for moisture content, one for the hydrometer test, and the last one for the sieve test. The minimum sample size should correspond to ASTM D 6913, as shown in the following table 2.4. The initial wet weight should be recorded. The sample to be sieved should be dried, weighed, and recorded, and the moisture content calculated. The test sample can now be washed over a No. 200 washing sieve till the water flowing through the sieve is clean. This process involves washing the clay and silt (fines) particles out of the sample. The washing process will also separate grain particles from each other. During the washing process, stir the soil, taking care not to manually force any of the particles through the bottom of the sieve. Carefully collect all the washed samples from the sieves and dry the washed samples in an oven. Once the sample is fully dried, weigh it and record it. The sieve analysis is performed by running the washed, weighted dry soil sample through a stack of sieves with the largest openings on the top, followed by the next smaller opening sieve until the smallest sieve is the No. 200 sieve, followed by the pan (no openings), figure 2.6. Generally, each progressing sieve opening is approximately one-half of the previous opening size. The sieve stack should be shaken for about 15 minutes. Each sieve should be carefully collected, weighted, and recorded. The soil collected in the pan should also be weighed and recorded.

TABLE 2.4 Minimum Sample Size for Sieve Analysis
Adapted from reference 4.

Diameter of largest particle	Minimum mass (g)
3/4 in. (19 mm)	1000
1 in. (25.4 mm)	2000
1 1/2 in. (38.1 mm)	3000
2 in. (50.8 mm)	4000
3 in. (76.2 mm)	5000

FIGURE 2.6 Sieve stack

Some common laboratory errors associated with the sieve analysis are as follows:

1. Failure to separate grains during the washing or shaking process,
2. Loss of material during the testing procedure,
3. Overloading the sieves,
4. Broken or distorted sieves,
5. Stacking the sieves in the incorrect order,
6. Mis-recording the results,
7. Inadequate shaking of the sieves,
8. Over-crushing the soil particles prior to washing.

2.4.1.1.2 Calculations

Calculate the percent finer by dry weight (percent passing) by obtaining the cumulative weight of dry soil retained on a given sieve and dividing the initial total dry weight of the soil prior to performing the wash. The following equation obtains the results in a percentage.

$$F = 100 - 100(R_{DS}/M_S)$$

F = percent finer by dry weight (percent passing)
R_{DS} = cumulative weight of soil retained on a given sieve
M_S = initial total dry weight of the soil prior to wash

The grain size distribution curve is created by plotting the percent passing along the y-axis and the grain diameter (mm) along the x-axis in log scale. See Figure 2.7. The diameter at a certain percent passing is denoted as D_x. For example, the diameter of the particles at 60% of the material passed is shown as D_{60}. Four diameters are important for additional identification purposes: D_{100}, D_{60}, D_{30}, and D_{10}. The maximum diameter of the soil is indicated by D_{100}.

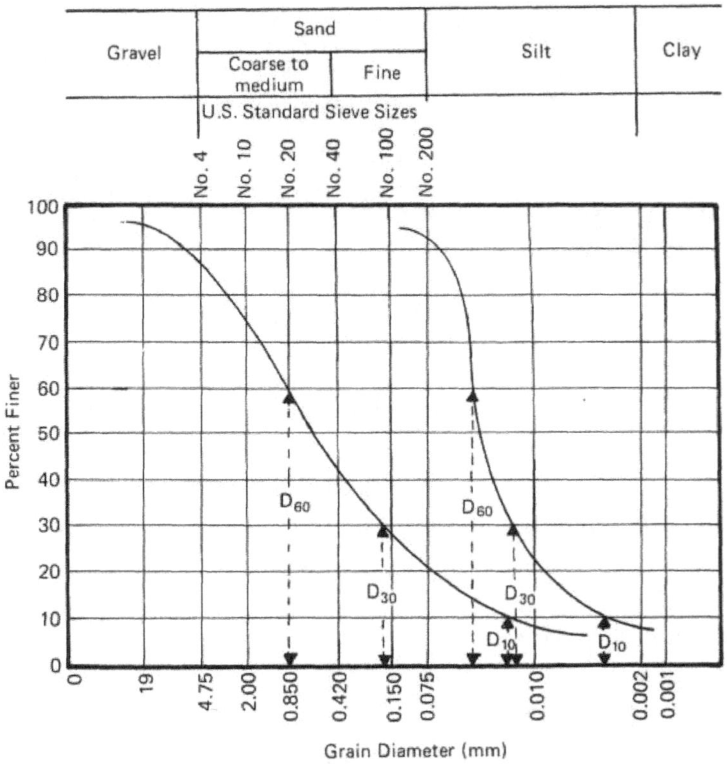

FIGURE 2.7 Grain size distribution curve
Adapted from reference 3.

Two coefficients used in the USCS method for classifying coarse grain soils are as follows:

Coefficient of uniformity (C_u) $C_u = D_{60}/D_{10}$
Coefficient of curvature (C_c) $C_c = (D_{30})^2/D_{60}D_{10}$

In the next section of the hydrometer analysis, we will see that the finer particle sizes are not determined if less than 5% of the soil passes the No. 200 sieve. It is common in the geotechnical field if less than 20% of the soils pass the No. 200 sieve the hydrometer analysis is not run unless more information about the fines is needed for engineering purposes. If the hydrometer analysis is not performed, the grain size distribution curve will end at the percentage passing the No. 200 sieve. If the hydrometer analysis is run, then the results of that test will be added to the curve corresponding to the same soil sample.

The particle size distribution curve shows the range of particle sizes present in the soil and the type of distribution of various particle sizes. This is demonstrated in Figure 2.8.

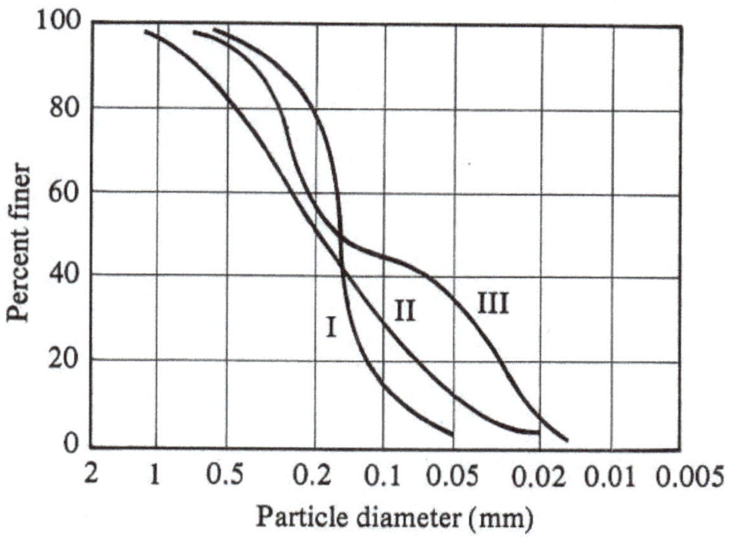

FIGURE 2.8 Types of Distribution Curves
Adapted from reference 1.

Curve I represent a type of soil in which most of the soil grains are the same size. This is called poorly graded soil. Curve II represents a soil in which the particle sizes are distributed over a wide range, which is termed well graded. A well-graded soil will have a uniformity coefficient greater than about 4 for gravel and 6 for sand and a coefficient of curvature between 1 and 3 for gravel and sand mix. A soil might have a combination of two or more uniformly graded fractions. Curve III represents such soil. This type of soil is termed gap-graded.[1]

2.4.1.1.3 EXAMPLE

We obtained soil sample #1 from a test pit three feet below grade near the existing flagpole. The pit was 20' south and 50' west from the flagpole. A washed sieve analysis is being requested to classify the soil sample. The dry sample and container weight before the wash was 1168.3 grams and 1065.0 grams after the wash. The container weight was 162.7 grams. The following sieves and the weight retained in the lab are as follows:

1-1/2"	0.0 g	#10	89.6 g
1"	88.7 g	#20	108.1 g
'3/4"	52.5 g	#40	132.4 g
'1/2"	108.5 g	#60	98.2 g
3/8"	90.6 g	#100	51.6 g
#4	146.3 g	#200	32.7 g
		Pan	1.1 g

See data sheet 2 below. The black ink shows the given data with the red representing the calculated data. Data Sheet 2 is found in Appendix J.

GRAIN SIZE ANALYSIS - MECHANICAL

Data Sheet 2

Project: Example 2 **Project No.:** AEC 2530

Location: Sample #1 @ 3' Pit 20'S & 50'W of Flagpole

Soil Description: Silty Sand

Tested By: Tom C. **Date of Test:** 6/4/2023

Total Soil		***Washed Soil***	
Wt. of dry sample + container (W_1)	1168.3 g	Wt. of washed dry sample + container (W_2)	1065.0 g
Container No.	C	Container No.	C
Wt. of container (W_c)	162.7 g	Wt. of container (W_c)	162.7 g
Wt. of dry sample (W_s) $W_s = W_1 - W_c$	1005.6 g	Wt. of washed dry sample (W_{wds}) $W_{wds} = W_2 - W_c$	902.3 g

Sieve No.	Diameter (mm)	Weight Retained	Cumlative Weight Retained	Cumulative % Retained	Cumulative % Passing
1-1/2"	37.500	0	0	0	100.0
1"	25.000	88.7	88.7	8.8	91.2
3/4"	19.000	52.5	141.2	14.0	86.0
1/2"	12.500	108.5	249.7	24.8	75.2
3/8"	9.500	90.6	340.3	33.8	66.2
#4	4.750	146.3	486.6	48.4	51.6
#10	2.000	89.6	576.2	57.3	42.7
#20	0.850	108.1	684.3	68.0	32.0
#40	0.425	132.4	816.7	81.2	18.8
#60	0.250	98.2	914.9	91.0	9.0
#100	0.150	51.6	966.5	96.1	3.9
#200	0.075	32.7	999.2	99.4	0.6
Pan	--	1.1	1000.3	99.5	0.5

Cumlative % Retained = 100 x Cumlative Wt. Retained/Ws

Cumlative % Passing = 100 - Cumlative % Retained

2.4.1.2 Hydrometer Analysis (ASTM D 7928)

Hydrometer analysis is the method to determine particle diameters of soils less than 0.075 mm better known as fines (silt and clay particles). This testing procedure is based on Stoke's law, which states the rate of settling suspended solids in a liquid solution. When a soil specimen is dispersed in water, the particles settle at different velocities depending on their shape, size, and weight. Stoke's law is based on a sphere in which fine soil is not always found in the shape of a sphere. For simplicity, the hydrometer analysis assumes the soils are a sphere. The idea of this test is that a larger (heavier) soil particle will fall faster through distilled water than a smaller (lighter) soil particle. The hydrometer readings are then correlated to the percent finer (percent passing), and the grain size distribution curve is plotted similarly to the method used in the sieve analysis.

2.4.1.2.1 Test Procedure

The sample set aside from the above procedure can now be tested. Three possible conditions determine how the hydrometer test should be conducted.

a. Low fine content
   If less than 5% of the soil passes the #200 sieve or if 5% to 15% passes the #200 sieve and the soils are non-plastic, a hydrometer analysis is normally not performed.

b. No large-size particles
   Typically, the largest size particle allowed in a hydrometer test is the soil particles passing the No. 10 sieve. Enough wet soil should be used to have about 50 g of soil particles if the particles are mostly fine and 100 g if they are mostly coarse-grained. The wet soil is placed in a pan and covered with distilled water.

c. Large-size particles
   The last condition is if the sample contains particles larger than passing the No. 10 sieve. In this case, the sample should be "wet sieved". Generally, the sample is wet sieved on the No. 10, No. 40, and No. 200 sieve. The hydrometer analysis can be performed on the material passing the No. 10, No. 40, or No. 200 sieve. In either case, the particles should be washed with distilled water, and all the samples should be collected by passing either the No. 10, No. 40, or No. 200 sieve, depending on which method you choose. [So, if the No. 40 sieve is chosen, you would only wet sieve the soil sample on the No. 40 sieve with distilled water while manually breaking up the particles with your fingertips, collect all the sample and water passing the No. 40 sieve, and discard all the material retained on the No. 40 sieve.] Wash approximately 50 g of wet soil through the sieve chosen if the particles are primarily fine and 100 g if it consists mostly of coarse-grained particles.

After the washing is complete, water may need to be evaporated so that there is less than 100 mL of soil-water solution in the pan. Add 5.0 g of hexametaphosphate to the soil and distilled water, mix thoroughly, and allow to soak overnight. The hexametaphosphate will prevent the clay-sized particles from flocculating during the test. Once the soaking period is completed, the soil-water-hexametaphosphate slurry is dispersed in accordance with ASTM. Pour the slurry into a 1000mL glass sedimentation cylinder. If clay clumps still exist, then additional mixing needs to be done. Once the slurry is transferred into the glass cylinder distilled water should be used to rinse all the slurry. Continue to add distilled water to the 1000 mL mark. Place the rubber stopper on the open end of the cylinder and shake for about 1 minute to complete the soil dispersion process. Upon completion of shaking, place the cylinder in an area where the cylinder will not be disturbed or subjected to vibrations. Take hydrometer readings at 1, 2, 5, 15, 30, 60, 250, and 1440 minutes. The hydrometer readings correspond to the top of the water meniscus. This reading is known as the actual hydrometer reading, R_s. Once the hydrometer reading is performed, the hydrometer is removed and placed in a cylinder of distilled water and cleaned by spinning the hydrometer in the cleaning cylinder. The temperature of the suspension should be determined after each hydrometer reading. It is best to use the 152H hydrometer since it is calibrated to record the actual mass of the soil particles and dispersing agent in suspension. After all the hydrometer readings are taken, pour the soil-water-hexametaphosphate solution into an evaporating dish. Rinse all the sediment from the test cylinder using distilled water. Place the evaporating dish into an oven set to 110° C. Weigh the evaporating dish with soil once the sample is fully dried.

Some common laboratory errors associated with the hydrometer analysis are as follows:

1. Oven drying the soils before the test,
2. Inadequate type or quantity of dispersing agent, resulting in flocculation of the soil particles,
3. Incomplete dispersion of soil,
4. Insufficient shaking of the cylinder at the start of the test,
5. Too much soil in suspension,
6. Disturbance of suspension during insertion or removal of hydrometer,
7. The hydrometer not kept clean,
8. Temperature variation during testing that was not recorded,
9. Mis-recording the results,
10. Loss of material during testing.

2.4.1.2.2 Calculations

The first step is to obtain the weight of the dry material M_s used in the hydrometer test. To account for the 5.0 g of sodium hexametaphosphate that was added to the soil-water slurry, subtract 5.0 g from M_3 to calculate the weight of dry soil M_s.

$$M_3 = M_2 - M_1$$
$$M_s = M_3 - 5.0\ g$$

M_1 = weight of the evaporating dish
M_2 = weight of evaporating dish with dry material
M_S = initial total dry weight of the dry soil

The hydrometer composite correction for a 152H hydrometer is as follows:

$$R_c = 13 - 0.4T$$

R_c = hydrometer composite correction
T = temperature (° C)

To determine the corrected reading, subtract the actual hydrometer reading from the hydrometer composite correction.

$$R = R_s - R_c$$

R = corrected hydrometer reading
R_s = actual hydrometer reading

The hydrometer is calibrated at a certain temperature (20° C) and a certain specific gravity (2.65). The following formula is used to calculate the correction factor for specific gravity:

$$\alpha = 1.0 + 0.2(2.65 - G_s)$$

α = correction factor
G_s = specific gravity of the soil

Unless the specific gravity of the soil has been tested, we make the following assumptions: $G_s = 2.65$ for predominately granular soil and $G_s = 2.70$ for predominately clay soil.

The equation for percent finer is calculated as follows:

$$F = 100R\alpha/M_s$$

F = percent finer remaining in solution

Adjusted percent finer is determined if the soil was first washed and separated on the No. 10, No. 40, or No. 200 sieve in test procedure c above. The adjusted percent finer is calculated by multiplying the value obtained from the percent finer remaining in the solution above (F) by the fraction of soil particles finer than the separation sieve. For example, the soil was washed on the No. 40, and the sieve analysis indicates that the soil is 75 percent finer than the No. 40 sieve. Therefore, the adjusted percent finer is equal to the percent finer remaining in solution (F) times 0.75.

The corresponding soil grain size can be calculated using Stoke's law, defined as

$$D = K(L/t)^{0.5}$$

D = soil grain size
K = correction factor for both viscosity of slurry and specific gravity of the soil.
The value K can be estimated from the following equation:

$$K = [13 + 1.15(24° C - T) + 4(2.65 - G_s)]/1000$$

T = temperature of solution (° C)
L = effective depth (cm) from the surface of solution to level at the density of solution is being measured by hydrometer as estimated by the following equation:

$$L = 16.3[1\text{-}(R_s/100)]$$

The soil grain size can be determined (mm) by inserting t, K, and L values.

2.4.1.2.3 EXAMPLE

A hydrometer analysis is performed on the soil sample from section 2.4.1.1.3. The sample is wet screened through a #40 sieve before performing the analysis; 18.8% of the soil passes the #40 sieve. The dry sample and container weight was 162.9 grams. The container weight was 108.2 grams. A specific gravity test on the soil will not be conducted, so we will assume a specific gravity of 2.65. The following lab data was obtained:

Time	Elapsed Time	Temp (°C)	Hyd. Reading
8:00 am	0.5 min.	20	29
8:01	1 min.	20	28
8:02	2 min.	20	27
8:05	5 min.	20	23
8:15	15 min.	20	21
8:30	30 min.	20	20
9:00	60 min.	20	17
12:10 pm	250 min.	20	14
8:00 am	1440 min.	20	12

See data sheet 3 below. The black ink shows the given data with the red representing the calculated data. Data Sheet 3 is found in Appendix J.

If the same sample is being tested, the results of the sieve and hydrometer analyses can be plotted on a grain size distribution curve. The results of examples 2.4.1.1.3 and 2.4.1.2.3 are from the same sample and are shown in Data Sheet 4 below. A blank Data Sheet 4 is included in Appendix J.

GRAIN SIZE ANALYSIS - HYDROMETER — **Data Sheet 3**

Project: Example #2 — **Project No.:** AEC 2530

Location: 20' S. & 50' W of Flagpole — **Bore No.:** — **Sample No.:** 1

Soil Description: Silty Sand — **Sample Depth:** 3.0'

Tested By: Tom C. — **Date of Test:** 6/5/2023

Hydrometer No.: 152 H — **Gs of solids:** 2.65 — **α =** 1

Dispersing agent Sodium Hexametaphosphate — **Amount:** 5 — **Wt. of soil (Ws):** 49.7

Soil washed on sieve? Yes (circled) | No — **If yes, Sieve No.** 10 | 40 (circled) | 200 — **% Passing on Sieve** 18.8

Total Weight of Material	
Wt. of dry material + container (W_1)	162.9
Container No.	C3
Wt. of container (W_c)	108.2
Wt. of dry material (W_3) $W_3 = W_1 - W_c$	54.7

NOTES: Assume G_s = 2.65 predominately granular soils.
Assume G_s = 2.70 predominately clayey soils.
Use adjusted % Finer only for washed sieve sample.

Wt. of dry soil (W_s) = W_3 - Dispersing agent = 49.7

Above equation is for a 152H hydrometer

Date	Time of Reading	Elapsed time (t) (min)	Temp (T) (°C)	Actual Hydrom. Reading (R_a)	Comp. Corr. Hydrom. (R_c)	Correct. Hydrom. (R)	% Finer (F)	Adjusted % F	Effective Depth (L) (cm)	Coor. Factor (K)	Particle Dia. (D) (mm)
6/5	8:00	0.5	20	29	5	24	48.29	9.08	11.57	0.0136	0.0654
6/5	8:01	1	20	28	5	23	46.28	8.70	11.74	0.0136	0.0466
6/5	8:02	2	20	27	5	22	44.27	8.32	11.90	0.0136	0.0332
6/5	8:05	5	20	23	5	18	36.22	6.81	12.55	0.0136	0.0215
6/5	8:15	15	20	21	5	16	32.19	6.05	12.88	0.0136	0.0126
6/5	8:30	30	20	20	5	15	30.18	5.67	13.04	0.0136	0.0090
6/5	9:00	60	20	17	5	12	24.14	4.54	13.53	0.0136	0.0065
6/5	12:10	250	20	14	5	9	18.11	3.40	14.02	0.0136	0.0032
6/5	8:00	1440	20	12	5	7	14.08	2.65	14.34	0.0136	0.0014

$R_c = 13 - 0.4T$ $R = R_a - R_c$ $F = 100R(\alpha)/W_s$ $\alpha = 1.0 + 0.2(2.65 - G_s)$ $D = K(L/t)^{1/2}$

Adjusted %F = F x % Passing on Sieve

$K = [13 + 0.15(24°C - T) + 4(2.65 - G_s)]/1000$ $L = 16.3(1 - R_a/100)$

GRAIN SIZE DISTRIBUTION

Data Sheet 4

Project: EXAMPLE 2 **Project No.:** AEC 2530

Location: SAMPLE #1 @ 3° - PIT - 20'S & 50'W OF FLAGPOLE

Soil Description: SILTY SAND

Tested By: Tom C. **Date of Test:** 6/5/23

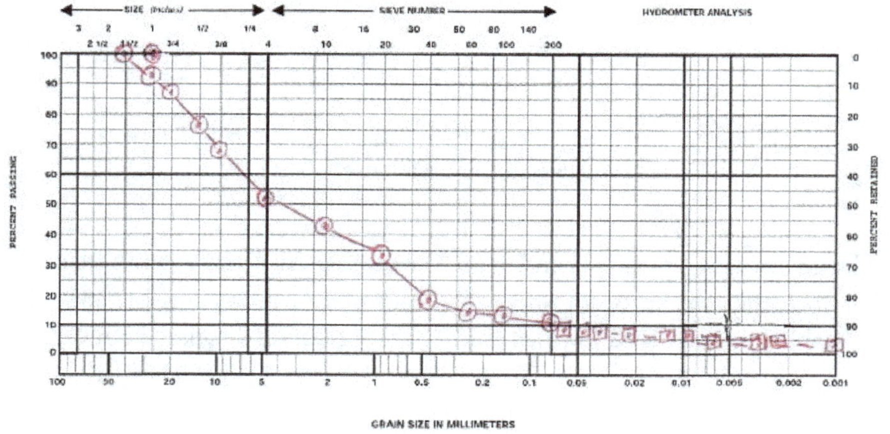

2.4.2 Soil Consistency and Atterberg Limits

Cohesive soils are generally defined by four states of consistency as introduced by a Swedish soil scientist, Albert Atterberg. The soil transfers from one state to the next with the addition or reduction of the water content in the soil. The soil is in a liquid state when the moisture content is very high. The soil flows when manipulated. As the water content decreases, the soil becomes plastic. The soil tends to stick to itself when worked. The percentage of water content where the soil changes from liquid to plastic is known as the liquid limit. With the continued drying of the soil, the soil changes to a semisolid state. The soil now becomes brittle as the soil is worked. The plastic limit is the water content where the soil changes from the plastic state to the semisolid state. Once the water content is very low the soil reaches the solid state. The soil now has insufficient water to change in volume. The water content where this occurs is the shrinkage limit. The four terms used in the Geotechnical Engineering field are the Atterberg Limits: liquid limit, plastic limit, shrinkage limit, and the plasticity index.

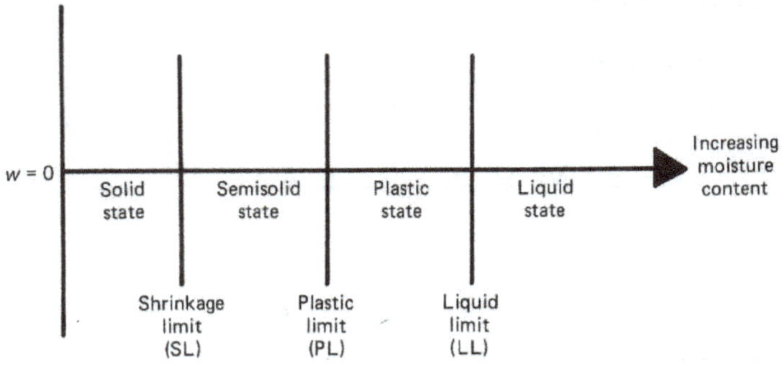

FIGURE 2.9 States of soil consistency based on Atterberg Limits
Adapted from reference 3.

Liquid limit, plastic limit, and plasticity index are the most used Atterberg Limits in the classification of soil; therefore, these test procedures will be discussed further. ASTM D 4318 designates the test procedure established by ASTM for the three Atterberg Limits discussed above.

2.4.2.1 Test Procedure

A soil sample is dried and broken apart in a manner that does not alter the individual particle size. The dried sample is passed through a No. 40 sieve, discarding all the material retained by the sieve. This soil sample can be one of the three samples that were separated during the sieve analysis. One of three conditions is possible with the soil sample, as follows:

1. Non-plastic soil (NP)
   Atterberg limit tests are not performed. Non-plastic soils can be identified if the sample cannot be rolled or molded even with the addition of water.
   Soil is classified as non-plastic if the plastic limit is equal to or greater than the liquid limit. This second alternative requires the Atterberg Limits to be performed.

2. High fines content
   If all the soil samples pass the No. 40 sieve, 150 to 200 g of wet soil should be thoroughly mixed with enough distilled water to bring it to a plastic state. The soil should then be sealed in a moisture-resistant container and allowed to sit overnight.

3. Intermediate condition
   The soil should be wet sieved on the No. 40 sieve. The particles passing the No. 40 sieve should be dried per ASTM until the soil reaches a plastic state. The liquid and plastic limit tests will need about 150 to 200 g of wet soil. The wet soil should be separated into two nearly equal parts.

2.4.2.1.1 Liquid Limit (LL)

The liquid limit device developed by Casagrande and described in the ASTM standards is shown in Figure 2.10. A soil paste is placed in the brass cup. The paste should be approximately 0.4 inches (10 mm) deep at the center, and the surface should be evenly leveled out. A grooving tool, as described by ASTM, is used to cut a groove from the top to the bottom of the soil paste. The grooving tool should cut through the soil and not displace the soil on both sides. The cup is raised and dropped 0.3937 inches (10mm) by turning the crank of the device. The liquid limit device should be calibrated and adjusted prior to commencing the test. The crank should be turned at a rate of 2 drops per second until the two halves come together for one-half inch (13 mm). A representative sample of the soil in the brass cup should be tested for moisture content. Cutting out the area where the two halves are touching is common. N refers to the number of blows the cup fell on the base to get the two halves touching one-half inch. Remove the remaining soil from the brass cup and mix it well with the other prepared soil. Either add water or let air dry to obtain a different moisture content. The addition of water will have a higher water content allowing the soil paste to flow more easily, producing a lower blow count. Follow this procedure to obtain four unique moisture contents that correspond between 15 to 35 blows. The moisture content and N (number of blows) are plotted on semi-logarithmic graphic paper. The number of blows should be plotted on the logarithmic axis of the log paper. A straight line should be drawn to connect the test results approximately. The moisture content corresponding to the N = 25 to the nearest whole number determines the liquid limit of the soil.

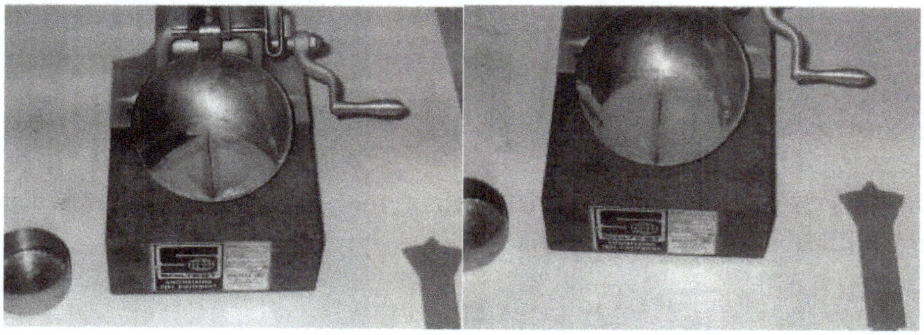

FIGURE 2.10 The left photo shows the Liquid Limit Device with the soil grooved prior to running the device. The right photo shows once the cup has been lifted and dropped by turning the crank with the flowing together.

2.4.2.1.2 Plastic Limit (PL)

The plastic limit is the water content at which the soil begins to break apart and crumble when rolled by hand into threads 1/8 inch (3 mm) in diameter. Obtain the second sample from the intermediate condition described above and air dry the soil until it no longer sticks to the fingers. The soil is then rolled by finger on a glass plate unit the soil crumbles see Figure 2.11. If the soil does not crumble at 1/8 inch diameter, mix the soil and keep rolling until it crumbles at the designated diameter. If the soil crumbles before the thread reaches 1/8 inch diameter, thoroughly mix the soil with a touch of water. Continue to roll the soil till the thread crumbles at the designated diameter. Collect two samples of a minimum of 2 g of wet soil when the thread starts crumbling at 1/8 inch diameter. Determine the water content of the two samples and verify the water content is similar, then average the two water contents. The average water content is recorded as the plastic limit to the nearest whole number.

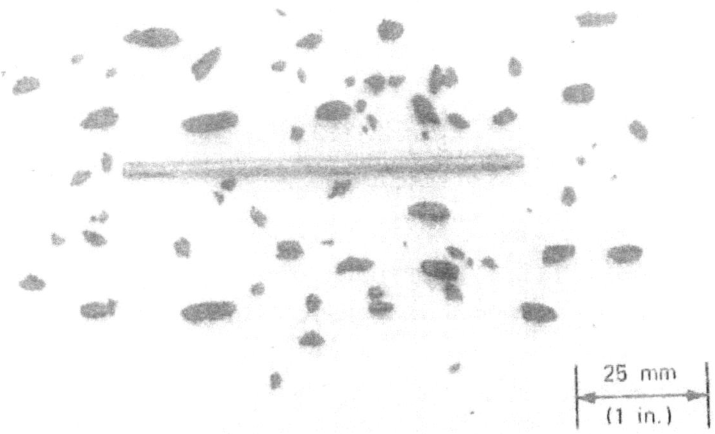

FIGURE 2.11 Plastic limit

2.4.2.1.3 Plasticity Index (PI)

The measure of a cohesive soil's plasticity is known as the plasticity index. It is calculated by subtracting the plastic limit from the liquid limit.

$$PI = LL - PL$$

The plasticity index is expressed as a whole number. The soil is considered non-plastic if the PI calculates as zero or a negative number.

Some common laboratory errors associated with the Atterberg Limits are as follows:

1. Nonrepresentative sample (the soil sample must be the same for both the liquid limit and the plastic limit),
2. Improperly prepared and cured samples,
3. Incorrect water content determination,
4. Mis-recording the results,
5. Loss of material during testing.

2.4.2.1.4 EXAMPLE

The same soil sample from examples 2.4.1.1.3 and 2.4.1.2.3 was also analyzed in the laboratory for its Atterberg Limits. See Data Sheet 5 below. The black ink shows the lab-determined data, with the red representing the calculated data. A blank Data Sheet 5 is found in Appendix J.

ATTERBERG LIMITS

Data Sheet 5

Project:	Example #2	**Project No.:**	AEC 2530
Location:	20' S & 50' W of Flagpole **Bore No.** Pit	**Sample No.:**	1
Soil Description:	Silty Sand	**Sample Depth:**	3.0'
Tested By:	Tom C.	**Date of Test:**	6/8/2023

Liquid Limit

Container No. (cup)	A1	C2	B4		
Wt. of cup + wet soil (W_1)	20.95	19.30	20.15		
Wt. of cup + dry soil (W_2)	18.80	17.63	18.60		
Wt. of cup (W_c)	10.84	10.92	11.33		
Wt. of dry soil (W_s) $W_s = W_2 - W_c$	7.96	6.71	7.27		
Wt. of water (W_w) $W_w = W_1 - W_2$	2.15	1.67	1.55		
Water content, ω% $\omega = 100W_w/W_s$	27.01	24.89	21.32		
No. of blows, N	19	24	29		

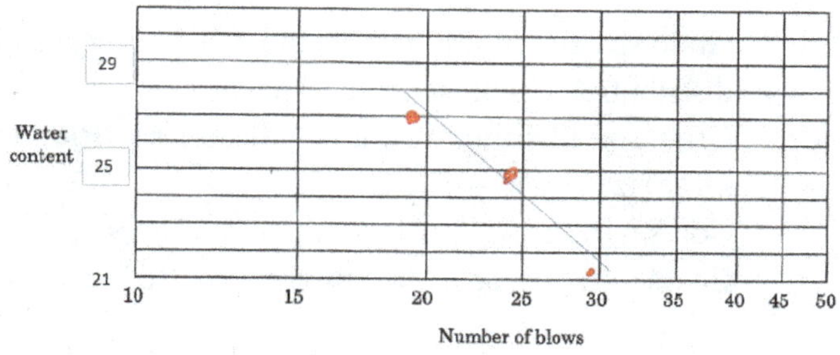

Plastic Limit

Container No. (cup)	B2	B3			
Wt. of cup + wet soil (W_1)	12.37	11.92			
Wt. of cup + dry soil (W_2)	12.19	11.77			
Wt. of cup (W_c)	11.25	10.98			
Wt. of dry soil (W_s) $W_s = W_2 - W_c$	0.94	0.79			
Wt. of water (W_w) $W_w = W_1 - W_2$	0.18	0.15			
Water content, ω% $\omega = 100W_w/W_s$	19.15	18.99			

Liquid Limit (LL) = 24	**Plastic Limit (PL) = 19**	**Plasticity Index (PI) = 5**

2.4.3 Classification

A soil classification system identifies different soils with similar properties into groups and subgroups. This means of classification allows the user to express the general characteristics of different soils to predict the behavior of influences such as loads, erosion, water, etc... The soil classification we will discuss has been developed for engineering purposes are based on simple index properties such as particle size distribution and plasticity. Earlier in this chapter, we discussed the three main classification systems: USC, AASHTO, and USDA. We will go into much more detail and learn how to identify the soils based on each system.

2.4.3.1 Unified Soil Classification System (USC) [ASTM D 2487]

Two main groups define soils: coarse-grained and fine-grained soils. The determining factor is the percentage passing the No. 200 sieve. When more than 50% of the soil is retained on the No. 200 sieve, the soil is classified as coarse-grained. When 50% or more of the soil passes the No. 200 sieve, the soil is noted as fine-grained.

2.4.3.1.1 Coarse-grained soils are further divided into gravel and sand, with the No. 4 sieve being the identifying separation point. More than 50% of the coarse fraction retained on the No. 4 sieve is then classified as gravel (G). When 50% or more of the coarse fraction passes the No. 4 sieve, the soils are classified as sand (S). The coarse-grained soils are then sub-grouped by the following identifiers:

W	Well graded	Grain size distribution curve indicated fairly even distribution of coarse-grained particles
P	Poorly graded	Grain size distribution curve indicated particles near the same size or gap-graded
M	Silt	Coarse-grained soil with some silt particles
C	Clay	Coarse-grained soil with some clay particles

Soil classification for engineering purposes is presented as ASTM D 2487. Note that gravel and sand may have dual classifications, such as SP-SC. The dual classification means the gradation meets the criteria for both soil types. The flow chart for classifying coarse-grained soils in ASTM D 2487 best describes this. Tables 2.5 and 2.6 list the coarse-grained soil classification using the Unified Classification System.

TABLE 2.5 Unified Soil Classification System(USC) – Group Symbols for Gravelly Soil

Group Symbol	Criteria
GW	Less than 5% passing No. 200 sieve; $C_u = D_{60}/D_{10}$ greater than or equal to 4; and $C_c = (D_{30})^2/(D_{10} \times D_{60})$ between 1 and 3
GP	Less than 5% passing No. 200 sieve; and not meeting both criteria for GW
GM	More than 12% passing No. 200 sieve; Atterberg's limits plot below "A" line (Figure 2.12) or plasticity index less than 4
GC	More than 12% passing No. 200 sieve; Atterberg's limits plot above "A" line (Figure 2.12) or plasticity index greater than 7
GC-GM	More than 12% passing No. 200 sieve; Atterberg's limits fall in hatched area marked CL-ML in Figure 2.12)
GW-GM	Percent passing No. 200 sieve is 5 to 12; and meets the criteria for GW and GM
GW-GC	Percent passing No. 200 sieve is 5 to 12; and meets the criteria for GW and GC
GP-GM	Percent passing No. 200 sieve is 5 to 12; and meets the criteria for GP and GM
GP-GC	Percent passing No. 200 sieve is 5 to 12; and meets the criteria for GP and GC

Group Symbol	Criteria
SW	Less than 5% passing No. 200 sieve; $C_u = D_{60}/D_{10}$ greater than or equal to 6; and $C_c = (D_{30})^2/(D_{10} \times D_{60})$ between 1 and 3
SP	Less than 5% passing No. 200 sieve; and not meeting both criteria for SW
SM	More than 12% passing No. 200 sieve; Atterberg's limits plot below "A" line (Figure 2.12) or plasticity index less than 4
SC	More than 12% passing No. 200 sieve; Atterberg's limits plot above "A" line (Figure 2.12) or plasticity index greater than 7
SC-SM	More than 12% passing No. 200 sieve; Atterberg's limits fall in hatched area marked CL-ML in Figure 2.12
SW-SM	Percent passing No. 200 sieve is 5 to 12; and meets the criteria for SW and SM
SW-SC	Percent passing No. 200 sieve is 5 to 12; and meets the criteria for SW and SC
SP-SM	Percent passing No. 200 sieve is 5 to 12; and meets the criteria for SP and SM
SP-SC	Percent passing No. 200 sieve is 5 to 12; and meets the criteria for SP and SC

TABLE 2.6 Unified Soil Classification System (USC) - Group Symbols for Sandy Soil
Adapted from reference 1.

2.4.3.1.2 **Fine-grained soils** are further divided into silt, clay, and organic as determined by plotting the Plasticity Index versus the Liquid Limit on the plasticity chart, Figure 2.12.

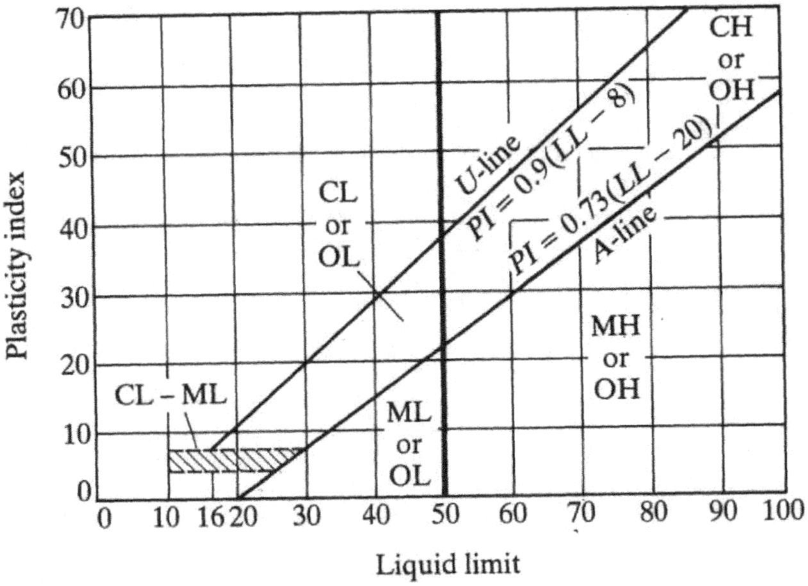

FIGURE 2.12 Plasticity Chart
Adapted from reference 1.

The plasticity chart has an A-line defined as PI = 0.73(LL-20). The Atterberg limits plotted on this chart for soil indicate a point above the A-line, and then the soils are classified as clay. If the point plotted is below the A-line, the soil is silt or organic. The fine-grained soils are then sub-grouped by the following identifiers:

L	Low plasticity	Liquid limit equal to or less than 50%
H	High plasticity	Liquid limit greater than 50%

The plasticity chart also has a U-line, which is defined as PI = 0.9(LL-8). The U-line is the upper limit of plasticity per Casagrande the creator of the plasticity chart. The laboratory results and calculations should be checked if a point is plotted above the U-line. In theory, no soil can have a plasticity higher than the U-line. The plasticity chart also has an area between the A-line and U-line where soil may be identified as a CL-ML. This means the soil sample tested will act as clay and silt with low plasticity. Another fine-grained soil is identified in Table 2.7, a highly organic soil commonly known as peat.

TABLE 2.7 Unified Soil Classification System (USC) – Group Symbols for Silty and Clayey Soil
Adapted from reference 1.

Group Symbol	Criteria
CL	Inorganic: LL < 50; PI >7; and plots on or above "A" line (Figure 2.12)
ML	Inorganic: LL < 50; PI < 4; and plots below "A" line (Figure 2.12)
OL	Organic; (LL - oven dried) / (LL- not dried) < 0.75; and LL < 50
CH	Inorganic: LL ≥ 50; and PI plots on or above "A" line (Figure 2.12)
MH	Inorganic: LL ≥ 50; and PI plots below "A" line (Figure 2.12)
OH	Organic; (LL - oven dried) / (LL- not dried) < 0.75; and LL ≥ 50
CL-ML	Inorganic; plot in the hatched zone in Figure 2.12
Pt	Peat, muck, and other highly organic soils

Other soil descriptors are commonly used to describe the soil in the field or the laboratory. We will not expound on these descriptors, but the student should be aware that they exist. Some additional descriptors are as follows:

1. Soil color
2. Soil structure
3. Soil texture
4. Clay consistency
5. Sand density condition
6. Soil moisture condition

2.4.3.1.3 EXAMPLE

A soil sample is tested in the laboratory with the following results:

LL = 30%
PL = 12%
Sieve analysis data:

Sieve Size	% Passing
3/8"	100.0
No. 4	86.5
No. 10	68.0
No. 40	37.9
No. 200	14.8

Classify the soil by the Unified Soil Classification System.

Solution:

1. Determine the percentage passing on the No. 200 sieve.
   a. Coarse-grained soils less than 50%
   b. Fine-grained soils 50% or more
   14.8% therefore coarse-grained soil
2. Coarse-grained soils – determine the percentage passing on the No. 4 sieve.
   a. Gravel soil less than 50%
   b. Sand soil 50% or more
   86.5% therefore sandy soil
3. Sandy soils - evaluate the percentage passing the No. 200 sieve.
   a. Less than 5% - clean sand
      i. SW or SP
   b. Greater than 12% - sand with fines
      i. SM or SC
   c. Between 5% and 12%
      i. SC-SM, SW-SM, SW-SC, SP-SM, or SP-SC
   14.8% therefore sand with fines, Table 2.6
4. Sand with fines – review plasticity chart
   a. Fines classify as ML or MH
      i. SM
   b. Fines classify as CL or CH
      i. SC

   PI = LL – PL = 30-12 = 18
   The plasticity chart shows CL or OL
   Results: Soil is a Clayey Sand or SC

2.4.3.1.4 EXAMPLE

A soil sample is tested in the laboratory with the following results:

LL = 61%
PL = 22%
Sieve analysis data:

Sieve Size	% Passing
3/8"	100.0
No. 4	100.0
No. 10	99.8
No. 40	91.0
No. 200	84.8

Classify the soil by the Unified Soil Classification System.

Solution:

1. Determine the percentage passing on the No. 200 sieve.
   a. Coarse-grained soils less than 50%
   b. Fine-grained soils 50% or more

   84.8% therefore fine-grained soil
2. Fine-grained soils – review plasticity chart.

   PI = LL – PL = 61-22 = 39

   The plasticity chart shows CH or OH, Figure 2.12

   Results: Soil is a <u>Clay with high plasticity</u> or <u>CH</u>

2.4.3.2 AASHTO Classification System (AASHTO) [ASTM D 3282] {AASHTO M145}

This classification was originally developed in 1929 by the Public Road Administration, which is now the Association of State Highway and Transportation Officials (AASHTO). It is provided to identify the best behavior of subgrade soil for constructing roads, especially the federal highway system.

The AASHTO classification system identifies granular materials as 35% or less of the total sample passing a No. 200 sieve. Granular materials are classified as A-1, A-2, and A-3, with classification into additional sub-grades for A-1 and A-2 soils. These soils are rated excellent to good for use as subgrade materials. Soils that have more than 35% passing the No. 200 sieve are classified as silt-clay materials. These materials are noted as A-4, A-5, A-6, and A-7. The A-7 materials have additional subgroups. Table 2.8 indicates the classification of the soils, including the subgrades.

TABLE 2.8 AASHTO Classification Criteria

General Classification	**Granular Materials** (35% or less of total sample passing No. 200)						
Group Classification	A-1			A-2			
	A-1-a	A-1-b	A-3	A-2-4	A-2-5	A-2-6	A-2-7
Sieve Analysis (percent passing)							
No. 10	50 max.						
No. 40	30 max.	50 max.	51 min.				
No. 200	15 max.	25 max.	10 max.	35 max.	35 max.	35 max.	35 max.
Atterberg Limits							
Liquid Limit				40 max.	41 min.	40 max.	41 min.
Plasticity Index	6 max.		NP	10 max.	10 max.	11 min.	11 min.
Unusual types of significant constituent materials	Stone fragments, Gravel and Sand		Fine Sand	Silty or Clayey Gravel and Sand			
General Subgrade Rating	Excellent to Good						

General Classification	**Silty-Clay Materials** (More than 35% of total sample passing No. 200)			
Group Classification	A-4	A-5	A-6	A-7 A-7-5 A-7-6
Sieve Analysis (percent passing)				
No. 10				
No. 40				
No. 200	36 min.	36 min.	36 min.	36 min.
Atterberg Limits				
Liquid Limit	40 max.	41 min.	40 max.	41 min.
Plasticity Index	10 max.	10 max.	11 min.	11 min.
Unusual types of significant constituent materials	Silty Soils		Clayey Soils	
General Subgrade Rating	Fair to Poor			

Plotting the plasticity index versus the liquid limit in Figure 2.13 can visually identify the soils for groups A-2, A-4, A-5, A-6, and A-7.
Adapted from reference 1.

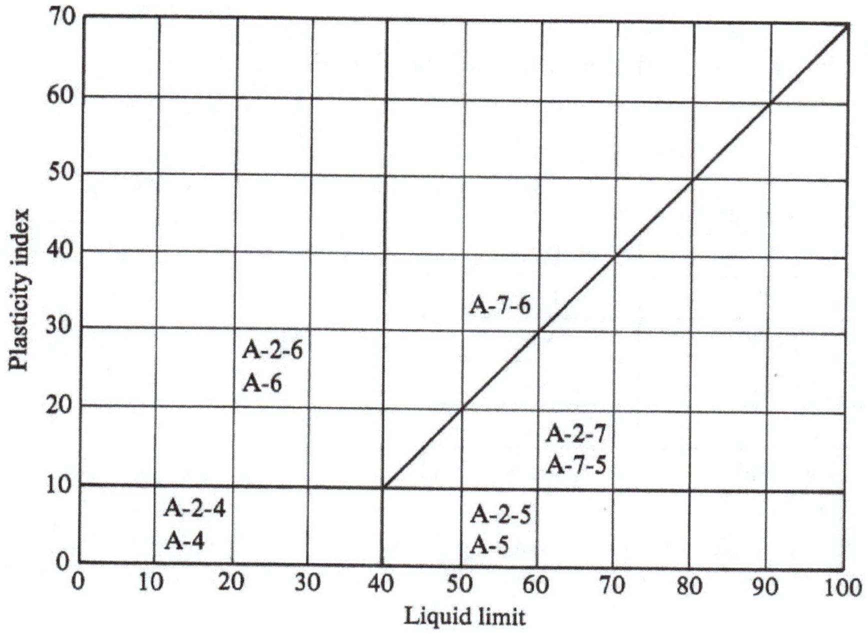

FIGURE 2.13 AASHTO classification chart for A-2, A-4, A-5, A-6, and A-7 soil
Adapted from reference 1.

Once a soil has been classified, it can be further described using a group index. The group index can be calculated using the following equation:

$$GI = (F-35) + [0.2+0.005(LL-40)] + 0.01(F-15)(PI-10)$$

GI = Group Index
F = percent of soil passing a No. 200 sieve
LL = liquid limit
PI = plasticity index

The group index should be rounded to the nearest whole number. It will be shown in parentheses after the grouping and subgroup. If the group index is calculated to be negative, use zero as the group index. The group index for soils belonging to A-1, A-2-4, A-2-5, and A-3 will always be zero.

The road-supporting characteristics of a subgrade may be assumed as an inverse ratio of its group index. A road subgrade having a group index of 0 indicates an excellent to good subgrade, providing good drainage and adequate bearing when properly compacted. A group index of 20 or greater indicates a very poor subgrade material due to impervious drainage and low bearing capacity.

2.4.3.2.1 EXAMPLE

A soil sample is tested in the laboratory with the following results:

LL = 30%
PL = 12%
Sieve analysis data:

Sieve Size	% Passing
3/8"	100.0
No. 4	86.5
No. 10	68.0
No. 40	37.9
No. 200	14.8

Classify the soil by the AASHTO Classification System.

Solution:

1. Determine the percentage passing on the No. 200 sieve.
   a. Granular materials 35% or less
   b. Silt-Clay materials of more than 35%

   14.8% therefore granular material

2. Granular material – evaluate PI.
   a. 6 max A-1
   b. NP A-3
   c. 10 max A-2-4 or A-2-5
   d. 11 min A-2-6 or A-2-7

   PI = LL – PL = 30-12 = 18
   therefore A-2-6 or A-2-7

3. A-2-6 or A-2-7 - Evaluate LL.
   a. 40 max
      i. A-2-6
   b. 41 min
      i. A-2-7

   LL = 30
   Results: A-2-6, Fig 2.13

2.4.3.2.2 EXAMPLE

A soil sample is tested in the laboratory with the following results:

LL = 61%
PL = 22%
Sieve analysis data:

Sieve Size	% Passing
3/8"	100.0
No. 4	100.0
No. 10	99.8
No. 40	91.0
No. 200	84.8

Classify the soil by the AASHTO Classification System.

Solution:

1. Determine the percentage passing on the No. 200 sieve.
   a. Granular materials 35% or less
   b. Silt-clay materials of more than 35%

   84.8% therefore, silty-clay material

2. Silty-clay material. Evaluate PI
   a. 10 max A-4 or A-5
   b. 11 min A-6 or A-7

   PI = LL – PL = 61-22 = 39
   Therefore, A-6 or A-7

3. A-6 or A-7. Evaluate LL
   a. 40 max A-6
   b. 41 min A-7

   61% therefore A-7
   LL versus PI chart shows A-7-6, Figure 2.13
   Results: The material is a <u>Clayey Soil</u>

Liu (1967) compared the AASHTO and USCS systems. His study's results are presented in Tables 2.9 and 2.10.

Soil group in Unified system	Comparable soil groups in AASHTO system		
	Most probable	Possible	Possible but improbable
GW	A-1-a	—	A-2-4, A-2-5, A-2-6, A-2-7
GP	A-1-a	A-1-b	A-3, A-2-4, A-2-5, A-2-6, A-2-7
GM	A-1-b, A-2-4, A-2-5, A-2-7	A-2-6	A-4, A-5, A-6, A-7-5, A-7-6, A-1-a
GC	A-2-6, A-2-7	A-2-4, A-6	A-4, A-7-6 A-7-5
SW	A-1-b	A-1-a	A-3, A-2-4, A-2-5, A-2-6, A-2-7
SP	A-3, A-1-b	A-1-a	A-2-4, A-2-5, A-2-6, A-2-7
SM	A-1-b, A-2-4, A-2-5, A-2-7	A-2-6, A-4, A-5	A-6, A-7-5, A-7-6, A-1-a
SC	A-2-6, A-2-7	A-2-4, A-6, A-4, A-7-6	A-7-5
ML	A-4, A-5	A-6, A-7-5	—
CL	A-6, A-7-6	A-4	—
OL	A-4, A-5	A-6, A-7-5, A-7-6	—
MH	A-7-5, A-5	—	A-7-6
CH	A-7-6	A-7-5	—
OH	A-7-5, A-5	—	A-7-6
Pt	—	—	—

[a] After T. K. Liu (1967)

TABLE 2.9 Comparison of USCS to AASHTO Classification System

Soil group in AASHTO system	Comparable soil groups in Unified system		
	Most probable	Possible	Possible but improbable
A-1-a	GW, GP	SW, SP	GM, SM
A-1-b	SW, SP GM, SM	GP	—
A-3	SP	—	SW, GP
A-2-4	GM, SM	GC, SC	GW, GP, SW, SP
A-2-5	GM, SM	—	GW, GP, SW, SP
A-2-6	GC, SC	GM, SM	GW, GP, SW, SP
A-2-7	GM, GC, SM, SC	—	GW, GP, SW, SP
A-4	ML, OL	CL, SM, SC	GM, GC
A-5	OH, MH, ML, OL	—	SM, GM
A-6	CL	ML, OL, SC	GC, GM, SM
A-7-5	OH, MH	ML, OL, CH	GM, SM, GC, SC
A-7-6	CH, CL	ML, OL, SC	OH, MH, GC, GM, SM

[a] After T. K. Liu (1967)

TABLE 2.10 Comparison of AASHTO to USC

Adapted from reference 1.

2.4.3.3 USDA Textural Soil Classification System (USDA)

Figure 2.14 presents the simple textural classification system per the U.S. Department of Agriculture (1975), where *percent sand* is defined as the percentage of soil particles between 2.0 and 0.050 mm in size, *percent silt* is defined as the percentage of soil particles between 0.050 and 0.002 mm, and *percent clay* is defined as the percentage of soil particles finer than 0.002 mm. Note that the definitions of sand and silt size particles differ from those used in the other classification systems. If a soil contains a significant soil fraction larger than 2.0 mm, the textural classification is preceded with the term *gravelly* for soil fragments up to 76 mm in diameter, *cobbly* for rock fragments between 76 and 250 mm, and *stony* or *boulder* for rock fragments larger than 250 mm.[4]

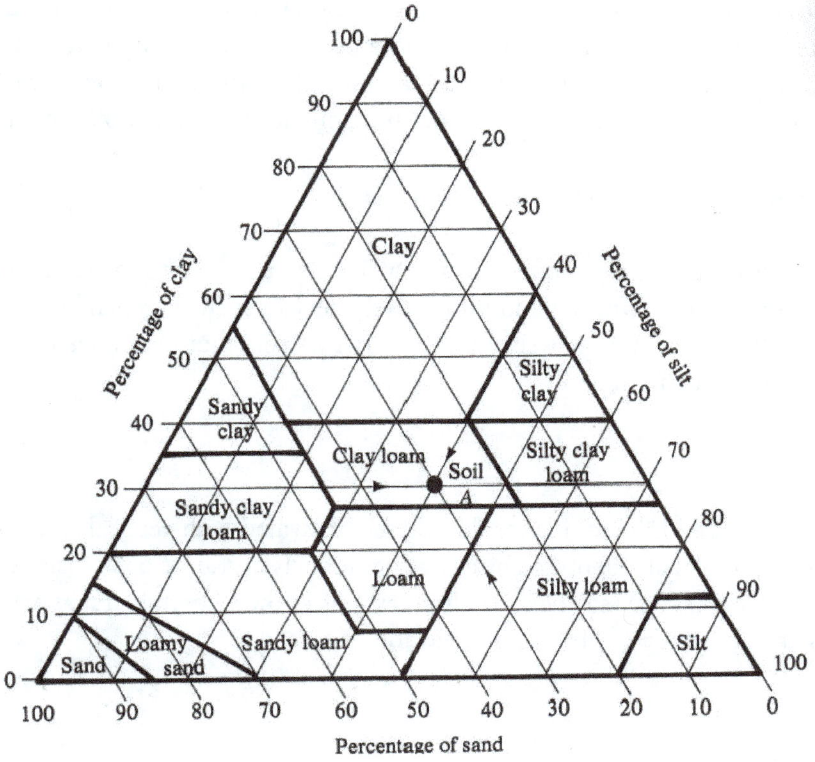

FIGURE 2.14 USDA Classification Chart
Adapted from reference 1.

The USDA also identifies soil by their Hydrologic Soil Groups. The Hydrologic Soil Groups are based on estimates of runoff potential. These soils are best identified by using the Web Soil Survey website, as described in the mapping section below. These soils are assigned to the following groups:

Group A. Soils have a high infiltration rate (low runoff potential) when thoroughly wet. These consist mainly of deep, well-drained to, excessively drained sands or gravelly sands. These soils have a high rate of water transmission. (Low runoff potential)

Group B. Soils have a moderate infiltration rate when thoroughly wet. They consist chiefly of moderately deep or deep, moderately well-drained or well-drained soils that have moderately fine texture to moderately coarse texture. These soils have a moderate rate of water transmission.

Group C. Soils have a slow infiltration rate when thoroughly wet. These consist chiefly of soils with a layer that impedes the downward movement of water or soils of moderately fine texture or fine texture. These soils have a slow rate of water transmission.

Group D. Soils have a very slow infiltration rate (high runoff potential) when thoroughly wet. These consist chiefly of clays with a high shrink-swell potential, soils with a high water table, soils that have a claypan or clay layer at or near the surface, and soils that are shallow over nearly impervious material. These soils have a very slow rate of water transmission. (High runoff potential)

2.4.3.4 OSHA Soil Classification System (OSHA)

OSHA classifies soils for the purpose of trench safety. The more stable the soil excavation, the steeper the slope in a trench can be without additional safety devices such as trench boxes. The following soil types are described on the OSHA website. The OSHA website should be consulted for additional safety information concerning soil.

Stable Rock

Stable Rock is natural solid mineral matter that can be excavated with vertical sides and remain intact while exposed. It is usually identified by a rock name such as granite or sandstone. Determining whether a deposit is of this type may be difficult unless it is known whether cracks exist and whether or not the cracks run into or away from the excavation.

Type A Soils

Type A Soils are cohesive soils with an unconfined compressive strength of 1.5 tons per square foot (tsf) (144 kPa) or greater. Examples of Type A cohesive soils are often clay, silty clay, sandy clay, clay loam, and, in some cases, silty clay loam and sandy clay loam. (No soil is Type A if it is fissured, is subject to vibration of any type, has previously been disturbed, is part of a sloped, a layered system where the layers dip into the excavation on a slope of 4 horizontal to 1 vertical (4H:1V) or greater or has seeping water.

Type B Soils

Type B Soils are cohesive soils with an unconfined compressive strength greater than 0.5 tsf (48 kPa) but less than 1.5 tsf (144 kPa). Examples of other Type B soils are angular gravel, silt, and silt loam; previously disturbed soils unless otherwise classified as Type C; soils that meet the unconfined compressive strength or cementation requirements of Type A soils but are fissured or subject to vibration; dry, unstable rock; and layered systems sloping into the trench at a slope less than 4H:1V (only if the material would be classified as a Type B soil).

Type C Soils

Type C Soils are cohesive soils with an unconfined compressive strength of 0.5 tsf (48 kPa) or less. Other Type C soils include granular soils such as gravel, sand, and loamy sand, submerged soil, soil from which water is freely seeping, and submerged rock that is not stable. Also included in this classification is material in a sloped, layered system where the layers dip into the excavation or have a slope of four horizontal to one vertical (4H:1V) or greater.

Multi-type soil

Layered Geological Strata. Where soils are configured in layers, i.e., where a layered geologic structure exists, the soil must be classified on the basis of the soil classification of the weakest soil layer. Each layer may be classified individually if a more stable layer lies below a less stable layer, i.e., where a Type C soil rests on top of stable rock.

2.4.4 Mapping

Throughout the United States, the U.S. Department of Agriculture (USDA) has conducted shallow soil (60") investigations and determined soil identification. This information has been compiled on the Web Soil Survey site www.websoilsurvey.com. Although the soil was generally mapped for agricultural purposes, a lot of soil information relates to engineering purposes. The information on this website is not recommended for use in the final design of projects. However, the information is very useful in preliminary reconnaissance and to identify what may be identified in a field site investigation.

The following overview of using the Web Soil Survey site was taken directly from the web page.

Here's a brief overview to get you started using Web Soil Survey:

1. Browse to the Web Soil Survey home page at the following link: Web Soil Survey Home Page.
2. To start the application, click the large round green button labeled **Start WSS**.
3. To work with Web Soil Survey interactively, jump to step 4 below. To download soil data for a full survey area or for State Soil Geographic Database (STATSGO2) data, click the **Download Soils Data** tab.
4. After the map refreshes, you can use the **Zoom In** map tool. Click on the **Zoom In** tool to put the map in **Zoom In** mode. Then either click on the map to zoom in or click and drag the map to zoom to the rectangle you've drawn. Zoom in as close as you need.

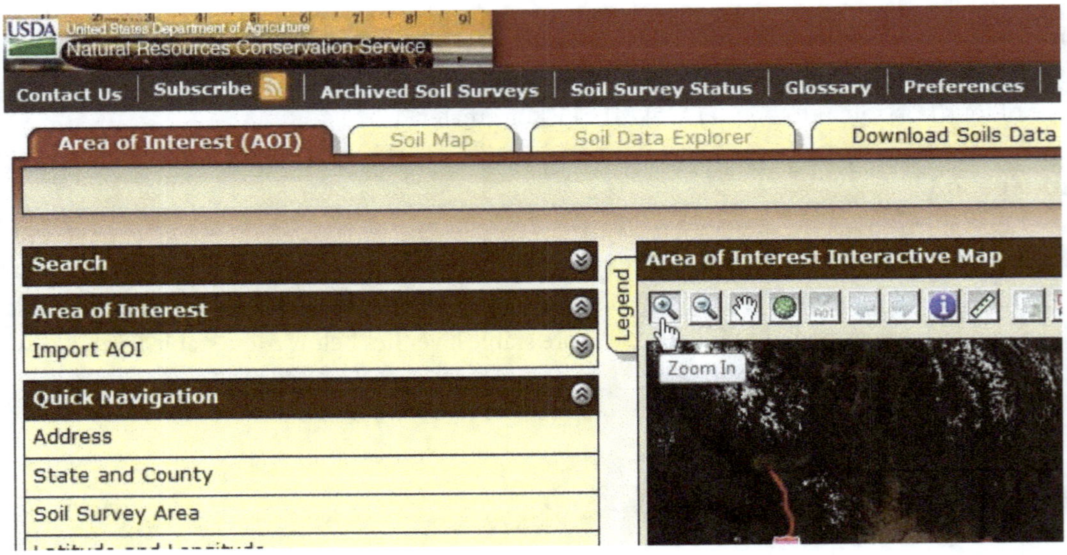

5. Or alternatively, on the left side of the browser window, in the **Quick Navigation** panel, click on one of the forms — **Address**, **State and County**, Soil Survey Area, etc. Enter the information, and click **View**.
6. Before you can view any soil data, you must define your Area of Interest or AOI. You can set your AOI by drawing a rectangle or a polygon on the map, or you can set your AOI to a whole Soil Survey Area. You can also create a multi-part AOI using the Create AOI from Shapefile or Create AOI from Zipped Shapefile options under the Import AOI menu. AOIs created using the **AOI Rectangle** and **AOI Polygon** tools are limited to a maximum of 100,000 acres, but Soil Survey Area AOIs are not. To set your AOI to a rectangle, click the **AOI Rectangle** map tool. Then click and hold the left mouse button, and drag a rectangle on the map. When you release the mouse button, the rectangle you've drawn will be set as your Area of Interest. To set your AOI to a polygon, click the **AOI Polygon** map tool. Then, click points on the map to define your AOI. To finish, either double-click the last point or hold down the Control key while clicking the last point. To set your AOI to a whole Soil Survey Area, in **Quick Navigation**, open the **Soil Survey Area** form. Choose a state and Soil Survey Area using the dropdowns. Then click **Set AOI**.
7. Once you have set your AOI, click the **Soil Map** tab to see the soil map and map unit information.
8. To create a printable document containing the map and information on the **Soil Map** tab, click the **Printable Version** button and then click the **View** button.
9. To run soil ratings or soil reports, click the **Soil Data Explorer** tab, then one of the inner tabs: **Suitabilities and Limitations for Use**, **Soil Properties and Qualities**, or **Soil Reports**
10. On the left side of the browser window, click the **Open All** button to expand all the folders or click an individual folder to list the items within it.
11. Click one of the items to open the form, then set options as desired, and click **View Rating** or **View Soil Report**. This will show the data in tabular form and, for the Ratings, in col-

or-coded map form. Click the Legend tab at the left side of the map to see a legend of the Rating values.

12. To create a printable version of the soil data, click **Printable Version** or **Add to Shopping Cart**: **Printable Version** generates a Portable Document Format (PDF) document containing the rating or report that you just ran. **Add to Shopping Cart** adds the report or map to the shopping cart. You can add multiple ratings and reports to the shopping cart and then create a Portable Document Format (PDF) document containing all the items you added to it. The AOI soil map, the list of map units, and their descriptions are added to the shopping cart by default. Once you're done adding content to the shopping cart, click the **Shopping Cart (Free)** tab and click the **Check Out** button. This will generate a single Portable Document Format (PDF) document containing all the items you added. By default, the **Soil Map** content is automatically included in your Portable Document Format (PDF) document. For best results, limit the number of items you add to the shopping cart to ten or fewer.
13. There is a **Search** function that you can use to search for where specific keywords occur in the application. The **Search** form is located in the upper left corner of each tab. Set your Area of Interest, then click **Search** and enter keywords, such as "hydraulic rating" or "KCl", and then click the **Search** button. The search results are links that navigate to the place in the application where the keywords were found.
14. After you have defined your AOI, you can also click the **Download Soils Data** link to download tabular and spatial soil data for your AOI. You will need Microsoft Access software to make use of the tabular data and ArcGIS software to make use of the spatial data.
15. Throughout Web Soil Survey, context-specific help is available by clicking the question mark icon. For more information, see "How can I get help using Web Soil Survey?" on the Frequently Asked Questions page, How can I get help using Web Soil Survey?
16. For assistance with specific soil data questions, you may click the Contact Us link. The Contact Us link is also located beneath the USDA logo in the upper-left corner of Web Soil Survey.

2.4.4.1 EXAMPLE

A Web Soil Survey report is provided in Appendix H.

2.5 Soil Compaction

Compaction is defined as the densification of the soil by mechanical means. This is performed by pressing the soil particles tightly together and replacing the air void with soil and/or water. Compaction is used to improve the structure of the soil during the construction process. Compacting the soil increases its density and improves its following characteristics. 1) Increases the soil's shear strength and bearing capacity. 2) Decreases the potential for settlement of the soil. 3) Decreases the permeability or the flow of water through the soil. 4) Reduces the potential for frost heave or swell. Compaction is a relatively inexpensive way to improve the properties of a given soil.

There are four basic factors that affect the compaction:

1. Soil type
2. Material gradation
3. Water content
4. Compaction effort (energy applied)

The relative compaction of the soil in the field is compared to the theoretical maximum compaction as determined in the laboratory. R. R. Proctor standardized a laboratory test that identifies the soil's theoretical maximum dry density at optimum moisture. This test is known as the Proctor Compaction Test. Two methods were used: the Standard Proctor and the Modified Proctor. The degree of compaction is measured by its dry unit weight. The dry unit weight can be calculated by dividing the wet unit weight by one plus the decimal moisture content as follows:

$$\gamma_d = \gamma/(1+\omega)$$

γ_d = dry unit weight
γ = wet unit weight
ω = moisture content

A soil sample given the same compaction effort will increase in dry unit weight as water is added. The water acts as a lubricant for the soil particles, making the sample denser. The dry unit weight continues to increase until the water starts replacing the soil particles in the structure, and then the dry density begins to decrease.

2.5.1 Proctor Compaction Test

The proctor compaction test is a laboratory method to identify a soil sample's maximum dry density and optimum moisture content. The test procedure is similar whether the Standard Proctor Method (ASTM D 698) or the Modified Proctor Method (ASTM D 1557) is performed. The Modified Proctor Method is conducted with more energy applied than the Standard Proctor Method. See Table 2.11. The preparation of the soil samples for both methods is the same. However, the Modified method places five soil layers versus three layers in the Standard method. The Modified method also uses a proctor hammer weighing 10 pounds with an 18" drop. The Standard method uses a proctor hammer weighing 5.5 pounds with a 12" drop.

TABLE 2.11 Summary of Compaction Test Specifications

Description		ASTM D698; AASHTO T99 Method A	Method B	Method C	ASTM D1557; AASHTO T180 Method A	Method B	Method C
Mold:							
Volume	ft3	1/30	1/30	1/13.33	1/30	1/30	1/13.33
	cm3	943.9	943.9	2124.3	943.9	943.9	2124.3
Height	in.	4.58	4.58	4.58	4.58	4.58	4.58
	mm	116.33	116.33	116.33	116.33	116.33	116.33
Diameter	in.	4	4	6	4	4	6
	mm	101.6	101.6	152.4	101.6	101.6	152.4
Weight of hammer	lb	5.5	5.5	5.5	10	10	10
	kg	2.5	2.5	2.5	4.54	4.54	4.54
Height of hammer drop	in.	12	12	12	18	18	18
	mm	304.8	304.8	304.8	457.2	457.2	457.2
Number of layers of soil		3	3	3	5	5	5
Number of blows per layer		25	25	56	25	25	56
Tested on soil fraction passing sieve		No. 4	3/8 in.	3/4 in.	No. 4	3/8 in.	3/4 in.

Adapted from reference 1.

2.5.1.1 Test Procedure

A representative soil sample should be obtained from the field. The sample should be obtained from the soil that will be compacted in the field. Variations in the soils may require multiple samples to be taken from the same project. The soil sample should be from 50 pounds to 200 pounds, depending on the maximum particle size. There are three methods used depending on the maximum particle size:

Method A	Material passing a No. 4 sieve
Method B	Material passing a 3/8" sieve
Method C	Material passing a 3/4" sieve

Two volumes of compaction molds are used depending on the maximum particle size found in the soil sample. Methods A and B use a mold that will produce 1/30 of a cubic foot with a 4" diameter, while Method C uses a 1/13.33 cubic foot with a 6" diameter mold. After the soil has been passed through the proper sieve, the soil should be dried till friable. The soil should be air dried but can also be dried in an oven not exceeding 140°F (60°C). A minimum of four samples of approximately 5 pounds will be separated into separate containers. Different amounts of water will be added to each container to create an approximate 2% difference in moisture content, with two samples on the dry side and two on the wet side of optimum moisture. The soil in each container should be thoroughly mixed and allowed to sit without losing moisture content. Start the compaction process after a specified curing period (minimum 12 hours for clays and a couple of hours for sands). This process will require adding the moisture-conditioned soil to the mold with the collar so that either 1/3 or 1/5 of the mold (without the collar) is filled once each layer is compacted, depending on whether the test is either a Standard or Modified Proctor. Each layer will be compacted by raising the hammer to the full stroke possible and allowing the hammer to drop freely onto the soil. Move the hammer around the mold until the appropriate number of blows per layer is completed. Add another layer to the mold, repeating the process until the appropriate number of lifts are performed. Remove the collar from the mold. The compacted soil should be slightly greater than the top of the mold. Carefully strike off the compacted soil till the top of the soil is even and level with the top of the mold. Weigh the mold with the soil so the wet density can be calculated. Remove the soil from the mold and obtain a moisture content of the soil. Repeat this process for the other three moisture-conditioned samples. With experience, you will be able to identify when the compacted soil has reached optimum moisture. This can also be determined by noting that the weight of the soil and mold is equal to or less than the previous moisture content once all the layers have been added and the excess soil has struck off.

Calculate the dry unit weight using the wet unit weight and moisture content. Plot the dry unit weight along the y-axis and the moisture content on the x-axis. A curve like Fig 2.15 should be formed. The peak point of the compaction curve is defined as the laboratory's maximum dry density, and the corresponding water content is known as the optimum moisture content. These are the values that should be used in the relative comparison with the field data while conducting field density testing. If the compaction curve does not form a distinct peak or all the points are located on one side of the curve, additional points need to be performed in the laboratory with the missing moisture content to identify the peak.

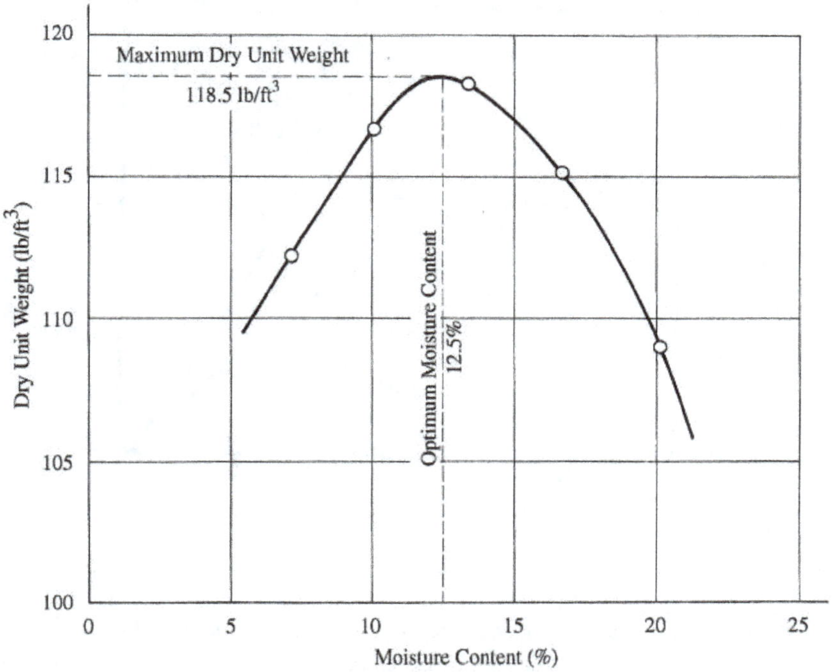

FIGURE 2.15 Compaction Test Curve
Adapted from reference 2.

The results for the same soil are different for the Standard Proctor and Modified Proctor Tests. Since the Modified Proctor Test is developed with greater compaction effort, we would anticipate the maximum dry density would also be greater. Figure 2.16 shows the results of the two tests on the same soils. Note that the maximum dry density is greater and the optimum moisture content is lower. This is caused by the greater compaction force that drives more soil particles into the void, allowing less space for water.

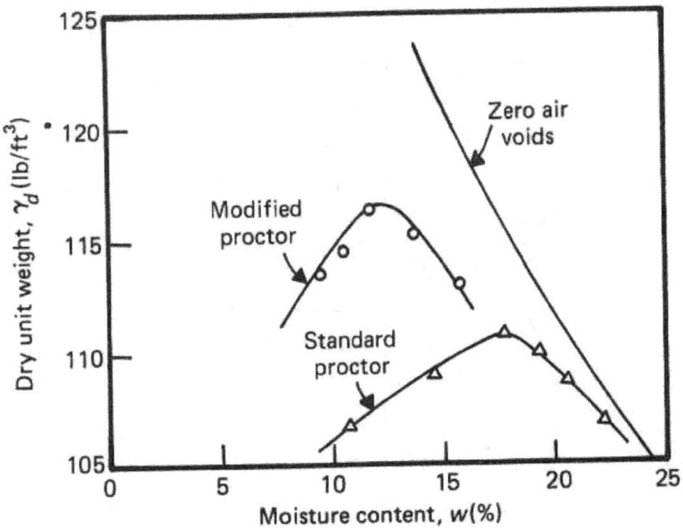

FIGURE 2.16 Standard and modified proctor test results for a silty sand
Adapted from reference 3.

Some common laboratory errors associated with the proctor tests are as follows:

1. Inadequate mixing of the water with the soil,
2. Reuse of the soil for multiple points on the compaction curve,
3. Not compacting the soil on a solid base,
4. Incorrect volume used for the compaction mold,
5. Not providing a full stroke with the compaction hammer,
6. Excessive variation in layers or total depth prior to trimming,
7. Water content not representative of the entire specimen at each point,
8. Mis-recording the results,
9. Loss of material during testing.

2.5.1.1.1 EXAMPLE

During the quality control testing for the over-lot grading of a new subdivision, a soil sample is obtained to conduct a standard proctor test for future nuclear density testing. The project name is Example #3, and it is the first sample taken on this project. The soil was obtained on the surface from Sunset Ridge Lot 1, Block 1. See data sheet 6 below. The black ink shows the data determined in the lab, and the red represents the calculated data. A blank Data Sheet 6 is found in Appendix J.

PROCTOR TEST

Data Sheet 6

Project:	Example #3	**Project No.:**	AEC 2530
Location:	Suset Ridge, L1, B1	**Bore No.:**	
		Sample No.:	1
Soil Description:	Silty Sand w/gravel	**Sample Depth:**	0'
Tested By:	Tom C.	**Date of Test:**	8/6/2023
Test Method:	Standard (circled) / Modified	**Procedure:**	A (circled) / B / C
Mold Diameter:	4" (circled) / 6"	**Mold Volume: (V_m)**	1/30 CF (circled) / 1/13 CF

Circle choice of boxes used above

Water Content

Sample No.	1	2	3	4	5
Container No. (cup)	A1	A6	A2	B5	
Wt. of cup + wet soil (W_1)	624.8	618.7	632.5	621.3	
Wt. of cup + dry soil (W_2)	585.3	570.9	575.8	557.0	
Wt. of cup (W_c)	103.4	102.8	103.3	104.3	
Wt. of dry soil (W_s) $W_s = W_2 - W_c$	481.9	468.1	472.5	452.7	
Wt. of water (W_w) $W_w = W_1 - W_2$	39.5	47.8	56.7	64.3	
Water content, ω% $\omega\% = 100W_w/W_s$	8.2	10.2	12.0	14.2	

Density

Wt. of soil + mold (W_s+W_m)	8.73	8.88	8.94	8.85	
Wt. of mold (W_m)	4.40	4.40	4.40	4.40	
Wt. of soil in mold (W_s)	4.33	4.48	4.54	4.45	
Wet Density (PCF) $\gamma_{wet} = W_s/V_m$	129.9	134.4	136.2	133.5	
Dry Density (PCF) $\gamma_d = \gamma_{wet}/(1+\omega)$	120.1	122	121.6	116.9	

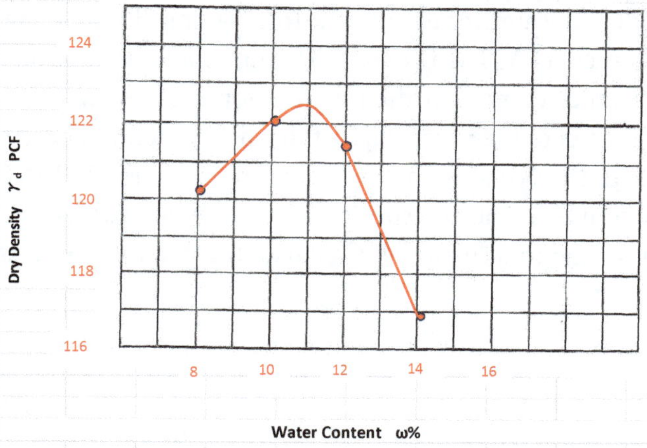

Optimum Moisture (%) = 11.1	**Maximum Dry Density (PCF) =** 122.4

2.5.1.2 Oversize Particles

Oversize particles are the gravel and cobble particles that exceed the maximum sieve used in the Proctor test described above. Table 2.12 shows the criteria for oversize particles in accordance with ASTM procedures.

TABLE 2.12 Oversize Particles In Accordance With ASTM Proctor Procedure

Method	Particle Size analysis	Are oversize particle corrections required?
Procedure A	5% or less retained on No. 4 sieve	No
	5% to 20% retained on No. 4 sieve	Yes
Procedure B	More than 20% retained on No. 4 sieve and 5% or less retained on 3/8 in. sieve	No
	More than 20% retained on No. 4 sieve and 5% to 20% retained on 3/8 in. sieve	Yes
Procedure C	More than 20% retained on 3/8 in sieve and 5% or less retained on 3/4 in. sieve	No
	More than 20% retained on 3/8 in sieve and 5% to 30% retained on 3/4 in. sieve	Yes
>30% Retained on 3/4 in. sieve	Cannot use ASTM procedures. No alternative options are provided.	

Note: This table is based on ASTM D 1557-98 and ASTM D 698-98 test criteria
Adapted from reference 4.

2.5.1.2.1 Procedure

Correction of unit weight and water content for the total sample. Prepare the sample in accordance with the test procedures such as ASTM D 698. Determine the weight of the finer and the oversize fraction. Determine the water content of the two fractions. Calculate the weight of the dry finer and oversize fractions. Calculate the percentage of dry finer fraction and oversize fraction. Determine the relative density of the oversize fraction in accordance with ASTM C 127. (It is a common practice in the industry to use 2.65 as the specific gravity of the oversized particles). Calculate the corrected water content of the total material. Calculate the corrected dry unit weight of the total material.

2.5.1.2.2 Calculations

Calculate the weight of the finer fraction.

$$W_{DF} = W_{MF}/(1+(\omega_F/100))$$

Calculate the weight of the oversized fraction.

$$W_{DC} = W_{MC}/(1+(\omega_C/100))$$

where:	W_{DF}	=	weight of the dry finer fraction
	W_{MF}	=	weight of the moist finer fraction
	ω_F	=	water content of the finer fraction
	W_{DC}	=	weight of the dry coarse fraction
	W_{MC}	=	weight of the moist coarse fraction
	ω_C	=	water content of the coarse fraction

Calculate the percentage of finer and oversize fractions.

$$P_F = 100\ W_{DF}/(W_{DF} + W_{DC})$$

$$P_C = 100\ W_{DC}/(W_{DF} + W_{DC})$$

where:	P_F	=	percentage of the finer fraction by weight
	P_C	=	percentage of the oversize fraction by weight

Calculate the corrected water content and corrected dry unit weight of the total material.

$$\omega_T = (\omega_F P_F + \omega_C P_C)$$

and

$$\gamma D_T = 100\gamma D_F G_M \gamma_\omega/(\gamma D_F P_C + G_M \gamma_\omega P_F)$$

where:	ω_T	=	corrected water content of both fractions expressed in %
	ω_F	=	water content of finer fraction expressed as a decimal
	ω_C	=	water content of the oversize fraction as a decimal
	γ_{DT}	=	corrected dry unit weight of both fractions
	G_M	=	relative specific density
	γ_{DF}	=	dry unit weight of the finer fraction
	γ_ω	=	unit weight of water

2.5.1.2.3 One-Point Proctor Test

The one-point proctor test is a quick method that can be used to obtain an estimate of the laboratory's maximum dry density and optimum moisture content. Prior to completing this test, you must have conducted several compaction curves on similar soils. The one-point Proctor requires conducting a single-point dry density test on soil that has a water content that is dry of optimum moisture content. A curve is drawn that mimics the curve shape of other similar soil compaction curves. The maximum dry density and optimum moisture are obtained by using the line of optimums drawn through the peak points of similar soil compaction curves. See Fig 2-10.

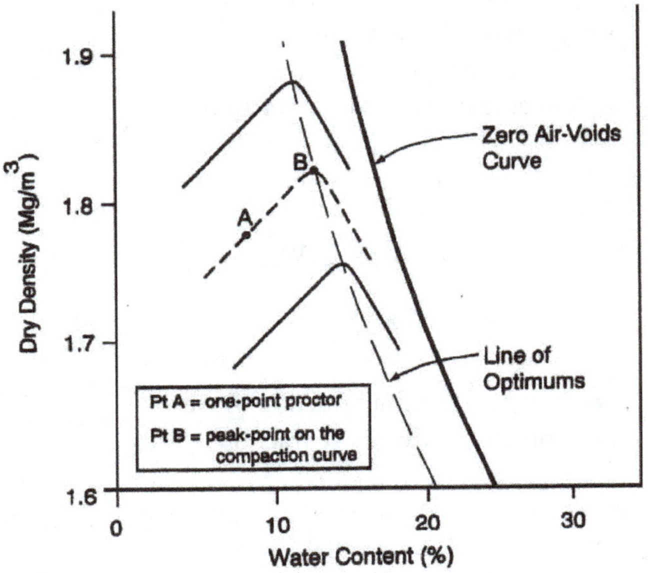

FIGURE 2.17 One-point proctor curve and procedure
Adapted from reference 4.

2.5.2 Field Compaction

Compaction of the soil in the field is an important part of any development. As previously stated, compaction is one of the simplest and least expensive means of improving the soil. This book will discuss surface compaction as opposed to deep foundation methods.

2.5.2.1 Construction Development

The development of raw land generally follows a prescribed method. Once the development has been engineered for the proposed use, the project design has been completed, and the necessary permits have been obtained, fieldwork commences.

A typical grading process would include the following:

1. **Clearing and grubbing.** This is the removal of vegetation and unwanted debris. It is important that the removed vegetation and debris are removed from the site so no structures are constructed on such material.
2. **Unsuitable soil removal.** Any soil found unsuitable for bearing structures during the geotechnical subsurface soil investigation is removed, stockpiled for other use in the development, or reworked to improve the soil characteristic function for use in the development. (In Colorado, where expansive soil is commonly found, an over-excavation process may be conducted as deep as ten feet below the lowest foundation level.) This involves the thorough mixing of water to optimum moisture or greater. Replacement of the moisture-conditioned soil in controlled compacted lifts.
3. **Scarifying and recompacting.** This allows a good bond between the in-place soil and the compacted fill.
4. **Cut and fill grading.** This process entails removing (cut) the soil in areas with a lower proposed elevation than the existing elevation to an area that needs additional (fill) material to meet the proposed finished elevation. The cut material is referred to as borrow material. The design elevations are commonly adjusted to provide proper drainage and meet adjacent properties' elevation. The ideal design will balance the cut with the fill, requiring no soil to be removed or brought onto the site. The earthwork contractor is normally required to provide the final grade to within a certain tolerance per the specifications or contract.
5. **Trench excavation.** The installation of the utilities (sewer, water, storm sewer, gas, elec-

FIGURE 2.18 Construction Development Upper
Top: Pipe installation
Middle: Overlot grading
Bottom:: House foundation excavation

tricity, telephone, cable, and others) generally requires a trench to be excavated in line and on grade per the approved plans. Once the utilities are installed, the trench is backfilled and compacted in lifts per the project or local jurisdiction specifications.

6. **Subgrade preparation.** Before the installation of roads or structures, the soil supporting the structure (which includes roads) is prepared that improves the soil characteristics for the support of the structure. The soil supporting the structure is referred to as the subgrade, sub-base, and/or base.
7. **Backfill.** Later in the development process, structures are being constructed. Soil may be added around the structure to help support its lateral movement. It is important that this soil is constructed by placing controlled backfill in layers per the specifications.

2.5.2.2 Field Compaction Methods

The type of field compaction equipment to perform the work will depend on the type of soil, the size of the project, the size of the area to be compacted, and the depth of the compaction needed. Compaction methods can be generalized in the following categories:

1. **Static weight or pressure.** This type of compaction equipment applies a relatively uniform pressure to the soil. This is accomplished by adding weight to equipment that is rolled over the soil. The weight could be soil in a front-end loader or scrapper using rubber-tired wheels or water in a smooth wheel roller. This method is most effective on clay and silt but will also work on gravel and sand.
2. **Kneading or mixing.** This equipment, such as a sheepsfoot roller or sheepsfoot wheel, consists of a drum with metal protruding "feet" attached. The idea is to allow the feet to knead and mix the soils while exerting significant pressure on the affected area beneath the "feet". This method is highly recommended for clay soils and is also effective with silts.
3. **Impact or sharp blow.** These devices compact the soil using a sharp blow. Walk-behind tampers or jumping jack compactors are common equipment. This equipment works on all soil types. Gravel and sand will need to be confined in a small area to prevent the soil from being displaced due to the energy provided. The walk-behind tamper is used in a smaller area, such as foundation footing, due to the number of passes it will require to meet the required compaction percentage. The tampers should also be used with thin layers of fill.
4. **Vibration or shaking.** This equipment imparts a shaking process to the soil to realign the soil particles. Rollers are commonly used as this equipment, as well as HED (high energy dissipaters) shakers. Gravel and sand are the ideal soils for this type of compaction. However, HED shakers are commonly used with silt and clay as well.

FIGURE 2.19 Field Compaction Methods: Upper Left: Static–loaded scrapper
Upper Right: kneading – sheepsfoot pinwheel
Lower Left: impact – jumping jack tamper
Lower Right: vibration – HED shaker head

It is common for the compaction methods listed above to be combined in different compaction equipment. For example, a sheepsfoot roller may have an option to vibrate. The key ingredient in soil compaction is providing the correct water content. Soil compaction is most easily reached when the soil is near optimum moisture content. The desired compaction percentage can often be obtained if the soils are drier than optimum, but more compaction effort will be required.

2.5.2.2.1 Field Density Control

Most projects will have specifications regarding soil compaction that the contractor must meet (see Appendix D.). These specifications will specify the minimum percent compaction and a range of moisture content in relation to the optimum moisture content determined by either ASTM or AASHTO. This requires field testing of the compacted soil at a specified frequency. Table No. 2.13 and 2.14 show the typical frequency of tests in many cities in Colorado for quality assurance and quality control, respectively.

TABLE NO. 2.13 Typical City Testing Frequency for Quality Assurance

GUIDELINE FOR QUALITY ASSURANCE SAMPLING AND TESTING

Sewer/Water Compaction	Moisture/Density Curve	One per soil type
	% Compaction % Moisture	Mainline: One test every 1000 L.F. Water Service: One test per 20 services Sewer Service: One test per 20 services Water Valve/Manhole: At Inspectors discretion. Test opposite sides
Storm Sewer Compaction	Moisture/Density Curve	One per soil type
	% Compaction	Mainline: One test every 1000 L.F.
	% Moisture	Lateral: One test per laterals
Structural Backfill	Moisture/Density Curve	One per soil type
	% Compaction % Moisture	1 per 100 C.Y., Minimum 1 per structure
	Gradation	1 per 1000 C.Y.
Embankment and Subgrade Soil Compaction	Moisture/Density Curve Gradation	One per soil type
	% Compaction % Moisture	One test per 2500 C.Y. for embankment fill. One test for each soil type for subgrade tests
Filter Material/Bed Course Material	Gradation	One per type
Aggregate Base Course	Moisture/Density Curve Gradation	One per type
	% Compaction % Moisture	One test every 1000 L.F., Staggered and on center line
Hot Bituminous Pavement	Design Check	Annual check
	Extraction/Gradation	One per project, depending on project size
	Superpave/Rice	One sample per project
	Density	One Core per 500 L.F.
Concrete sampling & testing for Sidewalks, Bike Paths	Slump Air Content Temperature Compressive Strength	One slump, air content, temp & 4 cyl. For each 1500 L.F. or fraction there of or at the Inspectors discretion (1 @ 7 days, 2 @ 28 days, 1 hold)
	Gradation	One per source
Concrete sampling & testing for Curb & Gutter	Slump Air Content Temperature Compressive Strength	One slump, air content, temp & 4 cyl. For each 1500 L.F. or fraction there of or at the Inspectors discretion (1 @ 7 days, 2 @ 28 days, 1 hold)
	Gradation	One per source
All other utilities and/or excavation in the R.O.W. or easements (Irrigation, Gas, Power, Cable T.V., Telephone line, etc.) Flowable fill may be used in lieu of compacting on-site material	Moisture/Density Curve	One per soil type
	% Compaction % Moisture	Mainline: One test every 1000 L.F. Servies: One test per 20 services

TABLE NO. 2.14 Typical City Testing Frequency for Quality Control

GUIDELINE FOR QUALITY CONTROL SAMPLING AND TESTING		
IDENTIFICATION	**TYPE OF TEST REQUIRED**	**MINIMUM SAMPLING/TESTING FREQUENCY**
Sewer/Water Compaction	Moisture/Density Curve	One per soil type
	% Compaction	Mainline: One test every 200 L.F., Every other 8" lift of backfill
	% Moisture	Water Service: One test every other 8" lift of backfill
		Sewer Service: One test every other 8" lift of backfill
		Water Valve/Manhole: One test each 8" lift. Test opposite sides
Storm Sewer Compaction	Moisture/Density Curve	One per soil type
	% Compaction	Mainline: One test every 200 L.F., Every other 8" lift of backfill
	% Moisture	Service: One test every other 8" lift of backfill
Structural Backfill	Moisture/Density Curve	One per soil type
	% Compaction	1 per 100 C.Y. or fraction there of, Minimum 1 per structure
	% Moisture	
	Gradation	1 per 200 C.Y. or fraction there of
Embankment and Subgrade Soil Compaction	Moisture/Density Curve	One per soil type
	Gradation	
	% Compaction	One test per 450 C.Y. or fraction there of, each lift placed shall be tested for embankment fill.
	% Moisture	One test for each 200 Lane Feet each lift (Not less than on test/day) Shall be tested for subgrade compaction
Filter Material/Bed Course Material	Gradation	One per type
Aggregate Base Course	See MGPEC - Item 7	
Hot Bituminous Pavement	See MGPEC - Item 9	
Concrete sampling & testing	Slump	One slump, air content, temp & 4 cyl. For each 1500 L.F. or fraction there of or at the Inspectors discretion (1 @ 7 days, 2 @ 28 days, 1 hold)
	Air Content	
	Temperature	
	Compressive Strength	
	Gradation	One per source
Concrete sampling & testing for Curb & Gutter	Slump	One slump, air content, temp & 4 cyl. For each 100 L.F. or fraction there of or at the Inspectors discretion (1 @ 7 days, 2 @ 28 days, 1 hold)
	Air Content	
	Temperature	
	Compressive Strength	
	Gradation	One per source
All other utilities and/or excavation in the R.O.W. or easements (Irrigation, Gas, Power, Cable T.V., Telephone line, etc.) Flowable fill may be used in lieu of compacting on-site material	Moisture/Density Curve	One per soil type
	% Compaction	Mainline: One test every 200 L.F. every other 8" lift of backfill
	% Moisture	Servies: One test every other 8" lift of backfill

There are several ways to determine the density and moisture content of the compacted soil in the field. There are destructive and non-destructive means to determine these properties. The destructive method to determine in-place density and moisture content requires the fill material to be disturbed, commonly removed from the site. The disturbed area is generally small and has no impact on the overall placement of the fill. The destructive tests take more time to identify the density, relative compaction percentage, and moisture content. A sample is obtained from the field, and weighting and drying of the sample are required to determine the density and moisture content. The three most common destructive tests are the sand cone method, balloon method, and drive cylinder method. This book will not get into each process since these methods are not commonly used anymore. The non-destructive tests involve placing a probe into the soil and obtaining a reading. There are two main means to obtain the reading with electricity resistance or nuclear radiation. The most common method is using a nuclear density gauge that reads the time for the radioactive particles to penetrate the soil and bounce back to the gauge. The next section will detail the nuclear density testing procedure since this is now the preferred way to obtain in-place nuclear density tests and moisture content.

2.5.2.2.1.1 Nuclear Density Testing (ASTM D 6938)

The nuclear density test method has become the common method to obtain in-place density and moisture content due to its reliable results and quick time in obtaining results in the field. The disadvantage of this testing method over the previous destructive methods is the cost of equipment and the necessity to obtain a federal or state license to work with a monitored substance (nuclear isotopes). To use the gauge, a pin is driven into the soil, allowing a retractable probe to be lowered into the ground. The gauge is activated for a certain amount of time to emit gamma rays through the soil. Some of the gamma rays are absorbed by the soil, and the remaining rays reach a detector in the gauge. Soil unit weight is directly proportional to the amount of radiation that reaches the detector. The nuclear count rate received is calibrated to determine the wet unit weight of the soil. The moisture content is determined by emitting alpha particles that bombard a beryllium target, releasing fast neutrons. The fast neutrons strike hydrogen atoms in the water captured by the soil and loss velocity. The detector counts the resulting low-velocity neutrons and calculates as moisture content. Be aware that moisture content may give false readings when additional hydrogen is available, such as in soils containing iron, boron, or cadmium, organic material is present, hydrocarbon-based paint such as damp proofing, or the gauge is surrounded by soil in the proximity of the gauge such as a narrow trench or hole. The gauge will calculate the dry unit weight, the wet unit weight, and the moisture content of the material being tested. See Figure 2.20.

FIGURE 2.20 Nuclear density gauge

The moisture content can be physically verified by obtaining a field sample beneath the nuclear density gauge and placing the sample in a plastic bag to retain the moisture. As presented earlier, the sample in the plastic bag can be tested in the laboratory to determine the moisture content. If the verification moisture content is significantly different from the field results, the dry density and percent compaction should be modified with the corrected moisture content as determined by the laboratory test results.

To determine the relative percent compaction, you take the in-place field dry density divided by the maximum dry density from the proctor you choose and multiply the results by 100 as follows:

$$RC\% = 100\gamma/\gamma max$$

This process appears simple until the soil technician is on a site with numerous proctors and varying types of soil. It is important not just to use a proctor that will provide the results that allow the field density test to pass the specifications but to obtain the proctor that best fits the soils encountered to the best of the technician's judgment. The first clue to determine the correct proctor is the visual soil description. The next clue is to feel the moisture content of the soil being field tested. The soil technician should have a good idea of the optimum moisture content for each soil. The soil technician should be able to guesstimate if the field soil is over, at, or under optimum moisture and by an approximate percentage. The technician can compare the field moisture content to the list of proctors for that job that match the soil description. If no existing proctor closely matches the field results, then a sample of the field soils should be obtained to run an additional proctor.

2.6 Soil Stabilization

The type of soil may not always behave in a manner that will allow the proposed construction. The soil may tend to swell, settle, erode, or slough. All these actions identify the soil as non-stable for the proposed construction. Adding or removing water in the soil greatly affects the soil's behavior. A fat clay will behave stable and not impact the construction until water is introduced to the clay. The water molecules adhere to the clay particles, and swell is experienced. The clay may also settle if the moisture is removed from the clay particles. Sand and silt are prone to settlement once water and a load are added to the soil, especially if the soil is not very consolidated. Erosion is common with clay, silt, and sand if water is introduced to the soil, causing the soil particles to separate and wash away. The land may slough or slide when water infiltrates and saturates a weaker soil with less friction angle, such as silt or clay.

We previously found that the soil strength increases and capacity to accept water decreases with the compaction of the soil. Sometimes this is not sufficient and additional soil stabilization is needed.

Soil stabilization can also be used to improve the properties of the natural soil by preloading the soil, adding special soil (mechanical stabilization), adding chemicals (chemical stabilization), or placing some kind of fabric materials (geosynthetics) in the soil. These means to achieve soil stabilization are discussed below.[2]

2.6.1 Preloading

Preloading is the addition of excess weight to the proposed construction site prior to the construction of the structures. The soil is consolidated prior to construction, thereby strengthening the soil and decreasing or reducing the potential for the structure to settle. The more weight and the more time the weight is on the soil, the more consolidation and best results. This method is effective with soft silty and clayey soils. The disadvantage is the amount of time needed and the cost to add and remove the weight to the site.

2.6.2 Mechanical Stabilization

Mechanical stabilization is a simple idea that allows the addition of soil material into the on-site soil to create a better blend or gradation to increase the strength of the soil. The mixture is mechanically mixed, wetted, and compacted. This could be the addition of gravel into fine sand or even gravel into silt or clay to increase the bearing capacity and allow the water to pass more freely without being captured by the soil.

Mechanical stabilization could also be the replacement of less desirable soils with soils that better fit the situation of the construction site. For instance, a construction site may have a high groundwater table that is still three feet below the bottom of the foundation. The soils at the foundation level are soft due to the moisture content of the soils and the proximity of the groundwater. Two feet of the soft soils are removed and replaced with angular aggregate with a minimum 1" diameter and few fines; the angular aggregate could even be crushed concrete. This will allow the structure to be supported on the aggregate, allowing the water to flow freely beneath the support soil without the concern of the material becoming saturated and losing strength.

2.6.3 Chemical Stabilization

Chemical stabilization requires the addition of chemicals, lime, cement, or fly ash into the soil, either altering the characteristic properties of the soil or mixing as a binder with the soil. The chemical/soil mixture can be mixed, wetted, and compacted to the depth specified by the Geotechnical Engineer. An alternative means to add chemical solution or slurry into the soil is injecting the solution with pressure (jetting) beneath the surface of the soil. This method is commonly requested when the soil needs improvement, but surface disturbance isn't desired. The spacing of the injections, the pressure, and the amount of solution are normally determined by the Geotechnical Engineer working with the contractor.

Portland cement or fly ash is commonly added to create a soil-cement mixture. The amount of cement or fly ash commonly ranges from 7% to 14% by weight for sandy to clayey soils, respectively. The cement or fly ash bonds with the soil particles, improving the strength of the soil. Once the mixture is placed and tilled into the soil, water is added to bring the soil-cement mixture near optimum moisture in accordance with the Standard or Modified Proctor, followed by compaction of the soil mixture. In Colorado, this is a common means of improving the strength beneath roadways and sidewalks.

Lime, calcium chloride, and some trade-name products are additives designed to alter the structure of clayey soil to reduce plasticity. This technique can be effective in reducing volume changes associated with some expansive clays.

FIGURE 2.21 Soil Stabilization Techniques
Top Left: Preloading, Top Right: Mechanical, Bottom: Chemical

2.6.4 Geosynthetics.

Geosynthetics are man-made materials made of fabric, fiberglass, or plastic. They may be used to stabilize and reinforce soil for erosion, slope stability, backfill for retaining walls, and reduce water flow through the soil, to name a few. Geosynthetics come in many forms designed for different uses and are known as geotextiles, geogrids, geonets, and geomembranes.

Geotextiles are woven fabric with very small openings. They are used with fine-grained soils to prevent the infiltration of the fine soils into a larger aggregate soil or a pipe designed for water flow. Strength can be achieved by placing the fabric over fine-grained weak soil and filling it with better fill or aggregate. Geotextiles are commonly used with other geosynthetics.

Geogrids are generally made of plastic and have openings ranging from 1" to 4". They increase the strength of the soils and are commonly used under roadways and parking lots. They also reinforce and reduce potential slip failure on retaining, block, and earth walls.

Geonets are like geogrids but have intersecting ribs. They are generally used for drainage purposes under roadways, landfills, dams, and retaining walls.

Geomembranes are thin plastic sheets installed to greatly reduce the flow of water into the soil beneath the membrane. They are commonly used as liners in landfills and lagoons.

FIGURE 2.22 Geosynthetic Soil Stabilization
Top: Geotextile Second: Geogrid Third: Geonet
Bottom: Geomembrane

2.7 Sampling

Sampling is critical to providing realistic results of the fieldwork being performed, yet it is rarely discussed in soil mechanic books. To produce results that best match what is occurring in the field, it is imperative that a representative sample is obtained. This section will discuss methods to gather samples that are as representative as possible. We will first discuss sampling to aid in the preparation of the subsurface soil investigation. This is not the primary intent of this book, but it is important to the overall evaluation of construction projects. Next, we will discuss the soil field sampling necessary for the quality control of the field construction work.

2.7.1 Subsurface Soil Investigation

Soil samples are obtained from the field for laboratory testing to develop recommendations for the subsurface soil investigation. The purpose of performing subsurface soil investigation is to identify the following:

- ✓ Soil strata
  - o Depth
  - o Thickness
  - o Variability
  - o Classification
  - o Engineering properties
    - Bearing capacity
    - Permeability
    - Swell potential
    - Settlement potential
- ✓ Rock strata
  - o Depth to rock
  - o Identification and classification
  - o Variability
  - o Engineering properties
    - Bearing capacity
    - Swell potential
    - Solubility
- ✓ Groundwater elevation
- ✓ Local conditions
  - o Potential dump site
  - o Uncontrolled fill
  - o Hazardous waste

- o Ravine filled
- o Previous pond location

Methods to obtain samples and collect information of the in-situ soil vary but here are a few of the common means:

1. Borings
    a. Continuous-flight augers
    b. Hollow-stem augers
    c. Circulating wash
    d. Rotary coring
    e. Percussion drilling
2. Test Pits
3. Test Trenches

FIGURE 2.23 Drill rig during subsurface soil investigation

The recovered samples are considered altered soil, disturbed samples, or undisturbed samples. The ***altered soil sample*** can be due to mixing more than one soil strata or contamination of more than one soil strata. These samples are not representative of the soil horizons and should not be tested and represented in the report. This can occur during the drilling process by allowing the upper soil to fall into the bottom of the borehole; the moisture content is altered by surface water or fluid entering the borehole, and the over-pushing or underdriving of the sampler to obtain the sample. ***A disturbed sample*** is a soil sample that does not represent the natural condition of the soil. These samples may be obtained by driving a split spoon sampler, auger cuttings, or an auger bucket. Disturbed samples can still provide valuable information for use in the subsurface soil investigation, such as moisture content and classification. Strength characteristics can also be determined if the soils will be used as fill and the samples are recompacted to desired moisture content and density. ***An undisturbed sample*** is obtained by slowly pushing a thin-walled tube through the soil. We understand the sample is still impacted, but it is considered undisturbed if the soil structure is not greatly affected during the sampling process.

The California Split Sampler is the sampler of choice for many Colorado Geotechnical firms. The samples obtained are a hybrid of disturbed and undisturbed. The sampler is driven with force into the soil, but the samples are collected in brass or steel liners with minimum disturbance. Figure 2.24 shows the sampler with the brass liners in place. Figure 2.25 shows the soils in the liners.

FIGURE 2.24 California sampler with liners

FIGURE 2.25 Brass liners with soil samples

Care must be taken during sampling, transporting, removal from the sampler, storage, and preparation of the sample in the laboratory to best represent the soil in the field. This is especially true when identifying the strength characteristics.

2.7.2 Field Sampling

To name a few tests, soil samples are obtained from the field for laboratory testing to determine the classification of the soil, gradation, proctor values, and R-value. Samples may be collected from the surface soil as fill is placed, from test pits, stockpiles, or conveyor belts. The key factor in collecting the sample is to best represent the soil used in the field construction. Identify what tests are to be performed and what type of construction is occurring on site. If a large development is commencing with the over lot, grading and compaction tests are required. Obtain samples of the soil as they are being placed by haulers or belly dump scrapers for the purpose of soil classification and proctors.

2.7.2.1 Surface or pit sampling

Collecting soil samples from the surface or pit it is recommended to take several samples from different areas but similar soil. Take sufficient samples to blend the samples together. Mix the samples thoroughly and split the appropriate amount to perform the test required. We will discuss splitting techniques later in this chapter.

2.7.2.2 Stockpile sampling

Sampling stockpiles also requires taking several samples from different locations of the stockpile. It is also important to sample different elevations of the stockpile as well. Think about what occurs when a stockpile is formed. The larger and heavier soil particles will fall from the top of the pile towards the bottom. To obtain a representative sample,, you will need to sample near the top, middle of the pile, and towards the bottom. Mix the samples together thoroughly and split the mixed samples to obtain the size sample needed for the required test.

2.7.2.3 Conveyor belt sampling

Two methods can be used for conveyor best sampling. The first method is performed while the conveyor belt is operating. Make sure the plant operator and the conveyor belt operator are aware you will be sampling. The other important item is to ensure you can safely sample while the belt moves. If you can safely sample the belt during operation, take a five-gallon bucket and pass it fully through the discharge stream from the conveyor belt. Continue this process until you have sufficient samples to perform the necessary tests. The second method requires the conveyor belt to be off at the time of sampling. A conveyor belt section has two lines scored with your sampling trowel or shovel. All the material is collected between the scored lines. It is important to collect all the material, including the fines. The fines may need to be brushed into a dustpan.

FIGURE 2.26 Stockpile sampling

FIGURE 2.27 Sampling conveyor belt

2.7.3 Splitting samples

After thorough mixing, a composite sample consisting of several samples may need to be split into the appropriate size for testing.

One method requires a piece of equipment, a splitter box, see Figure 2.28. The splitter box consists of openings directed in two directions. The openings must be larger than the nominal size of the material being split. Therefore, the sample is divided into half each time it runs through the splitter box. Each side of the sample will be collected into pans. One pan should be bagged up and not used any further in this splitting process. The other pan will be halved through the splitter until the appropriate size is obtained.

Another method, known as quartering, requires an area with a clean surface where the full sample can be spread out. The pile will then be separated into quarters with a clean division through the full sample. Two opposite quarters will be bagged and removed from further splitting. Care must be taken to include the fines when quartering and bagging. The remaining two quarters will be mixed and quartered again. This procedure will continue till the appropriate sample size is obtained for the test.

FIGURE 2.28 Splitter box

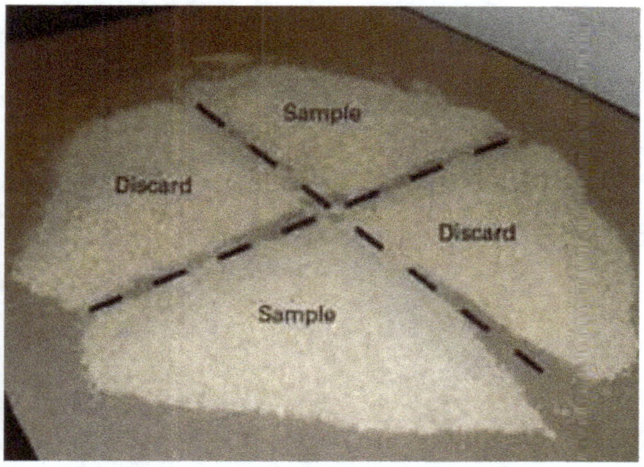

FIGURE 2.29 Quartering Technique

Problems – Chapter 2

1. Soil is formed by rock. True or False

2. How is soil formed? Choose all that apply.
   a. Weathering of the rock.
   b. Mineral composition causing chemical weathering.
   c. Change in temperature, mechanical weathering.
   d. All of the above.
   e. C only.

3. Why are soils the least predictable of all construction materials?

4. List the three phases of soils.

5. Which phase or phases make up the void volume? Choose all that apply.
   a. Solid volume.
   b. Water volume.
   c. Air volume.

6. Which case below describes saturated soil?
   a. The voids in the soil are partially filled with water.
   b. The voids in the soil are fully filled with water.
   c. The voids in the soil have no water.

7. What does USCS stand for?

8. Water content is the water weight divided by the sample's total wet weight. True or False

9. Using the swell-consolidation curve found in Appendix G, Geotechnical Exploration Report, Boring 5 at 4' to answer the following questions:

   a. Does this soil exhibit swell or consolidation?

   b. Approximately what percentage of the soil volume changes when water is added?

   c. Is the volume change positive or negative?

10. Using the swell-consolidation curve found below to answer the following questions:

    a. Does this soil exhibit swell or consolidation?

    b. Approximately what percentage of the soil volume changes when water is added?

    c. Is the volume change positive or negative?

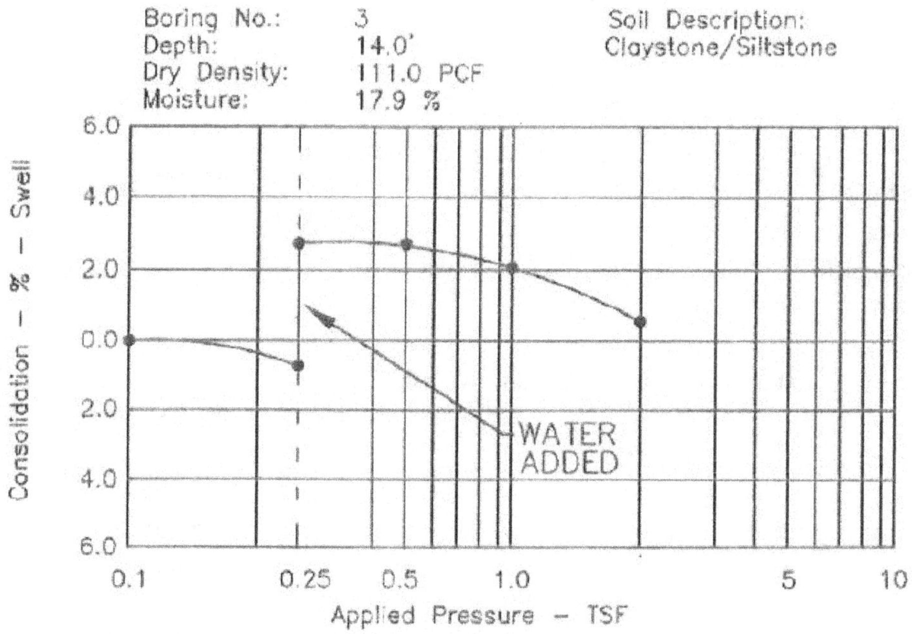

11. List four soil classification systems.

12. What is the most widely used soil classification system?

13. What soil classification system is most widely used in the development of the transportation system?

14. What two-grain size analysis tests are used to determine the size of the soil particles?
    a. Hydrometer Test
    b. Proctor Test
    c. Swell-Consolidation Test
    d. Sieve Analysis
    e. Atterberg Limits

15. Place the following sieves in the correct order from the largest opening above the smaller openings: ¾", #10, Pan, #100, 1", 3/8", #20, #200, #40, 1/2", ¼".

16. What is the largest size particle allowed in a hydrometer test?
    a. Passing the 1" sieve
    b. Passing the #40 sieve.
    c. Passing the #10 sieve.
    d. Passing the #200 sieve.

17. Name the four states of consistency in cohesive soils.

18. What is measured when the cohesive soils change from one state to another?

19. What is the limit when the cohesive soils change from the liquid state to the plastic state?
    a. Liquid Limit
    b. Plastic Limit
    c. Shrinkage Limit
    d. Plasticity Index

20. Complete the following sieve analysis.

GRAIN SIZE ANALYSIS - MECHANICAL

Project: ____________________ **Project No.:** __________

Location: ____________________

Soil Description: ____________________

Tested By: ____________________ **Date of Test:** __________

Total Soil		***Washed Soil***	
Wt. of dry sample + container (W_1)	766.8	Wt. of washed dry sample + container (W_2)	610.0
Container No.	AA	Container No.	AA
Wt. of container (W_c)	112.2	Wt. of container (W_c)	112.2
Wt. of dry sample (W_s) $W_s = W_1 - W_c$		Wt. of washed dry sample (W_{wds}) $W_{wds} = W_2 - W_c$	

Sieve No.	Diameter (mm)	Weight Retained	Cumlative Weight Retained	Cumulative % Retained	Cumulative % Passing
3/4"		0			
3/8"		35.2			
#4		30.5			
#10		21.6			
#20		49.5			
#40		102.6			
#60		89.1			
#100		95.6			
#200		60.4			
Pan		11.2			

Cumlative % Retained = 100 x Cumlative Wt. Retained/Ws

Cumlative % Passing = 100 - Cumlative % Retained

21. Complete the following hydrometer analysis.

GRAIN SIZE ANALYSIS - HYDROMETER

Project: Worksheet #3 **Project No.:** ______

Location: ______ **Bore No.** ______ **Sample No.:** ______

Soil Description: ______ **Sample Depth:** ______

Tested By: ______ **Date of Test:** ______

Hydrometer No.: 152 H **Gs of solids:** 2.65 assumed **α =** ______

Dispersing agent Sodium Hexametaphosphate **Amount:** 5.0 g **Wt. of soil (Ws):** ______

Soil washed on sieve? YES / No **If yes, Sieve No.** 10 / 40 / 200 **% Passing on Sieve** 24.3

Total Weight of Material	
Wt. of dry material + container (W_1)	162.9 g
Container No.	K7
Wt. of container (W_c)	108.4
Wt. of dry material (W_3) $W_3 = W_1 - W_c$	

NOTES: Assume G_s = 2.65 predominately granular soils.
Assume G_s = 2.70 predominately clayey soils.
Use adjusted % Finer only for washed sieve sample.

Wt. of dry soil (W_s) = W_3 - Dispersing agent = ______

Above equation is for a 152H hydrometer

Date	Time of Reading	Elapsed time (t) (min)	Temp (T) (°C)	Actual Hydrom. Reading (R_a)	Comp. Corr. Hydrom. (R_c)	Correct. Hydrom. (R)	% Finer (F)	Adjusted % F	Effective Depth (L) (cm)	Coor. Factor (K)	Particle Dia. (D) (mm)
		0.5	20	29							
		1	20	28							
		2	20	27							
		5	20	23							
		15	20	21							
		30	20	20							
		60	20	14							
		250	19	14							
		1440	19	12							

$Rc = 13 - 0.4T$ $R = R_a - R_c$ $F = 100R(\alpha)/W_s$ $\alpha = 1.0 + 0.2(2.65 - G_s)$ $D = K(L/t)^{1/2}$

Adjusted % Finer = F x % Passing Sieve

$K = [13 + 0.15(24°C - T) + 4(2.65 - G_s)]/1000$ $L = 16.3(1 - R_a/100)$

22. Graph the grain size analysis from problems 20 and 21 above.

GRAIN SIZE DISTRIBUTION

Project: ______________________ **Project No.:** __________

Location: ______________________

Soil Description: ______________________

Tested By: ______________________ **Date of Test:** __________

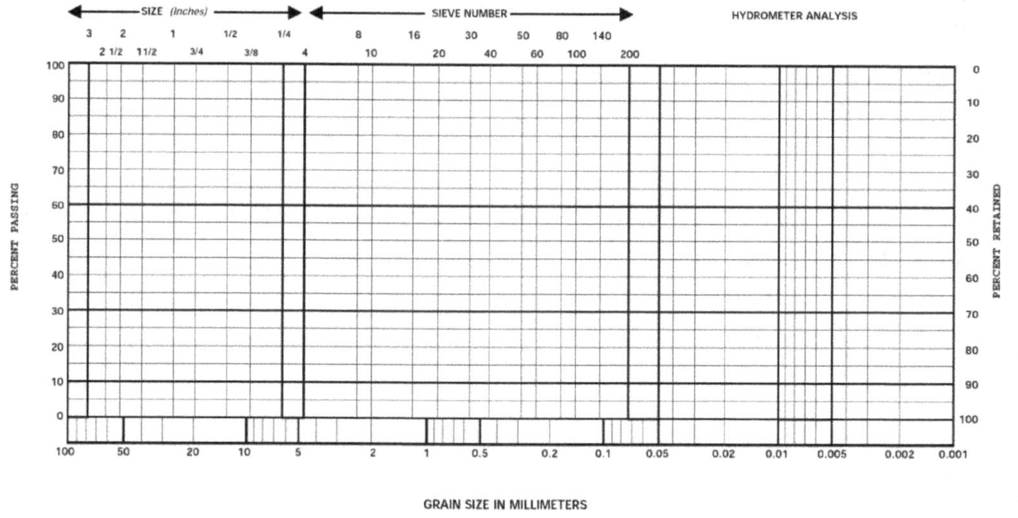

23. Identify the type of distribution curve generated in question 22 above.
    a. Well graded
    b. Poorly graded
    c. Gap graded

24. Calculate the uniformity coefficient and the coefficient of gradation from question 22.

25. When running an Atterberg limit test, what sieve size should the soils be passed through?

26. A liquid limit test results should be plotted on a curve with the moisture content versus N. What does N represent?

27. What N value represents the Liquid Limit for the soils tested?

28. What results need to be identified to calculate the Plasticity Index? List all that are applicable.
   a. % Finess
   b. Maximum dry density
   c. Plastic limit
   d. Grain size diameter
   e. Liquid limit
   f. All of the above.

29. The following are the results from the liquid limit and plastic limit tests for soil:

Number of blows (N)	Moisture content (%)
13	39.2
17	36.0
20	33.1
29	27.5

   Plastic Limit = 12.1%

   Draw the flow curve and obtain the liquid limit.
   What is the Plasticity Index of the soil?

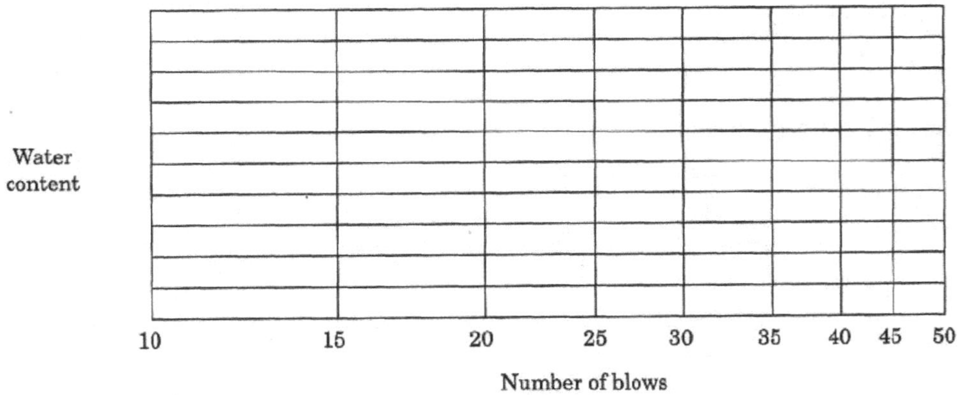

30. The following are the results from the liquid limit and plastic limit tests for a soil:

Number of blows (N)	Moisture content (%)
15	21.1
20	17.8
26	15.9

Plastic Limit = 16.1%

Draw the flow curve and obtain the liquid limit.

What is the Plasticity Index of the soil?

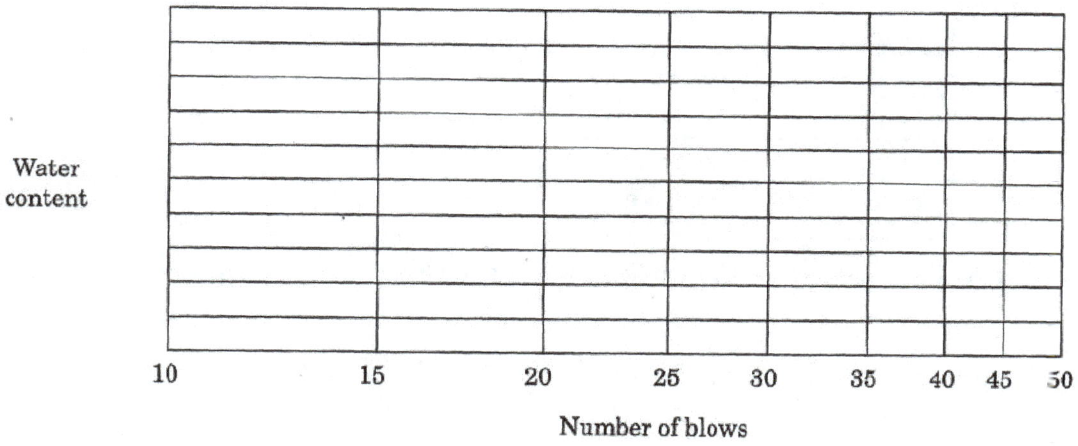

31. What are the two main groups of soil classification in the USCS?

32. What size sieve is used to determine the two groups of soil classification discussed above?

33. What size sieve is used to separate gravel from sand?

34. Using Table 2.6, identify the group symbol with more than 12 percent passing the No. 200 sieve and the Atterberg's limits plot above the "A" line and plasticity index greater than 7.

35. When the Liquid limit is greater than 50%, fine-grained soil is considered to be what subgroup?

36. When the Liquid limit is 50% a fine-grained soil is considered as what subgroup?

37. A soil sample is tested in the laboratory with the following results:

LL = 29%
PL = 17%

Sieve analysis data:

Sieve Size	**% Passing**
No. 10	92.2
No. 40	81.0
No. 200	51.8

Classify the soil using the Unified Soil Classification System. Give the group symbol and name.

38. A soil sample is tested in the laboratory with the following results:

LL = Not able to perform
PL = NP

Sieve analysis data:

Sieve Size	**% Passing**
No. 4	95
No. 10	62
No. 20	21
No. 40	11
No. 60	7
No. 100	5
No. 200	3

Classify the soil using the Unified Soil Classification System. Give the group symbol and name.

39. A soil sample is tested in the laboratory with the following results:

LL = 55%
PL = 28%

Sieve analysis data:

Sieve Size	% Passing
No. 10	100.0
No. 40	89.6
No. 200	77.7

Classify the soil using the Unified Soil Classification System. Give the group symbol and name.

40. In the AASHTO classification system, what percentage of passing and what sieve size identifies granular material?

41. Using the AASHTO classification system, identify the soil in problem 37 above.

42. Using the AASHTO classification system, identify the soil in problem 38 above.

43. Using the AASHTO classification system, identify the soil in problem 39 above.

44. Appendix H provides a sample report for a Web Soil Survey.

    a. What subdivision was this report prepared for?

    b. What was the total acres of the area of interest?

c. How many soil types were identified in the area of interest?

d. Fill out the following chart from that Web Soil Survey report:

	Soil Type #1	Soil Type #2	Soil Type #3
Map Unit			
Soil Type Name			
Slope			
Area			
Corrosion of Con-crete			
Corrosion of Steel			
Basements			
Limiting Factor			
Without Basements			
Limiting Factor			
pH (1 to 1 water)			
Hydrologic Soil Group			

45. What basic factors affect the compaction of the soil?
    a. Water Content
    b. Material Gradation
    c. Compaction Effort
    d. Soil Type
    e. All of the above

46. What is the dry density of the soil given a wet density of 135.5 pcf and a water content of 12.2%?

47. What two methods are used for Proctor Tests?

48. Method A Proctor Test uses what size diameter mold?

49. List the three soil fraction passing sieves for Method C, B, and A, respectively.

50. List three variables that differ between the Standard and Modified Proctor Tests.

51. Identify the following for a Method C, Modified Proctor Test:

   a. Volume of mold:

   b. Diameter of mold:

   c. Weight of hammer:

   d. Height of hammer drop:

   e. Number of layers of soil:

   f. Number of blows per layer:

   g. Soil fraction passing sieve:

52. In a proctor test curve, what does optimum moisture content identify?

53. A Standard Proctor will give a higher maximum dry unit weight with higher optimum moisture content. True or False

54. The results of a Standard Proctor test are given below. Determine this soil's maximum dry unit weight and optimum moisture content.

Volume of proctor mold (cf)	Weight of wet soil in mold (lb)	Moisture Content (%)
1/30	3.68	12.0
1/30	3.81	14.2
1/30	3.87	15.9
1/30	3.82	20.2
1/30	3.77	21.0

55. List at least five grading processes.

56. Complete the following Proctor Test giving the maximum dry unit weight and optimum moisture content.

PROCTOR TEST

Data Sheet 6

Project:	Black Canyon Road	**Project No.:**	AEC 2530
Location:	Sta 20 + 75, Centerline **Bore No.**	**Sample No.:**	22
Soil Description:	Sandy Silty Clay	**Sample Depth:**	0.0'
Tested By:		**Date of Test:**	
Test Method:	Standard / Modified	**Procedure:**	A / B / C
Mold Diameter:	4" / 6"	**Mold Volume: (V_m)**	1/30 CF / 1/13 CF

Water Content

Circle choice of boxes used above

Sample No.	1	2	3	4	5
Container No. (cup)	B2	A2	C5	C2	
Wt. of cup + wet soil (W_1)	164.12	180.97	180.69	178.02	g
Wt. of cup + dry soil (W_2)	144.36	156.78	154.02	149.70	g
Wt. of cup (W_c)	24.60	28.80	27.00	25.50	g
Wt. of dry soil (W_s) $W_s = W_2 - W_c$					
Wt. of water (W_w) $W_w = W_1 - W_2$					
Water content, ω% $\omega\% = 100W_w/W_s$					

Density

Wt. of soil + mold (W_s+W_m)	5.39	5.51	5.62	5.61	lbs
Wt. of mold (W_m)	1.25	1.25	1.25	1.25	lbs
Wt. of soil in mold (W_s)					lbs
Wet Density (PCF) $\gamma_{wet} = W_s/V_m$					
Dry Density (PCF) $\gamma_d = \gamma_{wet}/(1+\omega)$					

Dry Density γ_d PCF

Water Content ω%

Optimum Moisture (%) =	Maximum Dry Density (PCF) =

57. Match the following:

___Kneading or mixing compaction	a. loaded front end loader
___Impact or sharp blow	b. sheepsfoot wheel
___Vibration or shaking	c. jumping jack compactor
___Static weight or pressure	d. HED compactor

58. From the Typical City Testing Frequency, Table 2.14, complete the following questions?

    a. How often is a moisture/density curve required for an aggregate base course?

    b. How often are compaction tests required for stormwater installation?

59. From the soil specifications in Appendix D, complete the following questions?

    a. In the Earth Moving Section under compaction, density requirements:

        i. Which proctor method is required?

        ii. What is the minimum density for Driveways?

        iii. Where would you find the test intervals for this project?

    b. In the Trenching and Backfilling Section:

        i. What is the compaction and moisture requirements for the groundwater barriers?

60. What is the most common method to conduct field density tests?

61. List four types of means to achieve soil stabilization.

62. List four types of geosynthetics.

63. Answer the following questions From the Subsurface Soil Investigation in Appendix G.

    a. What are the recommendations for the exposed soil discussed in the Site Preparation Section?

    b. What are the recommendations for the fill soils discussed in the Site preparation Section?

    c. What type of foundation system was recommended?

    d. What was the recommended maximum net allowable bearing capacity for the above foundation system?

    e. How many bores were drilled?

f. Answer the following questions for Bore #4

   i. What was the total depth of the bore?

   ii. Was groundwater encountered? If yes, at what depth?

   iii. What depth was bedrock first logged?

   iv. What type of bedrock was first logged?

g. Answer the following questions for Bore #11

   i. What was the total depth of the bore?

   ii. Was groundwater encountered? If yes, at what depth?

   iii. What depth was bedrock first logged?

   iv. What type of bedrock was first logged?

h. Answer the following questions for Bore #18

   i. What was the total depth of the bore?

   ii. Was groundwater encountered? If yes, at what depth?

iii. What depth was bedrock first logged?

iv. What type of bedrock was first logged?

i. Answer the following questions by reviewing the Swell/Consolidation curves.

   i. Was swelling soil encountered?

   ii. Which curves indicated swelling soil?

   iii. What was the maximum swell pressure?

64. One sample point is generally accepted as a representative sample when sampling from a stockpile. True or False.

65. When quartering a sample, half of the sample is not used every time the soil is quartered. True or False.

3

3.1 Concrete

On the surface, concrete appears to be a simple material. However, it is very complex because it is manmade, thereby having a plethora of variables and purposes. Concrete's versatility, durability, and economy have made it the world's most used construction material. Concrete lends strength and support to structures, weight, and mass to dams, and durability and low permeability to ditches and drainage structures.

Concrete is found in nature in the form of cemented rock. This occurs when limestone is broken down into lime, water such as rain is added, and the mixture disperses into a sand or rock area. Man has taken this natural phenomenon and has modified the process to provide more consistency in the final product. Man has also learned that by adjusting the ingredients or adding additives, the final product can vary in strength, workability, initial set time, durability, permeability, and cost to produce, to name a few.

Concrete starts out in a plastic state when the concrete is freshly mixed. This allows the concrete components to be mixed, transported, placed, consolidated and finished prior to the initial set. The initial set generally occurs in approximately three hours unless an additive is mixed to increase or decrease the time. The initial set is the start of the phase change from a plastic material to a hardened concrete and can vary depending on environmental effects (air temperature, wind, and humidity) or additives designed to impact the set time and final strength.

3.1.1 Freshly Mixed Concrete

Freshly mixed concrete should be plastic or semifluid. All the sand and coarse aggregate grains are surrounded and held by suspension in the binding paste. Segregation of the materials is not desirable during this phase. Care is taken, especially during the placing of the freshly mixed concrete, to prevent segregation of the materials. A slump test is a means to measure the consistency of mixed concrete. The freshly mixed concrete will flow easily with a higher slump, flow sluggishly with a medium slump, and have a stiff consistency with a low slump concrete. Fleshly mixed concrete will generally be placed within a form system to control the flow of the concrete.

3.1.1.1 Mixing

Mixing the batched concrete requires effort and care to ensure the combined ingredients become homogeneous. The sequence of materials added and the timing of the water play important roles. The number and speed of revolutions are also critical to the final mixed concrete. Proper mixers will provide the material with an end-to-end material exchange in a rolling, folding, and kneading action.

3.1.1.2 Workability

Workability is the effort required to place, consolidate, and finish freshly mixed concrete. Concrete should be handled without segregation or bleeding (water present at the top surface). Excessive bleeding may weaken the top layer of the concrete. Entrained air can improve workability and the tendency of concrete to bleed.

3.1.1.3 Consolidation

Once placed in the forms, freshly mixed concrete should be consolidated. This will allow the aggregate and sand to fill in voids that may be formed by air pockets. Consolidation may be obtained by hitting the outside of the forms with hammers, rodding the wet concrete, or using a concrete vibrator. Care should be taken not to over-consolidate the concrete, as this can cause segregation of the materials.

3.1.1.4 Hydration and Setting Time

Hydration is A chemical reaction between the cement and the water. The hydration of the cement paste binds the sand and aggregates to make hardened concrete. The initial set is when enough hydration has occurred that the concrete behavior is altered by the phase change. Rigorously working the concrete, such as placing or consolidating once the initial set starts, will generally weaken the final strength of the concrete. The hydration process creates heat, which may be beneficial during cold weather pours but may be detrimental during massive pours, such as with dams, due to the undesired internal stresses created.

3.1.2 Hardened Concrete

Hardened Concrete is considered the final product of concrete and is the purpose of the construction of concrete. When we refer to concrete, we generally are discussing hardened concrete. We have already discussed several attributes of concrete, such as durability, low permeability, and weight. We have not discussed the aesthetics of this material since it can be placed in the form of any shape or size and allowed to harden. However, the greatest attribute of concrete is its strength characteristics. We will get into more detail on this subject later in this chapter. The characteristics of concrete are discussed below.

3.1.2.1 Moist Curing

Curing of the concrete continues if un-hydrated cement is present. The hydration of the cement requires the relative humidity above 80% and the temperature to be above freezing. Hydration and strength gain virtually stop if either environment is not present. Therefore, moisture curing the concrete after installation is advised to provide the quickest and most effective strength gain. Moisture curing may be provided for just a few days after placement to accelerate the curing or may be provided until design strength is met.

3.1.2.2 Drying of Concrete

Concrete does not harden or cure by drying. Concrete hardens due to the hydration of the cement in the concrete mixture. The drying of the concrete is a surface occurrence normally having little effect on the internal curing of the concrete. Sufficient water within the mix should hydrate with the cement if the concrete were designed correctly. However, this does take time. Concrete with a large surface area exposed directly to the sun may be prone to a drying effect on the surface. Concrete dusting or surface cracking is common evidence this has occurred. The surface may be treated with an additive or covered with material, such as burlap, and kept wet.

3.1.2.3 Strength

Engineers may design for several types of strength, such as flexural, tensile, torsional, and shear. However, compressive strength is the most common; the other strengths can be defined as a general ratio of compressive strength. Compressive strength is the maximum resistance of a load applied to crush the concrete. The design compressive strength is expressed in pounds per square inch (psi) at 28 days after being made. Water cement ratio and age are the principal factors affecting strength.

3.1.2.4 Unit Weight

The unit weight of concrete varies based on the relative density of the aggregate, the amount of air entrained, the amount of water, and the amount of cement. See table 3.1. Normal concrete has a unit weight of 140 to 150 pounds per cubic foot (pcf). Concrete containing reinforcing bars is generally assumed to be 150 pcf.

TABLE 3.1 Observed Average Unit Weight of Fresh Concrete

Maximum size of aggregate (inches)	Air content (%)	Water (pounds per cubic yard)	Cement (pounds per cubic yard)	Unit weight (pounds per cubic foot)*				
				Specific gravity of aggregate**				
				2.55	2.60	2.65	2.70	2.75
3/4	6	283	566	137	139	141	143	145
1-1/2	4.5	245	490	141	143	146	148	150
3	3.5	204	408	144	147	149	152	154
6	3.0	164	282	147	149	152	154	157

*Air-entrained concrete with indicated air content.

**On saturated surface dry basis.

Adapted from reference 9.

3.1.2.5 Resistance to Freezing and Thawing

Concrete is known for good durability and ability to resist exposure conditions. The most destructive weathering factor is freezing and thawing while the concrete is wet (not fully cured). Deterioration is caused by freezing of the water in the paste, the aggregate particles, or both. Two ways to reduce the effects of freezing and thawing is to reduce the water-cement ratio, therefore having less water in the mix to freeze and/or having more air in the concrete, giving an area to expand without increasing internal stress.

3.1.2.6 Permeability and Watertightness

Common concrete is considered impermeable but not watertight. Concrete reduces the flow of water or other substances through it even when pressure is applied, assuming the concrete is not cracked or contains rock pockets. The lower the water-cement ratio, the lower the permeability of the concrete. Concrete foundations with livable spaces below ground must have a damp-proofing material applied to the soil side of the concrete in accordance with the current International Residential Code (IRC).

3.1.2.7 Abrasion Resistance

Concrete is used in numerous areas subject to abrasion, such as pavements, floors, and hydraulic structures. The water-to-cement ratio impacts the resistance to abrasion since strong concrete tends to resist abrasion. Larger aggregates and steel-troweled surfaces are also more resistant.

3.1.2.8 Volume Stability

Concrete volume changes slightly due to environmental conditions (temperature, moisture, and humidity) and stress. Concrete that is allowed to dry will shrink. The amount of shrinkage depends on aggregate, concrete mass shape, curing method, degree of hydration, and time. Cement content has little to no effect on shrinkage on concrete when the cement content is between 5 and 8 bags per cubic yard.

3.1.2.9 Control of Cracking

Two basic causes of cracks in concrete are (1) stress due to applied loads and (2) stress due to drying shrinkage or temperature changes in restrained conditions. Drying shrinkage is an unavoidable property of concrete. We have all heard the saying, "Concrete is guaranteed to get hard and to crack."

Reinforcing steel or joints are placed to control the location and width of the crack. Figure 3.1 shows three basic joints used in concrete with slab-on-grade construction. Control joints are grooved, formed, or sawed into the surface of the concrete approximately one-quarter of the thickness. Isolation joints separate two types of concrete (slab and footing) with a pre-molded joint filler. Construction joints occur when a concrete pour stops one day and continues another day.

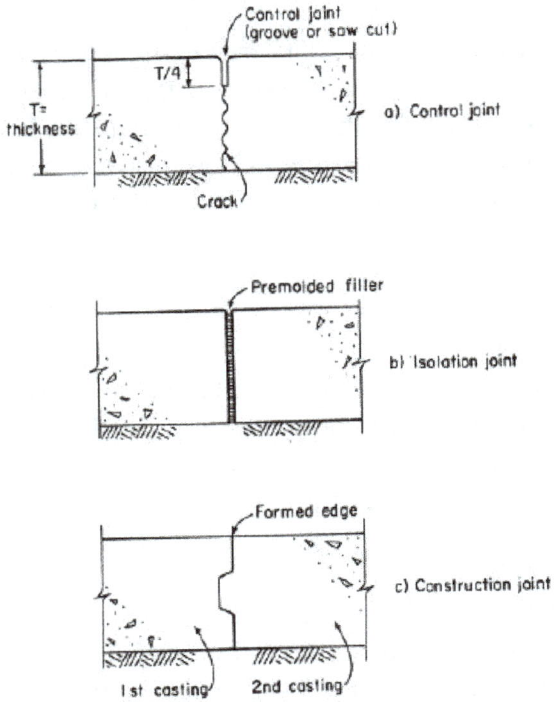

FIGURE 3.1 The three basic joints used in concrete slab-on-grade construction. Adapted from reference 9.

3.2 Ingredients

The basic ingredients of concrete are an inert material of various sizes and a paste that binds the material together. Concrete is primarily made of coarse and fine aggregate (the inert material), water and cement (the paste), and air. If proportioned properly, a good mix design will have enough cement and water to form a paste that will encapsulate all the coarse and fine aggregate. Some people misname concrete as "cement." As you can see, cement is just one ingredient required to make concrete.

3.2.1 Cement

Cements are divided into two classifications: hydraulic and non-hydraulic. Hydraulic cement can harden under water, while non-hydraulic cement requires air to harden and, therefore, cannot harden under water. This book will focus on hydraulic cement since it is the most common cement used in the construction industry.

Portland cement is the most commonly used cement. It must contain appropriate proportions of calcium, oxide, silica, alumina, and iron.[9]

Different types of Portland cement are manufactured to meet various normal physical and chemical requirements for specific purposes. The ASTM designation C 150 "Specifications for Portland Cement" provides eight Portland cement types. See Table 3.2.

TABLE 3.2 Types of Portland Cement

Type	Name	Use
I	Normal	General purpose cement.
IA	Normal, air entraining	General purpose cement with additional air.
II	Moderate	Precaution against moderate sulfate attack. Gains strength slower than Type I.
IIA	Moderate, air entraining	Moderate sulfate attack with additional air.
III	High early strength	Provides strength at earlier period.
IIIA	High early strength, air entraining	Provides strength at earlier period with additional air.
IV	Low heat of hydration	Minimizes the heat of hydration.
V	High sulfate resistance	Concrete exposed to severe sulfate content.

Masonry cement is designed for masonry construction. It meets the requirements of ASTM C 91, which classifies it as Type N, S, or M. Masonry cement is used as mortar, grout, and plaster. It is normally mixed with fine aggregate and does not contain coarse aggregate. This cement should not be used for making concrete.

Other types of cement are also manufactured, such as expansive cement, oil well cement, waterproof Portland cement, plastic cement, and regulated set cement, to name a few. This book will not discuss these cements since they are not commonly used in construction.

Pozzolans are a powder that will harden when combined with lime and water at ordinary temperatures. Pozzolans are not cement but are a replacement for a percentage of the cement due to the lower cost to produce than Portland cement. Pozzolans are normally substituted on a pound-for-pound or volume-for-volume basis with Portland cement. The most common synthetic pozzolans are fly ash. Fly ash is a waste product of coal fire power plants. Fly ash quality is regulated by ASTM C 618 "Fly Ash and Raw or Calcined Natural Pozzolans for Use in Portland Cement Concrete."

3.2.2 Water

Almost any drinking water can be used in the production of concrete. Drinking water with a noticeable taste, odor, or color should be tested for organic impurities. Some impurities affecting the concrete are dissolved chemicals, seawater, sugar, or algae. The effects on the concrete may be as follows:

1. Abnormal setting time
2. Decreased strength
3. Volume changes
4. Efflorescence
5. Discoloration
6. Corrosion of reinforcement.

3.2.3 Aggregates

The importance of using the right type and quality of aggregates cannot be overemphasized since fine and coarse aggregates generally occupy 60% to 75% of the concrete volume (70% to 85% by weight) and strongly influence the concrete's properties, mixture proportions, and economy. The most effective coarse aggregate is angular in shape; therefore, most coarse aggregate is crushed from larger rocks. Thin elongated aggregate can cause problems with pumping the concrete and should be reduced as much as possible, generally less than 15% of the weight of the total aggregate. More recently, recycled concrete has been used as a partial replacement for coarse aggregate. Normal weight aggregates should meet the requirements of ASTM C 33 "Specification for Concrete Aggregates." The grading specification in ASTM C 33 provides a range for each sieve size. This range has been proven to provide an optimum grading of coarse and fine aggregates in forming a dense concrete mixture. The aggregate should not have inorganic impurities, silt, clay, shale, iron oxide, coal, lignite, and soft particles. These substances may cause popouts by swelling due to absorbing water or freezing of the water present. Gypsum and anhydrite should be avoided because they may cause a sulfate attack on the concrete.

3.2.4 Chemical Admixtures

An admixture is a material other than water, aggregates, cement, and fiber reinforcement used as an ingredient to modify its freshly mixed, setting, or hardened properties. The main reasons for using admixtures are to reduce the amount of cement (reduce cost), alter concrete properties more effectively than other means, alter concrete properties in adverse weather conditions, and overcome certain emergencies during concreting operations.

During mixing, an **air entrainment admixture** is added to develop microscopic bubbles in the cement paste. Air entrainment reduces the potential for freezing and thawing effects and improves workability.

Water-reducing admixtures are used to increase the concrete's strength and workability and reduce cost. This admixture reduces the amount of water necessary to obtain the same slump, reducing the amount of cement while maintaining the same water-cement ratio.

Set-controlling admixtures are used to modify the setting time of the concrete. Retarding admixtures extends the set time, which is helpful during hot weather conditions. An accelerating admixture reduces the set time and accelerates early strength development, and this is beneficial for cold weather conditions.

The admixtures discussed above are the most commonly used to alter the properties of concrete. Other admixtures may assist the concrete with workability, bonding, damp proofing, permeability reduction, grouting, gas forming, color, corrosion inhibition, and pumping.

3.2.5 Fibers

Short fibers made from steel, plastic, glass, and synthetic polymers enhance cracking-related properties due to temperature expansion and contraction. Steel, plastic, and synthetic fibers significantly improve concrete flexural strength, impact strength, durability, and abrasion. Fibers are not required for common concrete and must be requested prior to batching.

3.3 Strength

The strength and cost of the product are the main purposes for using concrete over numerous other products in the construction industry. The strength of concrete is taken as an important index of its general quality. The most common test of hardened concrete is determining its strength. Although the test results of strength require time (design strength at 28 days), the results provide essential information to disclose any trends in concrete quality and allow adjustments to be made to future concrete. The results of strength tests can be used as a qualitative indicator of other properties of hardened concrete. Generally, strong concrete will be more impermeable, better able to withstand severe exposure, and more wear resistant. However, strong concrete may have greater shrinkage and be more susceptible to cracking. Concrete is the strongest when a compressive load is applied compared to flexural, tensile, torsional, or shear loads.

3.3.1 Compressive Strength

The compressive strength of concrete is a primary physical property and is typically used in design calculations for bridges, buildings, and other structures. Most general-use concrete has a compressive strength of 3000 to 5000 psi.[3]

1. Compressive strength testing is common, and specimens are generally obtained from the following means:
2. **Laboratory testing**. Concrete cylinders are made in the laboratory for the purpose of researching different effects of variations in materials or storage, control tests in reference to concurrent field tests, or evaluating variations in mix designs. ASTM C 192, Method of Making and Curing Concrete Compression and Flexure Test Specimens in the Laboratory, is the ASTM designation for this test.
3. **Field testing**. Cylinders are made to represent the concrete delivered to the construction site. The compressive strength results may help the contractor determine when forms may be removed or when a structure may be put in service. These results also verify and control the quality of the concrete being delivered to the construction site. The tests are conducted under ASTM C 31, Method of Making and Curing Concrete Compression and Flexure Test Specimens in the Field.
4. **Field coring testing**. Core samples of field-placed concrete structures are obtained and tested in accordance with ASTM C 42, Methods of Obtaining and Testing Drilled Cores and Sawed Beams of Concrete. This test is not commonly performed unless the field sample cylinders do not meet design strength or when investigations are made on old structures that may not have compressive strength testing on file.

FIGURE 3.2 Concrete cylinder after compressive strength testing. Note the hardened paste around the aggregate and how some of the aggregates broke during the testing.

3.3.2 Flexural Strength

Flexural tension is most commonly developed in beams and slabs because of loads, temperature changes, shrinkage, and moisture changes. Figure 3.3 depicts how flexural tension is developed in structural members when a load is applied. The top portion of concrete exhibits compression when sufficient load is applied to make the concrete member start to bend. While the top portion is in a compression loading, the lower part of the member is in tension. Designers will place reinforcing steel in the concrete to allow the tension to be transferred to the steel, which is capable of handling greater tension than the concrete. The amount of reinforcement steel and the location of the steel in the concrete are both important to ensure the concrete member will react as designed.

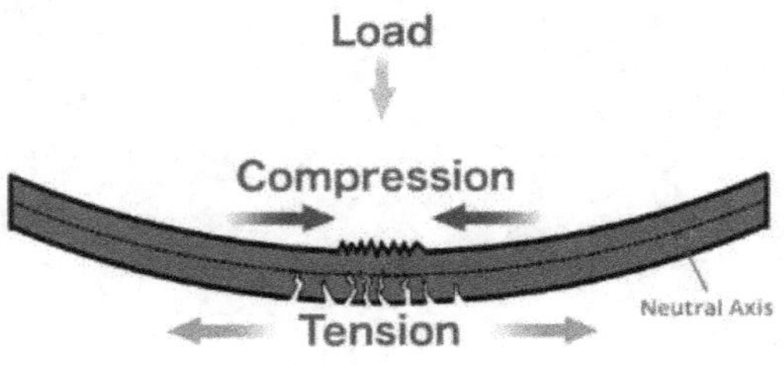

FIGURE 3.3 Flexural tension developed on the lower side of the beam.

Unreinforced concrete is normally tested by ASTM C 78, Test for Flexural Strength of Concrete Using Simple Beam with Third-Point Loading. Flexural strength testing is common for airport runways, concrete paved roads, and larger slabs. Compressive strength can be used as an index of flexural strength. The flexural strength or modulus of rupture of normal-weight concrete is about 15 to 20 percent of the compressive strength and is often calculated by 7.5 to 10 times the square root of the compressive strength.[9]

3.3.3 Tensile Strength

Axial tension tests on concrete are not easily performed. It is difficult to apply a true axial load by gripping the ends of the specimen without creating high local stresses. Therefore, the flexural strength tests discussed above are generally used. The tensile strength of common concrete is roughly 8 to 12 percent of the compressive strength and is normally estimated as 5 to 7.5 times the square root of the compressive strength.

3.3.4 Shearing Strength

The shearing strength of concrete is a very important property of the material, as it is the primary determining factor in the compressive strength of short columns.[3] The shear strength of normal concrete is around 20% of the compressive strength.

3.3.5 Torsional Strength

The application to a concrete specimen of torsion alone produces pure shearing stresses on certain planes. Pure shear acting on a plane does not occur in actual structures. Failure under these conditions occurs in tension rather than shear. Therefore, the strength of concrete subjected to torsion is related to its tensile strength.[3]

3.4 Proportioning Concrete Mixtures

The objective in designing concrete mixtures is to determine the most economical and practical combination of readily available materials to produce concrete that will satisfy the performance requirements under particular conditions. To fulfill this objective, a properly proportioned concrete mix should possess these qualities:

1. Acceptable workability of freshly mixed concrete
2. Durability, strength, and uniform appearance of hardened concrete
3. Economy

Understanding the basic principles of mixture design is as important as the actual calculations. The above qualities can be obtained only in concrete production with the proper selection of materials, mixture characteristics, and proper proportioning.[9]

Conducting numerous trail batch samples before submitting a final concrete mixture for approval is highly recommended. This will allow the opportunity to test the concrete mixture, identify any shortcomings, and make necessary adjustments as needed prior to actual construction. Depending on the construction project, sufficient time to adjust mixes that may not perform as required is not available. For instance, the compressive strength is not fully defined per ACI (American Concrete Institute) for 28 days. Numerous concrete pours may have occurred with a faulty mix until the first concrete cylinders identify the mix does not meet the design strength required. An Engineer now needs to evaluate if the placed concrete will work for the proposed structure or if the concrete needs to be removed and replaced.

This book is not intended to provide all the information needed to design a concrete mix. However, it should provide sufficient information to allow an educated review of a mix and an intelligent discussion of the mix design with the designer. *Design and Control of Concrete Mixtures*, generated by the Portland Cement Association, is an excellent manual if additional information is sought.

3.4.1 Water-cement ratio

As we have previously seen, the water-cement ratio has the greatest impact on the strength, durability, permeability, wear resistance, and workability of all the ingredients in the concrete mixture. Compressive strength is inversely related to the water-cement ratio in the normal range of strength concrete. This means the lower the water-cement ratio, within reason, produces the highest strength concrete. There are a few key criteria for this thought. First, there must be sufficient paste to surround all the aggregate. Second there must be sufficient water to fully hydrate the cement/binder in the mix. Third, the curing environment must allow full hydration to occur. And lastly time is required to allow the concrete an opportunity to harden sufficiently to accept the load applied. Experience has demonstrated considerable information about water-cement ratios. Tables 3.3, 3.4, and 3.5 summarize some of the available data.

TABLE 3.3 Maximum Water-Cement Ratios for Exposed Conditions

Exposure condition	Maximum water-cement ratio by weight for normal-weight concrete
Concrete protected from exposure to freezing and thawing or application of deicer chemicals	Select water-cement ratio on the basis of strength, workability, and finishing needs
Concrete intended to be watertight:	
a. Concrete exposed to fresh water	0.50
b. Concrete exposed to brackish water or seawater	0.45
Concrete exposed to freezing and thawing in a moist condition.	
a. Curbs, gutter, guardrails, or thin sections	0.45
b. Other elements	0.50
c. In the presence of deicing chemicals	0.45
For corrosion protection for reinforced concrete exposed to deicing salts, brackish water, seawater, or spray from these sources	0.40*

*If the minimum concrete cover required by ACE 318 Section 7.7 is increased by 1/2", the water-cement ratio may be increased to 0.45.

Adapted from reference 9.

TABLE 3.4 Typical Relationship Between Water-Cement Ratio and Compressive Strength of Concrete

Compressive Strength at 28 days (psi)	Water-cement ratio by weight	
	Non air-entrained concrete	Air-entrained concrete
6000	0.41	-
5000	0.48	0.40
4000	0.57	0.48
3000	0.68	0.59
2000	0.82	0.74

Adapted from reference 9.

TABLE 3.5 Maximum Permissible Water-Cement Ratios for Concrete When Strength Data from Field Experience or Trial Mixtures Are Not Available

Compressive Strength at 28 days (psi)	Water-cement ratio by weight	
	Non air-entrained concrete	Air-entrained concrete
2500	0.67	0.54
3000	0.58	0.46
3500	0.51	0.4
4000	0.44	0.35
4500	0.38	Field Data Reg.

Adapted from reference 9.

3.4.2 Cement

Cement is generally described or specified per sack. A sack contains 94 pounds and is assumed to be 1 cubic foot in volume. Project specifications commonly require a minimum cement content and a maximum water-cement ratio. Minimum cement requirements serve to ensure satisfactory durability and finish ability, improve the wear resistance of slabs, and provide a suitable appearance on vertical surfaces. The minimum cement requirement may exceed the amount of cement required to meet the strength specification. For place ability, finish ability, abrasion resistance, and durability in flatwork, the quantity of cement should not be less than shown in Table 3.6.

Propositioning should minimize the amount of cement for economy, since cement is normally the most expensive ingredient in the concrete mixture. However, the quality of the concrete mixture should not be compromised in the process of reducing the amount of cement.

TABLE 3.6 Minimum Cement Requirements for Normal Weight Concrete Used in Flatwork

Maximum size of aggregate (In.)	Cement (lb per cubic yard)
1-1/2	470
1	520
3/4	540
1/2	590
3/8	610

Adapted from reference 9.

3.4.3 *Water*

The actual weight of water is important to reduce the water-cement ratio. We just learned that the amount of cement is generally reduced, if possible, to assist in the cost of the concrete mixture. Therefore, the amount of water must also be reduced to ensure the maximum water-cement ratio is not exceeded. The quality of the concrete mixture characteristics remains an essential concern with the reduction of water. The amount of water is influenced by nearly every ingredient in the mixture. Coarse aggregate may affect the amount of water by the quantity, size, and shape of the aggregate. Coarse aggregate does not tend to absorb water; therefore, the more coarse aggregate and the larger the aggregate in the mix, the less water is required. The shape of the coarse aggregate affects the amount of water since angular surface aggregate has more surface area for water to adhere to. This requires more water than the rounded aggregate. Fine aggregate can absorb water, then increasing the amount of water with more fine aggregate in the mix. Weather conditions play a large role in water content with fine aggregate. A rainy or humid day will increase the moisture content of the fine aggregate, while a hot, sunny day will decrease the moisture content. The batch plant will check the moisture content of the fine aggregate stockpile every day or at least when the weather conditions have changed. Admixtures are designed to alter the amount of water needed in the concrete mixture and have been discussed in the *ingredients* section of this chapter.

To better understand how the amount of water can alter a mix, adding 10 pounds of water per cubic yard of concrete can increase the slump by 1 inch. This will be further discussed in construction procedures.

TABLE 3.7 Recommended Slumps for Various Types of Construction

Concrete construction	Slump (In.)	
	Maximum	Minimun
Reinforced foundation walls and footings	3	1
Plain footings, caissons, and substructure walls	3	1
Beams and reinforced walls	4	1
Building columns	4	1
Pavements and slabs	3	1
Mass concrete	2	1

Adapted from reference 9.

3.4.4 Aggregate Blend

The aggregate blend influences the proportioning of concrete mixtures because it affects the fresh concrete's workability and the mix's economics. Coarse aggregates should be sized as large as practical for the job conditions. The largest size aggregate depends on the size and shape of the concrete member to be cast and the amount, size, and distribution of reinforcing steel in the member. The largest size coarse aggregate should not exceed one-fifth the narrowest dimension between sides of form nor three-fourths the clear space between individual reinforcing bars or wire, bundles of bars, or prestressing tendons or ducts. The maximum size should not exceed one-third of the slab thickness for unreinforced slabs on the ground. It is also a good practice to limit aggregate size to not more than three-fourths of the clear space between the reinforcement and the forms.[9]

3.5 Construction Procedures

This section is intended to give a general description of different construction practices. It is not intended to dictate or direct how a contractor practices their trade in the field but to give the student an idea of some construction practices. It is also important for the student to understand that he or she should never direct the contractor on how to do their work beyond what is stated in the plans and specifications. The author of this book may add observations from forty years of field experience.

3.5.1 Batching

Batching is the process of weighing or volumetrically measuring and introducing the ingredients into the mixer to produce a batch of concrete. The ingredients must be accurately measured to generate a consistent and uniform product. Most specifications require that batching be done by weight since this method is more accurate.

3.5.2 Mixing

Mixing the ingredients occurs as the ingredients are batched. All concrete should be thoroughly mixed until the aggregate particles are coated with cement paste and a homogenous mixture is obtained. The mixture is uniform in appearance, with all ingredients evenly distributed. The ingredients are added to a mixing drum that contains mixing blades that carry the ingredients partway around the drum, are turned over, and drop to the bottom of the drum.

Fresh concrete that is left to agitate in the mixer drum tends to stiffen before the initial set develops. Such concrete may be used if, upon remixing, it becomes sufficiently plastic to be compacted in the forms. A small amount of water may be added to remix the concrete, providing the following conditions are met:

1. Maximum allowable water-cement ratio is not exceeded.
2. Maximum allowable slump is not exceeded.
3. Maximum allowable mixing and agitating time or drum revolutions are not exceeded.
4. Concrete is remixed for at least half the minimum required mixing time or number of revolutions.

3.5.3 Transporting

This section discusses transporting the fresh concrete from the batch plant to the construction site. The key to properly transporting the concrete is to prevent the segregation of the ingredients, assuring the product arrives in good quality and uniformity. The ready-mixed concrete truck (concrete truck) is the most common means to transport fresh concrete. An advantage of this truck is the ability to mix the concrete while being transported. These trucks can either slowly agitate or spin at a faster speed to mix the concrete. The trucks generally carry a water tank that can add water to the mix and a chute to discharge the concrete once on site.

FIGURE 3.4 Concrete Truck

FIGURE 3.5 Concrete Truck discharging into pumper hopper

3.5.4 Placing

The proper placement of concrete in the forms is an important factor in obtaining durable concrete structures. The forms should be accurately set, clean tight, adequately braced, free of debris and accumulated water prior to placement of the fresh concrete. Furthermore, any reinforcing steel should be clean, free of loose rust or mill scale, properly placed, and secured when concrete is placed. To obtain uniform concrete, it should be placed in the forms evenly and without segregation of materials. The concrete should be placed as near as possible to its final position. Excessive movement of the concrete provides additional opportunities for segregation. Consolidation of the concrete can be spaded, forms hammered, or vibrated. Care should be taken not to segregate the ingredients by over-vibrating.

Depositing the concrete into the forms from the concrete trucks is provided by numerous means, methods, and equipment. The fresh concrete can be deposited into the forms directly from the chutes on the concrete trucks. This is common for curb, gutter, and other concrete easily and safely driven next to. The concrete trucks may also be stopped and discharged into a wheelbarrow, power buggy, bucket and crane, conveyor belt, truck-mounted pump and boom, or screed machine. It depends on the type of concrete poured and the accessibility to the final placement of the concrete. As a preferred placement method, truck-mounted pump and boom systems are becoming commonplace on construction sites.

FIGURE 3.6 Several means of placing concrete. Top left – Mixer chute discharging into a drainage structure below grade. Top right – Mixer chute discharging into sidewalk forms at grade. Bottom left – Mixer discharging into a clamshell bucket controlled by a backhoe. Bottom Right – Mixer discharging into a pumper.

3.5.5 Finishing

The surface must be finished once the concrete has been deposited and consolidated. Several methods can be used to finish the concrete: steel trowel, wooden float, spading, brooming, burlap, or grinding tool.

Concrete slabs and sidewalks require edging along all edge forms, isolation, and construction joints. Edging densifies and compacts the concrete next to the form, making it more durable and less vulnerable to spalling and chipping. Proper jointing practices can eliminate unsightly random cracks. Jointing can be conducted on the fresh concrete during the edging or can be provided on the hardened concrete by sawing.

FIGURE 3.7 Finishing Concrete Left – Leveling fresh concrete, Center – gas powered steel trowel, Right – finished basement concrete slab.

3.5.6 Curing

As previously described, concrete sets and hardens due to a chemical reaction between the cement and the water. The process continues as long as water is present. We have also discussed that temperature is another important factor in curing. If the temperature is too high, undue cracking will occur. If the temperature is too low, the curing process will not continue. Curing aims to provide the best opportunity to meet the concrete's full potential strength. This can be achieved by preventing moisture loss from the concrete and maintaining a favorable concrete temperature for a definite period.

3.5.6.1 Hot Weather

The main concern with hot weather pours is the loss of water in the mix, either by the temperature of the concrete, high winds, or evaporation due to low humidity and direct sunlight. The water loss in the mix will decrease the slump and workability and increase the setting rate. Water can be added to the mix, keeping in mind that this will reduce the water-cement ratio, negatively affecting the final product. The increase in the concrete temperature impacts the increase of setting, thereby prompting early curing.

Hot weather curing may be reduced by one or more of the following means:

1. Lower the temperature of the concrete materials before mixing. Water is the easiest ingredient to cool. Substituting ice for an equivalent amount of water is sometimes used. Aggregate can be shaded from the direct sun and/or sprinkled with cool water, allowing the water temperature and the effects of evaporation to cool the aggregate.
2. Site preparation prior to placement of concrete. Concrete mixers should be kept as cool as possible. This could include shading the mixers while not in use, painting the equipment white to reflect the sunshine, and periodically wetting the equipment with cool water. On-site, the forms, reinforcing steel, and subgrade should be sprinkled with cool water just before the concrete is placed. This process cools the contact surface and surrounding air and increases its relative humidity.
3. the concrete should be transported and placed as quickly as practicable. Sufficient labor and equipment must be available at the jobsite to quickly place and finish the concrete to reduce concrete temperature increases. Prolonged mixing should be avoided.
4. Reduce plastic shrinkage cracking on the surface, especially on horizontal surfaces. The following simple precautions listed below can minimize the possibility of plastic shrinkage cracking:
   a. Reduce the temperature of the concrete during mixing as described above.
   b. Moisten the subgrade and forms prior to placing the concrete.
   c. Erect temporary windbreaks to reduce wind velocity over the concrete surface.
   d. Erect temporary sunshades to reduce direct sunlight.
   e. Protect the concrete with temporary coverings such as polyethylene sheeting or apply an evaporation retarder immediately after screeding.
   f. Reduce delays during construction.
   g. Protect the concrete immediately after final finishing to minimize evaporation. Use a fog spray to raise the relative humidity of the ambient air. Fogging should be continued until a suitable curing material, such as a curing compound, wet burlap, or curing paper, can be applied.

3.5.6.2 Cold Weather

Cold weather is defined by ACI 306 as "a period when for more than 3 successive days the mean daily temperature drops below 40° F." Normal concrete practices can be resumed once the ambient temperature is above 50° F for more than half a day. Preparations should be made to protect the concrete and the location where the concrete is placed. This may include enclosures to act as windbreaks as well as areas that can be heated by portable heaters. Insulation and blankets should be available to maintain the concrete temperature once the concrete is finished. Forms, reinforcing steel, and embedded fixtures must be clear of snow and ice at the time concrete is placed.

Freshly mixed concrete must be protected against the disruptive effects of freezing until the degree of saturation of the concrete has been sufficiently reduced by the process of hydration. The time at which this reduction is accomplished corresponds roughly to the time required to attain a compressive strength of 500 psi. At normal temperatures, this occurs within the first 24 hours after placement. A significant reduction of ultimate strength occurs if the concrete is frozen prior to this period.[9] Low temperatures retard the rate at which the hydration of cement occurs, therefore hindering the hardening and strength gain of concrete.

Several of the following precautions can be provided to assist in the curing of cold-weather concrete:

1. Cover the finished concrete with insulation or blankets. Not only does this help keep the cold from the forms, but it also assists in the capture of the heat of hydration as the chemical process occurs as the cement reacts with water to form a hard and stable paste.
2. Special cold weather concrete mixtures.
    a. Use type III cement, high early strength to allow the concrete to obtain strength more quickly than normal cement.
    b. Add additional Portland cement (100 to 200 pounds per cubic yard). Provides additional strength to the concrete mixture.
    c. Chemical accelerators such as calcium chloride added to the mixture in small dosages can accelerate the setting time.
3. Air-entrained concrete is desirable during cold weather conditions because it absorbs internal stresses caused by ice formation.
4. Increase concrete temperature as mixed. Frozen aggregate, especially fine aggregate, should not be added to the batch mix because the fine aggregate may be frozen together and not allow the paste to surround the aggregate properly. Hot water can be used to offset the cold temperatures from the aggregate in the mix. However, the cold aggregate and hot water should be added, mixed, and checked to make sure the combined temperature is appreciably less than 100° F before adding the cement. Cement added to a temperature greater than 100° F may cause quick or flash set of the concrete.
5. Reduce the time between mixing and placing the concrete. Concrete mixers have a large steel surface area that gets cold quickly. Maintaining the concrete temperature will be easier if the concrete is not in the mixer truck for any additional time.
6. Do not place fresh concrete on frozen ground.
7. The site of the concrete placement can be enclosed to provide a windbreak, and the air around the concrete forms can be heated with portable heaters. The most efficient heaters produce steam, ensuring the heated air does not dry out the concrete.
8. Allow the forms to remain on the concrete longer than normal weather conditions stripping. The forms serve to maintain and distribute heat more evenly.

3.6 Sampling and Testing

Satisfactory concrete construction and performance requires concrete possessing specific properties. Quality control and acceptance testing are indispensable parts of the construction process to ensure that these properties are obtained. Experience and sound judgment must be relied on in evaluating tests and assessing their significance in the ultimate performance of the concrete.[9]

Quality control is the testing of the material generally conducted by a representative of the product manufacturer. It is common for quality control testing to be conducted in the plant or location where the concrete is being batched and mixed prior to allowing the concrete to be sent to the job site. Quality assurance testing is normally performed by an individual not associated with the production of the product. Assurance testing is commonly performed by an independent third-party company on-site to verify that the delivered product meets the specifications for the project.

An important factor in concrete sampling and testing is the frequency of concrete being tested. The project specifications often provide the frequency required for quality assurance testing. Jurisdictions may also specify the frequency of testing. The more frequent testing schedule should be adhered to should a conflict arise between specifications. An example of city specifications is included in Table No. 2.13. The City Specifications are valid for all work performed in the city's right-of-way. Building Code Requirements for Reinforced Concrete (ACI 318) specifies the frequency of strength tests of each class of concrete place each day.

All the ingredients in the concrete mix require quality control testing. We have previously discussed the testing of grading in the aggregate. Much of the testing, such as for Portland cement, is conducted by the manufacturer of the cement and will provide the results upon request to the ready-mix concrete company. It is important that the industry personnel know this information is available upon request to the concrete producer. This book will focus on the quality assurance testing conducted in the field, as listed below. The procedural checklists listed in each test section are taken directly from the Concrete Field Testing Technician – Grade I ACI Certification Technician Workbook.

3.6.1 Sampling

A critical part of sampling fresh concrete in the field is obtaining a representative sample. ASTM C 172 gives the standard procedures for obtaining and handling samples. This procedure also discusses the time limits regarding slump, air content, and the beginning of the molding of test specimens. Samples taken for the purpose of slump, air content, and strength testing should be collected in two or more portions taken from the middle of the batched load. It is important that the sample being tested is representative of all the water and admixtures being added to the batch and placed in the forms. It is common for field crews to want the concrete tests at the start of the discharge from the truck. That is allowed for the assistance of the field crew information, generally slump, but additional sampling will be required for the recorded result in reference to slump, air, and cylinders. Another common practice in the field by the concrete crew is to add water once you have sampled the concrete. The sample obtained is no longer representative of the concrete being placed in the forms. Therefore, a new sample should be obtained with the concrete containing the additional water.

The location of the sample taken often comes into question. The intent of obtaining a sample to test slump, air content, and strength is to verify that the batch plant has delivered concrete that meets the specifications for the project. ASTM C 172 in Note 2 states: "Sampling should normally be performed as the concrete is delivered from the mixer to the conveying vehicle used to transport the concrete to the forms; however, specifications may require other points of sampling, such as the discharge of a concrete pump." This allows the sampling from another point besides the concrete mixer but should be called out in the specifications. Field experience has shown that the concrete from the mixer may have a slump that exceeds the maximum slump allowed in the specifications as it is being discharged into the pumper. However, the slump discharged from the pump hose at the point of delivery into the forms may be within specification tolerances. This will be discussed further in the slump testing section.

The ACI procedural checklist is as follows:

1. Obtain a representative sample source (i.e. revolving drum truck mixer):
    a. Ensure that all water has been added.
    b. Sample the concrete at two or more regularly spaced intervals during discharge of the middle portion of the batch.
    c. Repeatedly pass the receptacle through the entire discharge stream or completely divert the discharge stream into the sampling container.
    d. Transport samples to the place of testing. Combine and remix to form a composite sample.
    e. Obtain composite sample within 15 minutes from the start of sampling.
2. Start tests for the slump, temperature, and air content within 5 minutes after obtaining the final portion of the composite sample.
3. Start molding specimens within 15 minutes after fabricating the composite sample.
4. Protect the sample against rapid evaporation and contamination.

FIGURE 3.8 Sampling Concrete Left – Sampling from chute, Right – Concrete sample in bucket.

3.6.2 Temperature

Concrete temperature is one of the most important factors influencing concrete quality, time of set, and strength. Concrete temperature also affects the performance of chemical admixtures, air-entraining admixtures, pozzolanic materials, and other additives and admixtures. The temperature of concrete is used to indicate the type of curing and protection that will be needed, as well as the length of time the curing and protection should be maintained.

Fresh concrete testing should be performed per ASTM C 1064, Standard Test Method for Temperature of Freshly Mixed Hydraulic-Cement Concrete. The ACI procedural checklist is as follows:

1. Place the temperature measuring device (TMD) in the concrete so that the sensing portion is submerged a minimum of 3 inches.
2. Gently press concrete around the TMD so that ambient (outside) air temperature does not influence the measured temperature.
3. Leave the TMD in the concrete for at least 2 minutes but not more than 5 minutes.
4. Read and record the fresh concrete temperature to the nearest 1° F. Do not remove the TMD from the concrete when reading the temperature.

FIGURE 3.9 Thermometer in sample bucket.

3.6.3 Slump

The slump test, ASTM C 143, Standard Method for Slump of Hydraulic-Cement Concrete, is the most common method used to measure the consistency of the concrete. This is a measure of the relative fluidity or mobility of the concrete mixture. Slump does not measure the water content or the workability of the concrete. Nearly all concrete specifications state the range of slump required or a not-to-exceed slump. The important concern with concrete should be the water-cement ratio, not the slump. However, if the slump specified is not to exceed 4 inches and the slump test indicates a slump of 8 inches the contractor is well within their rights to reject the concrete load because the water-cement ratio is most likely compromised.

In our earlier discussion concerning slump and pumped concrete during the sampling section. The maximum slump specified may be 4 inches, and the concrete discharged from the mixer to the pump is determined to be 5 inches. A slump is also obtained from the pump hose discharge during the same load tested earlier, and the slump may only be 3.5 inches; this is common. The problem that occurs is that the pumper needs a sufficient flow (slump) to prevent the concrete from plugging the pump hose during delivery and some slump is lost during the pumping procedure. If the specifications did not state that the sampling of the concrete could be obtained at the pump hose discharge, the concrete is out of specification. An unexperienced inspector or superintendent may reject this concrete load due to not meeting the slump specification at the mixer discharge point. The water-cement ratio has not changed based on batched weights and may be within specifications. It is my opinion that slump is a general indicator of the water-cement ratio, and rejecting concrete loads just due to the slump not meeting the specifications should be done cautiously and with more information than just slump data.

The ACI procedural checklist is as follows:

1. Dampen the slump mold.
2. Place the mold on a rigid, flat, level, moist, nonabsorbent surface, free of vibration, that is large enough to contain all the slumped concrete.
3. Hold the mold firmly in place during filling and cleaning by standing on the two-foot pieces on either side of the mold. A base plate with clamps is also acceptable.
4. Using a scoop, fill the mold in three layers, moving the scoop around the perimeter of the mold opening to ensure an even distribution of the concrete. For the first layer:
   a. Fill the mold to approximately 1/3 of its volume (2-5/8 inches).
   b. With the rounded end of the tamping rod, rod the layer 25 times throughout its depth. Uniformly distribute the strokes over the cross-section of the layer. Incline the rod slightly, starting near the perimeter, and progress with vertical strokes spirally toward the center.
5. For the second layer:
   a. Fill the mold to approximately 2/3 of its volume (6-1/8 inches).

b. Rod the layer 25 times uniformly over the cross-section, with the rounded end of the rod. Rod through the layer and into the layer below approximately 1 inch. Uniformly distribute the strokes over the cross-section of the layer.

6. For the third layer:
   a. Heap concrete above the top of the mold.
   b. Rod the layer 25 times uniformly over the cross-section, with the rounded end of the rod. Rod through the layer and into the layer below approximately 1 inch. Uniformly distribute the strokes over the cross-section of the layer.
   c. Should rodding the layer result in the concrete falling below the top of the mold, add concrete to keep an excess above the mold. Continue the rodding count from the value reached before concrete was added to the mold.
7. Use the tamping rod to strike off the top surface of the concrete in a screeding and rolling motion.
8. Maintain downward pressure while removing any concrete that collected around the base of the mold during strike-off.
9. Immediately remove the mold by raising it in a steady, vertical direction. There should be no lateral or torsional motion of the mold while lifting. Lift the mold off the concrete 12 inches. The cone should be fully lifted from the sample in 5 seconds ±2 seconds.
10. Complete the slump test from the start of filling the mold through the removal of the mold in 2-1/2 minutes.
11. If falling away or shearing off of concrete from one side or portion of the mass occurs, disregard the test and make a new test on another portion of the sample.
12. Immediately measure the slump. This is the vertical distance between the top of the cone and the displaced original center of the specimen's top surface.
13. Report the slump to the nearest ¼ inch.

FIGURE 3.10 Slump testing Left – pulling slump cone, Right – concrete slump once cone pulled.

3.6.4 *Density (Unit Weight)*

The density test is a valuable tool to control the quality of freshly mixed concrete. Once the concrete mixture is designed, one of the characteristics determined is density. The density will not vary much if the concrete mixture is produced as designed. Therefore, a change in the concrete density during construction indicates a change in the mixture. A lower density may indicate a change in the specific gravity of the aggregate, a higher air content, a higher water content, a change in the proportions of ingredients, or a lower cement content. A lower density in the fresh concrete mixture will normally produce lower strengths and a reduction of other desirable qualities in the concrete. If the reduction of the concrete density was due to an increase in air content, the concrete will be more durable in its resistance to cycles of freezing and thawing, but the strength, abrasion resistance, resistance to chemical attack, shrinkage, and cracking qualities will be adversely affected.

The ACI procedural checklist for ASTM C 138, Standard Test Method for Density (Unit Weight), Yield, and Air Content (Gravimetric) is as follows:

1. Dampen the interior of the measure mold and remove any standing water from the bottom of the measure.
2. Determine the empty measure's mass (lb) and place it on a flat, level, firm surface.
3. Using a scoop, place the concrete in the measure in three layers of approximately equal volume, moving the scoop around the perimeter of the measure opening to ensure an even distribution of the concrete. For the first layer:
   a. Fill the measure to approximately 1/3 of its volume.
   b. Rod the layer 25 times, uniformly over the cross-section, with the rounded end of the rod. Rod the layer throughout its depth using care not to damage the bottom of the measure.
   c. Tap the sides of the measure 10 to 15 times with the mallet to close any voids left by the tamping rod and release any large air bubbles.
4. For the second layer:
   a. Fill the measure to approximately 2/3 of its volume.
   b. Rod the layer 25 times, uniformly over the cross-section, with the rounded end of the rod. Rod through the layer and into the layer below approximately 1 inch.
   c. Tap the sides of the measure 10 to 15 times with the mallet to close any voids left by the tamping rod and release any large air bubbles.
5. For the third layer:
   a. Add material so as to avoid overfilling.
   b. Rod the layer 25 times, uniformly over the cross-section, with the rounded end of the rod. Rod through the layer and into the layer below approximately 1 inch.
   c. Tap the sides of the measure 10 to 15 times with the mallet to close any voids left by the tamping rod and release any large air bubbles.

6. After consolidating the third layer, an approximate (1/8 inch) of excess concrete above the top of the measure is considered optimal. Representative material can be added or removed as needed prior to strike-off.
7. Strike off the top surface of the concrete and finish it smoothly with the flat strike-off plate, leaving the measure just level full:
    a. Press the strike-off plate on the top surface of the measure to cover 2/3 of the surface. Withdraw the plate with a sawing motion to finish only the area originally covered.
    b. Place the plate on top of the measure to cover the original 2/3 of the surface. Advance it with vertical pressure and a sawing motion to cover the whole surface of the measure. Continue to advance the plate until it slides completely off the measure.
    c. Incline the plate and perform several strokes with the edge of the plate to produce a smooth finish.
8. Completely clean the measure's exterior and determine the measure's mass filled with concrete.
9. Report the concrete's density (unit weight) to the nearest 0.1 pcf.

FIGURE 3.11 Unit weight bucket being weighted.

3.6.5 Yield

Yield is the volume of the batch in cubic feet. The mix design will determine the expected batch yield. An over-yield during construction is generally produced by a lower density of the concrete mix. This means that the required cement content for a cubic yard is now diluted to produce a greater volume of concrete, therefore reducing the strength and other qualities.

Yield is provided by calculating the amount of materials (aggregate, cement, and water, including water contained in the fine aggregate as well as the water added on site) divided by the density of the concrete.

3.6.6 Air Content

The Pressure Method of testing air entrainment, ASTM C 231, can be used to determine the air content of normal and heavyweight concretes. This method should not be used with highly porous aggregates used in lightweight concrete. ASTM C 173, the Standard Test Method for Air Content of Freshly Mixed Concrete by the Volumetric Method, should be used with highly porous aggregates. This method is noted but will not be further discussed in this book since this test is not commonly used, except for lightweight concrete and certification testing for the ACI Concrete Field Testing Technician – Grade I.

Air entrainment is necessary for concrete that will be exposed to cycles of freezing and thawing and to deicing chemicals. The microscopic entrained air voids provide a source of internal pressure relief within the concrete to accommodate the pressures that develop as ice crystals form in the pores and capillaries of the concrete. Without the proper air content in the paste portion of the concrete, normal concrete that is exposed to cycles of freezing and thawing will scale and/or spall, resulting in a durability failure of the concrete surface.[10]

The ACI procedural checklist for ASTM C 231, Standard Test Method for Air Content of Freshly Mixed Concrete by the Pressure Method (Type B meter only) is as follows:

1. Dampen the interior of the measure mold and remove any standing water from the bottom of the measure.
2. Using a scoop, place the concrete in the measure in three layers of approximately equal volume, moving the scoop around the perimeter of the measure opening to ensure an even distribution of the concrete. For the first layer:
   a. Fill the measure to approximately 1/3 of its volume.
   b. Rod the layer 25 times, uniformly over the cross-section, with the rounded end of the rod. Rod the layer throughout its depth using care not to damage the bottom of the measure.
   c. Tap the sides of the measure 10 to 15 times with the mallet to close any voids left by the tamping rod and release any large air bubbles.
3. For the second layer:
   a. Fill the measure to approximately 2/3 of its volume.
   b. Rod the layer 25 times, uniformly over the cross-section, with the rounded end of the rod. Rod through the layer and into the layer below approximately 1 inch.
   c. Tap the sides of the measure 10 to 15 times with the mallet to close any voids left by the tamping rod and release any large air bubbles.

4. For the third layer:
    a. Add concrete in a manner to avoid excessive overfilling.
    b. Rod the layer 25 times, uniformly over the cross-section, with the rounded end of the rod. Rod through the layer and into the layer below approximately 1 inch.
    c. Tap the sides of the measure 10 to 15 times with the mallet to close any voids left by the tamping rod and release any large air bubbles.
5. Strike off the top surface of the concrete:
    a. If using a strike-off plate:
        i. Press the strike-off plate on the top surface of the measure to cover 2/3 of the surface. Withdraw the plate with a sawing motion to finish only the area originally covered.
        ii. Place the plate on top of the measure to cover the original 2/3 of the surface. Advance it with vertical pressure and a sawing motion to cover the whole surface of the measure. Continue to advance the plate until it slides completely off the measure.
        iii. Incline the plate and perform several strokes with the edge of the plate to produce a smooth finish.
    b. If using a strike-off bar:
        i. Strike off the top surface by sliding the strike-off bar across the top flange or rim of the measuring bowl with a sawing motion until the bowl is level full.
6. Thoroughly clean the flange/rim of the measuring bowl and cover assembly.
7. Clamp the cover assembly to the measuring bowl, ensuring a pressure-tight seal.
8. Close the main air valve between the air chamber and the measuring bowl. Open both petcocks on the cover.
9. Use a syringe to inject water through one petcock until water emerges from the opposite petcock. Jar the meter gently until all air is expelled from the opposite petcock.
10. Close the air bleeder valve and pump air into the chamber until the dial gauge's hand is on the initial pressure line. Allow a few seconds for the compressed air to cool.
11. Stabilize the gauge hand at the initial pressure line by bleeding, pumping, and lightly tapping the gauge by hand.
12. Close both petcocks.
13. Open the main air valve between the air chamber and the measuring bowl. Tap the sides of the measuring bowl smartly with the mallet. Lightly tap the pressure gauge by hand to stabilize the gauge hand.
14. Read the air percentage on the pressure gauge's dial.
15. Release the main air valve. Release the pressure in the bowl by opening both petcocks before removing the cover.
16. Report the air content to the nearest 0.1% using the aggregate correction factor (or to the nearest ½ scale division if the gauge reading exceeds 8%).

FIGURE 3.12 Air Content testing Left – adding water during test, Right – Dial showing air content in concrete.

3.6.7 Making Concrete Test Specimens

Concrete is generally desired and specified based on strength test results. Strength is normally what is used as a basis for acceptance of concrete. ASTM C 31 is a means to ensure the strength results are reliable and can be reproduced by different testing individuals while producing similar results. ASTM C 31 details the procedure for molding and curing concrete cylinders and beams. It is important to follow the standard in the making of the specimens and in the curing as well. The strength results may be unreliable and questionable if the standards and procedures are not followed.

The ACI procedural checklist for ASTM C 31, Standard Practice for Making and Curing Concrete Test Specimens in the Field, is as follows:

1. Use a scoop to place concrete in the mold. Use care to distribute the material evenly around the perimeter of the mold.
2. For the first layer:
    a. Fill the mold with the appropriate quantity of concrete: approximately ½ the volume for the 4" x 8" mold or 1/3 the volume for the 6" x 12" mold.
    b. Rod the layer 25 times, uniformly over the cross-section, with the rounded end of the rod. Rod the layer throughout its depth using care not to damage the bottom of the mold.
    c. Tap the outsides of the mold lightly 10 to 15 times with the mallet to close any holes left by rodding and to release any large air bubbles that may have been trapped. Use an open hand to tap cylinder molds that are susceptible to denting or other permanent distortion if tapped with a mallet.
3. For the second layer:
    a. Fill the mold with the appropriate quantity of concrete: fill a 4" x 8" mold to the top or to approximately 2/3 the volume for a 6" x 12" mold.

   b. Rod the layer 25 times, uniformly over the cross-section, with the rounded end of the rod. Rod through the layer and into the layer below approximately 1 inch.
   c. Tap the outsides of the mold lightly 10 to 15 times with the mallet to close any holes left by rodding and to release any large air bubbles that may been trapped. Use an open hand to tap cylinder molds that are susceptible to denting or other permanent distortion if tapped with a mallet.
4. For the third layer (6" x 12" specimens):
   a. Add an amount of concrete that will fill the mold after consolidation.
   b. Rod the layer 25 times, uniformly over the cross-section, with the rounded end of the rod. Rod through the layer into the layer below approximately 1 inch.
   c. Tap the outsides of the mold lightly 10 to 15 times with the mallet to close any holes left by rodding and to release any large air bubbles that may have been trapped. Use an open hand to tap cylinder molds that are susceptible to denting or other permanent distortion if tapped with a mallet.
5. During consolidation of the top layer (after the second layer with 4" x 8" specimens), adjust the concrete level of underfilled and overfilled molds, if necessary.
6. Strike off any excess concrete with the tamping rod, a handheld float, or a trowel to produce a flat, even surface.
7. Provide protection to prevent sample moisture loss.
8. Mark the specimens to identify the concrete they represent.
9. Move the specimens to an initial curing place for storage and prevent being disturbed.

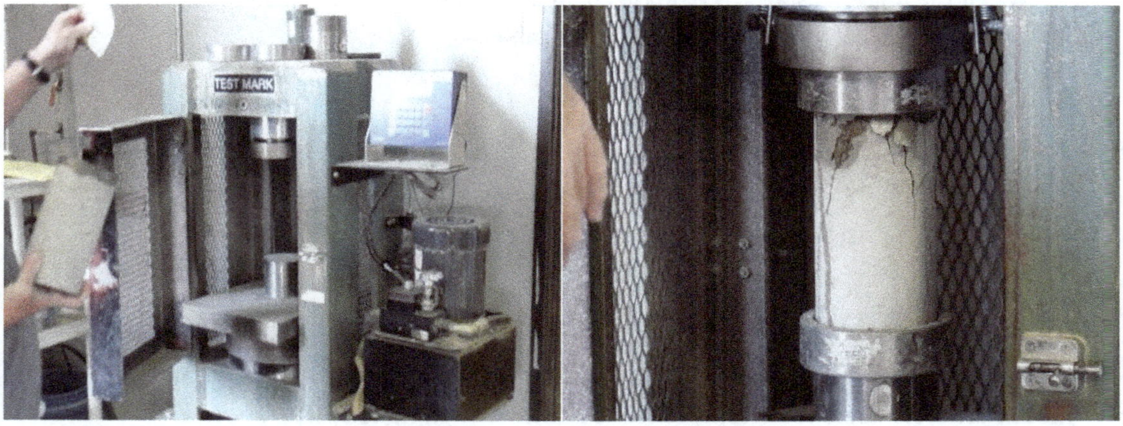

FIGURE 3.13 Concrete test specimens Left – 4" x 8" specimen about to be tested for a compressive test, Right – Concrete specimen during compression testing.

Problems – Chapter 3

1. Concrete is the world's most used construction material. True or False

2. How does concrete help in the construction industry?
   a. Strength.
   b. Weight.
   c. Durability.
   d. Low permeability.
   e. All of the above.

3. List at least four reasons why it is helpful that concrete starts out in a plastic state.

4. What does an initial set mean in terms of concrete?
   a. The first songs played at a concert.
   b. When a ship first starts to sail.
   c. When the material starts, the phase changes from plastic to hardened.
   d. The first time you place silverware on the table.
   e. All of the above.

5. What influences the initial set?
   a. Air temperature.
   b. Wind.
   c. Humidity.
   d. Additives.
   e. All of the above.

6. Should concrete becomes a homogeneous mixture when mixed. True or False

7. Name at least three items that play an important role in mixing concrete.

8. What should be avoided to prevent surface bleeding?

9. How does consolidation affect concrete?

10. The following is involved in the hydration of the cement. List all that apply.
    a. Is a chemical reaction between cement and water.
    b. Providing lots of water for the body.
    c. Binds the sand and aggregate.
    d. Initial set of the concrete.
    e. Produces chilling to the concrete.

11. Hardened concrete is normally referred to as concrete. True or False

12. What is the greatest attribute of concrete?

13. What is required for the hydration of concrete to continue to occur? List all that apply.
    a. Un-hydrated cement is present.
    b. Sufficient water in the mix to react to the cement.
    c. Temperature below freezing.
    d. Relative humidity above 80%.
    e. All of the above.

14. Concrete hardens by drying. True or False

15. What is the most common strength designed for concrete? List all that apply.
    a. Flexural
    b. Tensile
    c. Torsional
    d. Compressive
    e. Shear
    b. All of the above

16. In what manner is the load applied in compressive strength? List all that apply.
    a. Bending
    b. Pulling
    c. Twisting
    d. Crushing
    e. Sliding
    f. All of the above

17. What units are used to express compressive strength?

18. How many days are required to achieve design compressive strength?
    a. 1
    b. 3
    c. 28
    d. 365

19. What is the observed average unit weight of fresh concrete with a maximum aggregate size of ¾" and a specific gravity of 2.65?

20. What are the two ways to reduce freezing and thawing deterioration of concrete?

21. Causes of cracks in concrete are stress due to applied loads and stress due to drying shrinkage or temperature changes in restrained conditions. True or False?

22. What are the three main ingredients to make concrete?

23. What is the difference between hydraulic and non-hydraulic cement?

24. Portland cement is the most common cement used. True or False?

25. Match the following types of Portland Cement. Each line will have a letter and number.

___High early strength	a. general purpose	1. IV
___High sulfate resistance	b. general purpose w/air	2. IIIA
___Moderate, air entraining	c. moderate sulfate, slow strength	3. I
___Normal	d. moderate sulfate w/air	4. IA
___Low heat of hydration	e. quicker strength	5. V
___Moderate	f. quicker strength w/air	6. III
___High early strength, air entraining	g. reduces the heat of hydration	7. IIA
___Normal, air-entraining	h. severe sulfate	8. II

26. What water impurities may have a negative effect on concrete? List all that apply.
    a. Dissolved chemicals
    b. Algae
    c. Drinking water
    d. Seawater
    e. Sugar
    f. All of the above

27. Aggregates generally make up what range of percentage of the concrete weight?

28. What properties may be altered by adding admixtures to the concrete mix? List all that are applicable.
    a. Set control
    b. color
    c. Air entrainment
    d. Damp proofing
    e. Water reduction
    f. All of the above.

29. The required compressive strength will be found in the project specifications. True or False?

30. In Appendix E, Section 0330005 – Civil Cast-In-Place Concrete, what is the maximum water/cement ratio?

31. List three qualities a concrete mixture should possess.

32. Compressive strength is inversely related to the water-cement ratio. This means more water to cement will give higher compressive strength. True or False?

33. List the following ideas that are needed to design the best water-cement ratio. Circle all that apply.
    a. You must use the correct mixture of aggregate.
    b. There must be sufficient paste to surround the aggregate.
    c. There must be sufficient water to fully hydrate the cement.
    d. The curing environment must allow full hydration to occur.
    e. Time is required to allow the hardening to sufficiently accept the load.
    f. All of the above

34. Using Tables 3.3, 3.4, and 3.5 to answer the following questions:

    a. What is the maximum water-cement ratio for exposed curb and gutter?

    b. What compressive strength would you expect with a water-cement ratio of 0.48 in an air-entrained concrete?

    c. What is the maximum water-cement ratio for 3000 psi concrete using air-entrained concrete when no historical data is available?

35. How many pounds of cement is in a six-sack mix?

36. What is the minimum number of pounds of cement for normal-weight concrete used in flatwork? How many sacks is this?

37. What is the minimum to maximum slump recommended for reinforced foundation walls and footings?

38. Name three items that would limit the largest size aggregate.

39. The largest size aggregate should not exceed _______ the clear space between individual reinforcing bars.

40. Concrete that has stiffened in the mixer drum can have water added as long as which conditions are met.
    a. Maximum allowable water-cement ratio is not exceeded
    b. Maximum allowable slump is not exceeded
    c. Maximum allowable mixing and agitating time or drum revolutions are not exceeded
    d. Concrete is remixed for at least half the minimum required mixing time or number of revolutions.
    e. All of the above.

41. Choose all that apply when concrete is being placed in forms. The forms should be:
    a. Accurately set
    b. Clean
    c. Adequately braced
    d. Set by noon
    e. Free of debris
    f. Free of water
    g. All of the above

42. Reinforcing steel should be clean, free of loose rust, property placed and secured when concrete is placed. True or False.

43. Excessive movement of the concrete may cause what problem?

44. List at least four methods to finish concrete.

45. What is the main concern with hot weather concrete pours?

46. Loss of water in the mix will decrease __________ and ____________ while increasing the ________________________.

47. Hot weather curing may be reduced by? List all that apply.
    a. Lower the temperature of the materials before mixing.
    b. Keep concrete mixers as cool as possible.
    c. Transport and place the concrete as quickly as possible.
    d. None of the above.

48. What simple precautions can minimize the possibility of plastic shrinkage cracking on horizontal surfaces? List all that apply.
    a. Reduce the temperature of the concrete.
    b. Moisten the subgrade and forms prior to placing concrete.
    c. Protect the placed concrete from wind and sunlight.
    d. Prevent delays during construction.
    e. All of the above.

49. In accordance with ACI 306, what defines cold weather concrete placement?
    a. Temperature drops below 40°F.
    b. Two successive days below 40°F.
    c. Three successive days below 40°F.
    d. Five successive days below 40°F.

50. What precautions should be taken in cold weather concrete placement? List all that apply.
    a. Add ice to the concrete mixture.
    b. Clear the snow and ice from the forms.
    c. Proceed as normal conditions.
    d. Provide heaters and windbreak shelters.
    e. All of the above.

51. Concrete can be placed on frozen ground. True or False?

52. Who generally conducts Quality Control?

53. Quality Assurance testing is commonly provided by an independent third-party company. True or False?

54. List the Quality Assurance tests and frequency for the following concrete test from Table No. 2.13.

    a. Curb and Gutter

    b. Sidewalks

55. While sampling, it is not important to obtain representative samples. True or False?

56. Slump measures water content in the concrete. True or False?

57. When is sampling from the end of the pumper discharge hose acceptable? List all that apply.
   a. When additional water is added.
   b. The slump exceeds the specifications.
   c. The project specifications require it.
   d. The air temperature is below 40°F.
   e. None of the above.

58. The time to start the procedure is within how many minutes of generating the composite sample. Match the following: more than one answer can be used.

____Slump test	a. 3 minutes
____Temperature check	b. 5 minutes
____Air content test	c. 10 minutes
____Molding specimens	d. 15

59. Which comments are true while checking the concrete temperature? List all that apply.
   a. Submerge the thermometer at least 3 inches into the sample.
   b. Try to minimize the influence of the ambient temperature.
   c. Leave the thermometer in the concrete for at least 1 minute and not more than 5 minutes.
   d. ead the temperature to the nearest 1°F.

60. The slump cone is filled in ____ layers based on volume.

61. How many rod strokes per lift are required while doing a slump test?

62. Tap the side of the slump cone with the rubber mallet after each lift. True or False?

63. How much time is allowed for lifting the slump cone?
    a. 2 seconds to 5 seconds
    b. 5 seconds ±3 seconds
    c. 5 seconds ±2 seconds
    d. 3 seconds to 10 seconds
    e. None of the above

64. Tap the side of the unit weight measure with the rubber mallet after each lift. True or False?

65. The unit weight of the concrete should be reported to the nearest __________?

66. Thoroughly clean the flange/rim of the measuring bowl. True or false?

67. The air content of the concrete should be reported to the nearest __________?

68. What is the ASTM standard for making and curing concrete test specimens?

69. How many layers are required in making 4" x 8" concrete test specimens?

4

4.1 Asphalt

Asphaltic materials have been utilized since 3500 B. C. primarily as adhesives, waterproofing agents, and mortars for brick walls. Asphalt material is native and occurs when petroleum rises to the earth's crust and evaporates the volatile oils. With the invention of the automobile, smooth all-weather surfaces came into demand. The processes for producing asphalt cement from crude petroleum products grew, leading to modern asphalt cement's development.

This book will focus on the design, production, construction, and testing of asphalt pavement used for roadways. Asphalt pavement is also known as bituminous pavement. Bitumen can be found in nature or manufactured and is soluble in carbon disulfide. Another common term is "hot-mix asphalt" or HMA. This includes many different types of mixtures of aggregate and asphalt cement produced at an elevated temperature.

4.2 Properties

The most suitable bituminous pavement produces a stable, durable, flexible, and skid-resistant surface at a minimum cost with adequate aggregate. We will examine the properties that describe asphaltic pavements.

4.2.1 Stability

Stability is related to strength and refers to the ability of the pavement to resist deformation under the application of loads. The stability of the pavement is affected by aggregate interlocking developed among aggregate particles and the cohesiveness supplied by the asphalt cement. A crushed angular particle shape with hard aggregate and just enough asphalt cement to coat the aggregate particles to provide good adhesiveness will maximize the stability of the pavement.

4.2.2 Durability

Durability refers to the resistance of the pavement to disintegration under traffic loads. That means during the summer heat, the pavement will remain smooth and serviceable without rutting, and during the winter cold, the pavement will not crack or unravel. The pavement will also resist the forces of freezing and thawing. Durability is best met by coating all the aggregate and filling all the voids between the aggregate. Therefore, water is not able to enter the asphalt pavement. Maximizing durability compromises stability and skid resistance by increasing the asphalt cement to reduce voids in the pavement section.

4.2.3 Flexibility

Flexibility refers to a pavement's ability to withstand deflection and bending without cracking. Using an open-graded aggregate mixture will maximize flexibility. However, stability will then be impacted.

4.2.4 Skid Resistance

Skid Resistance is a form of stability that, if not met, causes slippery pavement. Pavement bleeding occurs if too few voids are left in the mix, and compaction occurs under traffic loads. This is because the asphalt is forced to the surface. Most specifications require 2 to 3 percent air voids to prevent surface bleeding. The second slippery surface is due to aggregate polishing once the pavement receives continuous wear from traffic and the surface aggregate becomes polished and slick. The best way to resolve this situation is to use relatively hard aggregate to prevent aggregate breakdown.

4.3 Ingredients

The ingredients for asphaltic concrete are similar to concrete using a binder and a mixture of aggregate to produce a product that produces strength. The main difference is the binder, asphaltic cement is used instead of Portland cement with asphalt. The design process becomes much more complicated when the binder is adjusted to alter the cure time, temperature, and expense. These adjustments can be made by the addition of liquid asphalt (cutbacks), road oils, or asphalt emulsions into the binder, to name a few. This book will not get into the details, but the construction personnel should know these adjustments and note what asphalt has been provided in the field from the delivery ticket.

4.3.1 *Asphaltic Cement*

Asphaltic cement is obtained from the refining of crude oil. Three methods are used to grade asphaltic cement: the penetration grading system and the viscosity grading system, which have been widely used. The performance grading (PG) system was developed under the Strategic Highway Research Program (SHRP). Since asphaltic cement is considered a viscoelastic material, behaving as elastic solids at low temperatures and viscous liquids at high temperatures, it is important that temperature is defined in the grading systems.

4.3.1.1 Penetration Grading System

The penetration grading system is determined by an empirical test, ASTM D 946 and AASHTO M 20. Asphaltic binders are classified into grades according to the penetration of a specific size needle into the unaged asphalt binder with a temperature of 77°F. The penetration is measured in units of 0.1 millimeter or deci-millimeter. Thus, an 85-100 binder would have a minimum penetration value of 85 deci-millimeters and a maximum penetration value of 100 deci-meters. The softest 200-300 penetration grade asphalt is moderately firm at room temperature, as noted by gentle finger pressure indents on the surface of the sample. The hardest 40-50 penetration grade asphalt would only permit a slight thumbprint under firm pressure at room temperature.

The penetration grading system does provide a means to evaluate the stiffness of an asphalt binder, but it does have some limitations. First, the test measures the binder's viscous and elastic behavior. The test results must be correlated with pavement performance to understand the value of the results. Second, the penetration grading system only measured asphalt stiffness at an intermediate temperature. The binder may react differently at different temperatures. Finally, the tests identified the penetration of short-term aging. There are no provisions to identify the effects of long-term aging of asphalt stiffness after being on pavement for a number of years.

4.3.1.2 Viscosity Grading System

The viscosity grading system in the 1970s began to be used more frequently than the penetration grading system. ASTM D 3381 and AASHTO M 226 establish using absolute viscosity at 140°F as the principal physical property for grading. 140°F was chosen as an approximation of the maximum temperature of asphalt pavement surfaces in service. The standards also have a viscosity of 275°F as an approximation of the laydown temperature of HMA. Thus, the specification addresses asphalt binder consistency during the laydown process of HMA, when workability is an issue, and consistency at summer pavement temperatures when stiffness is needed.

The viscosity grading system is divided into two systems. The **AC (Asphalt Cement) system** consists of AC-2.5, AC-5, AC-10, AC-20, AC-40, and sometimes AC-30 grades. The numerical values designation denotes the viscosity of the asphalt binder in hundreds of poises at 140°F with an allowable tolerance of ± 20 poises. Therefore, an AC-10 would be expected to have a viscosity of approximately 1000 poises but would include asphalt binders with a viscosity between 800 to 1200 poises at 140°F. The AR (Aged Residue) system consists of AR-1000, AR-2000, AR-4000, AR-8000, and AR-16000 grades. The numerical values denote the viscosity in poises at 140°F with an allowable tolerance on viscosity of ± 25 percent. The viscosity is measured after the asphalt has been subjected to the rolling thin film oven test. Therefore, the AR-4000 grade would include asphalt binders with a viscosity of 3000 to 5000 poises at 140°F after the rolling thin film oven test.

The viscosity graded system specifications were an improvement over the penetration graded specification, but it still had some of the same limitations. The viscosity graded system determined the stiffness at the higher mixing temperatures and typical high pavement temperatures. The stiffness of the asphalt binder at cold temperatures could only be inferred from the intermediate temperature data. Another issue with the viscosity graded system is it did not provide provisions for long-term aging of asphalt binders stiffness after the pavement is in service for several years. The viscosity graded system does not represent asphalt binders with polymer modification well. The viscosity of a modified asphalt binder may not provide sufficient information about the asphalt binder's physical properties to judge its performance potential properly.

4.3.1.3 Performance Grading System

The Performance Grading System differs from the other two methods by testing the asphalt binders in conditions that simulate critical stages during the binder's life cycle. The first stage represents the effects of transport, storage, and handling on the binder. The second stage represents the binder during mix production and construction. The third stage simulates years of service aging in the pavement. The test methods used to characterize PG Asphalt Binder Specifications are described in AASHTO M 320.

The advantage of the PG asphalt binder system is that the tests are conducted at the full range of temperatures expected during the production and service life of the asphalt pavement. Therefore, this grading system should provide greater performance for hot and humid areas as well as cold and dry areas.

4.3.2 Aggregates

The performance of hot mix asphalt (HMA) is greatly impacted by the aggregate. Aggregate is primarily responsible for the load-supporting capacity of pavement. Asphalt paving mixtures generally consist of 90-95 percent by weight or 75-85 percent by volume of aggregate. Aggregate refers to hard, inert mineral material used for mixing in graduated particles or fragments.

Aggregates are evaluated and tested for several properties. Aggregate material chosen for asphalt pavement depends on the availability, the cost of the material, and the type of construction proposed, as well as the following properties:

- Maximum particle size and grading
- Particle shape and surface texture
- Toughness
- Pore structure, porosity, and absorption
- Specific gravity
- Freeze-thaw and weathering resistance
- Deleterious substances
- Affinity for water or asphalt
- Cleanliness

Reclaimed asphalt pavement (RAP) may be used in the mix to offset a portion of the aggregate. This reduces the cost of aggregate and recycles the asphalt pavement that needs to be removed and replaced due to pavement failure.

4.3.2.1 Maximum Particle Size and Grading

Maximum particle size is defined in AASHTO MP 323 as "one sieve size larger than the first sieve to retain more than 10 percent of the material". For effective paving results, the lift thickness of HMA should be at least 2 ½ times the maximum aggregate size or 4 times the nominal maximum aggregate size in the mix.

The grading or gradation of an aggregate is the distribution of the various particle sizes. The procedure for grading aggregates was discussed in Chapter 2. Table 3.6 in TAH (The Asphalt Handbook[16]) shows common gradations for HMA, as discussed in ASTM D 3515.

4.3.2.2 Particle Shape and Surface Texture

Particle shape refers to the aggregate particles' shape (angular, round, cubic, flat, or elongated). Particle shapes influence the workability and strength characteristics of the HMA as well as the compaction effort needed to meet the required density. Irregular or angular particles, such as crushed stone and gravel tend to interlock when compacted and resist displacement. Round particles do not also tend to interlock and will react more like ball bearings. Flat or elongated particles tend to break during compaction, leaving aggregate particles with uncoated faces, therefore decreasing strength. Flat and elongated particles are more likely to segregate from the mix.

Surface texture refers to the smoothness of the surface. It is the most important characteristic contributing to frictional resistance and also affecting the workability, compaction, and strength of the asphalt paving mixture. A rough or pitted surface will allow the binder to adhere to the aggregate much better than a smooth surface, impacting the ability to resist rutting. However, a rough texture will require a greater compaction effort.

4.3.2.3 Toughness

An aggregate's toughness is measured by its ability to resist wear and degradation. The manufacture, placement, compaction, and abrasion under traffic loads influence the toughness of the aggregate. The surface aggregate is more susceptible to surface abrasion losses than lower layers of HMA due to the traffic loads directly wearing on the surface. A desirable characteristic in HMA aggregates is a low percentage of abrasion loss, which is better described as a high resistance to wear.

4.3.2.4 Pore Structure, Porosity, and Absorption

The characteristics and configuration of voids in aggregate particles are known as pore structures. Pore structure can be considered impermeable if the pores are encapsulated within the aggregate particles, while permeable pores are open to the surface.

Porosity is the total volume of voids in the aggregate particle. Lower density, specific gravity, and lower strength are characteristics of higher porosity in the aggregate. Higher porosity will mean higher absorption if the pores are permeable.

Absorption is the intake of liquid into the aggregate particle pores. Aggregate with more permeable voids will absorb more asphalt binder. Therefore, the porosity of the aggregate needs to be known in the design of the asphalt mixture to ensure more asphalt binder is added to meet the coating and absorption of the aggregate. AASHTO T 85 (ASTM C 127) and AASHTO T 84 (ASTM C 128) are commonly used to determine the adsorption of coarse and fine aggregate, respectively.

4.3.2.5 Specific Gravity

Specific Gravity is the ratio of the weight of a unit volume of material to the weight of the same volume of water at 73°F. The specific gravity of aggregate is primarily used in mix volumetric calculation for void determination. Bulk-specific gravity is used in the computation for adjusting quantities of the aggregate components that are to be used in the HMA mix because of the differing specific gravities of various aggregates.

4.3.2.6 Freeze-Thaw and Weathering Resistance

The aggregate's ability to resist degradation caused by freezing and thawing cycles when the aggregate is in a saturated condition is known as freeze-thaw resistance. Internal expansive forces develop in the pores of the aggregate each time the absorbed water transforms into a frozen condition. An aggregate's ability to resist the effects of wetting and drying cycles and freezing and thawing cycles is called the aggregate durability or soundness property. The durability or soundness of the aggregate is mostly a concern in stockpiles prior to HMA production. Once HMA production is provided the aggregate soundness is less impacted due to the moisture within the aggregate being driven out by heat and the coating of the aggregate by the asphalt binder.

4.3.2.7 Deleterious Substances

Any substance in the aggregate that diminishes or is detrimental to the HMA properties is considered a deleterious substance. Deleterious materials typically include clay and other surface coatings on aggregate particles, clay lumps, soft or weak particles, and organic materials.

4.3.2.8 Cleanliness

Surface coating of the aggregate is one way of describing aggregate cleanliness. Aggregate coating can affect the bond of the binder to the aggregate.

4.4 Mix Design

The hot mix asphalt (HMA) design procedure is more of a trial and test process than what we find with the concrete mix design. In the asphalt mix design, asphalt cement and aggregate are blended together in different proportions in the laboratory. The mixes are evaluated using standard criteria set to identify the appropriate mixture. The optimum mixture will ensure a balance between the following physical properties:

- Resistance to permanent deformation
- Fatigue resistance
- Low-temperature cracking
- Moisture resistance

- Durability
- Skid resistance
- Workability

The goal is to select the mix design that most closely meets the needs requested. No single asphalt mix can achieve all these properties. There are three mix design procedures that will be briefly discussed: the Marshall Method, the Hveem Method, and the Superpave Method. Regardless of the design procedure used, the design of an HMA mix will consist of the following three steps:

- Selection of the type and gradation of the aggregate
- Selection of the type and grade of the asphalt binder
- Selection of the amount of asphalt binder to satisfy the project-specific requirements.

4.4.1 Marshall Method

The Marshall Method was originally designed by the Army Corps of Engineers for the design of military airfields. The goal was to have a procedure that used simple portable laboratory equipment that could be used to design pavement. The procedure was later modified by the Corps of Engineers to assist in the design of HMA for highway pavement. The procedure is similar to conducting a proctor. The asphalt and steel mold is heated to a set temperature, the asphalt is added to the mold in layers, and a drop hammer is used to compact the asphalt in the mold. Several samples are molded, providing an average density for the asphalt mix.

4.4.2 Hveem Method

The Hveem Method was developed by the California Division of Highways. This method was extensively used in the western United States. The process is similar to the Marshall Method, except the compaction was provided by a device called a Hveem kneading compactor. The kneading compactor also consisted of a drop hammer but then additional pressure was applied to the HMA sample. The intent was to be more representative of the field compaction procedure created by the steel and pneumatic-tired rollers. A laboratory testing device known as the Hveem Stabilometer was created to evaluate the ability of the HMA to resist the shear forces applied by traffic.

4.4.3 Superpave Method

The Superpave Method was developed by the Federal Highway Administration. This method was sponsored under the Strategic Highway Research Program (SHRP). The program was designed to significantly improve the design and construction of the nation's highway infrastructure, which was being affected by the increase in traffic. The result of the research was a new mix design system called Superpave (SUperior PERforming Asphalt PAVEments).

The Superpave Method includes both a volumetric design procedure and performance tests of the resulting mixes from the volumetric design. Due to the cost and specialization of the performance testing equipment, not many testing facilities have the equipment required to conduct laboratory performance tests. Therefore, the Superpave method design is generally provided as a volumetric mix design procedure. The design procedure weighs heavily on the voids in the mineral aggregate (VMA). The targeted air voids in the mix are four percent (4%). The laboratory mixes are compacted by a Superpave gyratory device. This equipment places the hot mix in a heated mold like the previously discussed methods. The mold with HMA included is set at a set angle from the level and rotated at a given revolutions per minute with a known force applied from above. The number of revolutions or gyrations depends on the anticipated traffic load in ESAL (Equivalent Single Axial Loading). Table 4.1 shows the gyratory compaction effort range conducted in the design of the mix.

TABLE 4.1 Superpave Gyratory Compaction Effort

Design ESALs (million)	Compaction Parameters			Roadway Applications
	$N_{initial}$	N_{design}	$N_{maximum}$	
< 0.3	6	50	75	Local roads, county roads, city streets. Low truck traffic.
0.3 to 3	7	75	115	Collector roads. Medium traffic city streets and county
3 to 30	8	100	160	State Highways, U.S. Highways. Heavy traffic city and county roads.
> 30	9	125	205	U.S. Interstate Highway.

Adapted from reference 16.

4.5 Construction Procedures

The installation of hot mixed asphalt depends on several factors. The site preparation is as important or more important than the design of the mixed asphalt and the installation practices of the HMA. Poor workmanship or improper procedures of any of these processes may cause failure in the asphalt mat. Full-depth asphalt pavement requires the subgrade to be properly moisturized and compacted. A proof roll (driving a heavily loaded vehicle such as a fully loaded water truck over the subgrade identifying deflection in the support soils) is helpful for evaluating the subgrade condition as well as obtaining compaction tests. For the best results, the new HMA placed on a granular base coarse will require the base to be stable, the surface dry, and solid (no deflection during a proof roll). A new mat of HMA placed on top of an existing asphalt layer requires the existing layer to be level (potholes filled), cracks sealed, and the surface cleaned before a tack coat is placed to provide a bond between the two asphalt layers.

4.5.1 Site Preparation

Preloading is the addition of excess weight on the proposed construction site prior to the construction of the structures. The soil is consolidated before construction, thereby strengthening the soil and decreasing or reducing the potential for settling the structure. The more weight and the more time with the weight on the soil will yield the most consolidation and best results. This method is effective with soft silty and clayey soils. The disadvantage is the amount of time needed and the cost to add and remove the weight to the site.

4.5.1.1 Subgrade

The subgrade soils should meet all applicable requirements for moisture content, density, structural support, and smoothness whether the subgrade is supporting the pavement directly or the base coarse. Once these requirements have been met, a proof roll should be performed to ensure the soil can support the weight of the haul traffic. If distortion of the subgrade is observed, means must be taken to stabilize the soil. Stabilization methods were discussed in Chapter 3.

It is important the subgrade is shaped to prevent water accumulation beneath the proposed asphalt or base coarse. The water beneath the HMA mat may weaken the soil strength as well as place additional expansion-contraction either due to swelling soils or freeze/thaw effects.

4.5.1.2 Base Coarse

If the asphalt layer is to be constructed on a granular base coarse layer, the base material should meet all the requirements, including gradation, Atterberg Limits, moisture content, density, structural strength, and smoothness. Proof rolling of the base should be conducted to ensure deflection and indention do not occur during the haul traffic. If deflection or indentions of the base occur, the base should be reworked until stability occurs. A cement-base treated base coarse has been found to be effective when additional stability is needed.

A prime asphalt coat may be sprayed on the base coarse prior to placement of the HMA mat. The prime coat acts as a temporary waterproofing layer that protects the base coarse from absorbing excess moisture during rain before paving. The prime coat also assists the bond between the base and the HMA layer.

4.5.1.3 Existing Asphalt Surface

The extent of preparation on an existing asphalt pavement depends on its condition. At a minimum, failed areas should be removed and replaced, potholes patched, cracks cleaned out and sealed, and ruts either filled or milled and filled.

4.5.1.3.1 Replacement and Patching

Removal and replacement should occur in areas where severe load-related distress has occurred. All HMA and granular base materials should be excavated to a depth where the underlying soils are stable or can be stabilized. Proper subsurface drainage should be installed as necessary. The granular base material, stabilized base course layers, or HMA should be placed and compacted in lifts to bring the supporting soil to the same level as the surrounding good pavement layers. If the area is large, then the HMA should be placed with a paver and compacted with large rollers.

Localized failed areas should be patched properly. The failed areas should be cut back to sound pavement and squared. Loose material and water should be removed. A tack coat should be applied to the sides and bottom of the hole, and the mix should be placed in the hole and compacted to specifications. If the hole is deeper than four inches (4"), the HMA should be placed in more than one layer, compacting each layer properly.

4.5.1.3.2 Crack Filling

Crack filling is conducted to minimize the amount of moisture that will get into the pavement section as well as the underlying materials supporting the asphalt. Large areas with cracked pavement, especially those with pattern cracking, must be patched or replaced. Before installing the crack sealing material, cracks should be cleaned and dried, generally blown out with air. If the crack is narrow, less than 3/8 inch in width, the crack width should be increased to allow the crack sealing material to penetrate the asphalt section. The sealant should be placed slightly lower than the surrounding pavement surface.

With numerous small cracks, such as reflective cracking, surface treatment (seal coat or slurry seal) should be considered.

4.5.1.3.3 Leveling Courses

Leveling courses are used to improve the rideability of the pavement and sometimes to improve the drainage from the pavement surface. Since more asphalt is placed in low areas such as ruts, a phenomenon known as differential compaction may occur. Differential compaction is created due to thicker asphalt sections compacting more than areas with thinner mix. Multiple courses may be needed over a pavement section that is badly out of shape before a smooth surface is obtained. A rule of thumb is that conventional asphalt mixes will compact approximately ¼ inch per 1 inch of screeded placed asphalt. Therefore, if the contractor desires 1 inch of compacted asphalt, 1 ¼ inch of screed asphalt should be placed prior to compaction. Another rule of thumb is that one layer after compaction will remove approximately 80 percent of a low spot. Two layers, each compacted separately, will remove approximately 95 percent of the low spot.

4.5.1.3.4 Milling

Milling or cord planning may be used to remove the high areas of the existing surface instead of installing a leveling course. It is also common to mill the existing asphalt before installing an overlay to ensure the final surface elevation does not get too high. An added benefit of milling, if the surface is properly cleaned, is the milled surface can provide an improved bond between the new and old layers, reducing the possibility of slippage between the two layers of asphalt. It is important to clean the milled section due to the dust created in the milling process before installing the tack coat.

4.5.1.4 Existing Concrete Surface

Placing an HMA pavement over a Portland cement concrete (PCC) pavement is like placing a layer over the asphalt; the surface should be properly prepared. Any severely distressed areas in the concrete slab should be cut out, moved, and properly replaced with PCC or HMA. The underlying subgrade and base course should be reviewed and corrected if needed. Milling the PCC slab is commonly used prior to installing the HMA mat.

4.5.2 Hot Mixed Asphalt (HMA)

Construction with Hot-Mixed Asphalt (HMA) is similar to construction with concrete. Generally, the raw materials are brought into a batch area, where they are blended together. The main difference between the two materials is that HMA is heated to a desirable hot temperature, as the name implies. Once batched, the HMA is transported to a site for placement and compaction.

4.5.2.1 Batching

A hot mix asphalt batch plant aims to blend aggregate and cement together at an elevated temperature to produce a homogeneous asphalt paving mixture. It is important the batch plant can accurately weigh the different products to ensure the mix meets the required specifications for the project. The elevated temperature is different for each product: the asphalt cement should be stored at a constant temperature between 300°F and 350°F; the aggregate will be run through a drying drum providing a discharge temperature between 290°F and 325°F; and the RAP temperature will vary depending on the moisture content of the reclaimed material and the percentage of RAP used in the replacement of the aggregate, temperature ranging from 250°F up to 820°F in the drying drum will be required to generate a mix discharge temperature of 220°F to 280°F. Drying drums are normally designed to tolerate a maximum of 500°F aggregate temperature. An extended amount of time above this temperature may cause warping of the drum shell, increasing maintenance costs. The temperature of the RAP can be reduced by using a smaller percentage of RAP and RAP with lower moisture content.

4.5.2.2 Transporting

Transporting the HMA mix from the plant to the construction site must be done quickly and efficiently with minimal change in the characteristics of the mix and without segregation of the mix aggregate. Three types of trucks are usually used to transport the HMA: end-dump, bottom or belly dump, or live bottom. An end dump truck backs up to the delivery point and raises the truck bed allowing the HMA to flow out the back of the truck. The end dump truck is most efficient when it discharges directly into the hopper of the paver. However, the end dump can also discharge the HMA as a stockpile on site where a front-end loader may distribute the asphalt or place it into the paver's hopper. The bottom or belly dump truck discharges the HMA from under the truck while the truck continues forward. The windrow of asphalt that is left from the belly dump truck will need intermediate equipment, such as a front-end loader or material transfer vehicle, to load the paver. The live bottom truck discharges the HMA from the back of the truck via a conveyor belt without the need to raise the truck bed. The asphalt from the live bottom truck can be placed directly into the paver and discharged into a stockpile or windrow.

4.5.2.3 Placing

The ideal placement of the HMA is with a paver. The paver's main purpose is to place the HMA to the desired width, grade, cross slope, and thickness to produce a uniform mat texture. Pavers are self-propelled with a hopper to accept the HMA from the transport vehicle or intermediate loading equipment, slat conveyors to move the HMA to the augers without segregation, augers to remix the HMA to continue a homogeneous mix, with final placement at the screed to control the thickness of the asphalt layer and provide the initial texture and density. Pavers can be loaded directly by the end dump truck, live bottom truck, front-end loader, or a material transfer vehicle. The HMA may also be distributed by a front-end loader for areas that don't allow access to the paver or when the paver breaks down.

FIGURE 4.1 Paver

FIGURE 4.2 Hopper to Paver

FIGURE 4.3 Skid Loader

FIGURE 4.4 Skid Loader loading paver

4.5.2.4 Compacting

Compaction is the most important factor in the performance of an HMA pavement. Adequate compaction of the mix increases fatigue life, decreases permanent deformation (rutting), reduces oxidation or aging, decreases moisture damage, increases strength and stability, and decreases low-temperature cracking.[19] Compaction is the procedure used to compress and reduce the volume of the asphalt in the field. Compaction reduces air voids and increases the unit weight or density of the asphalt.

Several factors impact the ability to meet the HMA's compaction requirements in the field. Some of these factors are that the mix delivered in the field is different from the design mix, the thickness of the layer, asphalt temperature, air temperature, the compaction equipment used, and the timing of the compaction equipment used.

At the time of laydown, the higher the asphalt mix temperature, up to a certain temperature, the easier it is to compact. If the initial temperature is too high, the mix may be considered "tender," and the asphalt binder viscosity is in an elastic state. Once the temperature decreases, the asphalt binder viscosity increases, allowing the asphalt layer to be compacted. More compaction effort will be required if the asphalt temperature is too low. In fact, if the asphalt temperature is too low, the required density may never be achieved. Asphalt mixes are usually manufactured at temperatures between 270°F and 325°F. Depending on the environmental conditions and haul distance, the asphalt mix can decrease the temperature from 5°F to 25°F while being transported to the laydown site. The important temperature to monitor is the mix temperature as it comes out from under the paver screed.

Three primary types of self-propelled compaction equipment are used to provide compaction to the asphalt mat: static steel wheel rollers, pneumatic tire rollers, and vibratory steel wheel rollers. The static steel wheel roller normally weighs from 3 to 14 tons, which can be adjusted either by adding water inside the wheel or external weights above the center of the wheel. The pneumatic tire rollers provide the compaction as a function of the wheel load of the machine, the tire pressure, the tire design (tire size and ply rating), and the depth of penetration of the tires into the mix. The greater the contact pressure between the tire and the mix, the greater the compaction effort applied by the roller. Vibratory steel wheel rollers can be either a single-drum vibratory roller or a double-drum vibratory roller. The rollers can be operated in a static mode (vibration turned off), with one drum vibrating or both drums vibrating. These rollers provide both a static weight and dynamic (impact) force.

FIGURE 4.5 Steel Wheel Roller

FIGURE 4.6 Pneumatic Roller

FIGURE 4.7 Combo Roller

There are several different variations that are necessary to determine the most effective compaction effort. The thickness of the mat, the aggregate size, environmental conditions, the amount of asphalt proposed to be placed in a day, and the equipment available. Generally, the superintendent and the quality control technician for the paving company will establish a rolling pattern in the field. As stated earlier, each asphalt mat may be approached differently, but generally, the compaction is provided by the following procedure: After the asphalt mat has been laid by the paving machine, the asphalt compaction process is started, known as the break-down process, with a vibratory steel wheel roller while the asphalt temperature is very hot. Once the asphalt is allowed to cool some but still hot, an intermediate compaction is provided with a pneumatic tire roller. The final compaction is conducted with a static steel wheel roller. The quality control technician will provide test strips to identify the best number of passes with each roller and at what temperature each roller should perform its compaction passes.

FIGURE 4.8 Small Vibrator Roller

FIGURE 4.9 Walk behind compactor

4.6 Sampling and Testing

Sampling is critical to providing realistic results from the fieldwork being performed, yet it is rarely discussed in soil mechanic books. To produce results that best match what is occurring in the field, it is imperative that a representative sample is obtained. This section will discuss methods to gather samples that are as representative as possible. This book will touch on just the sampling of the HMA once it is mixed and ready to be installed in the field. Sampling is a crucial part of the overall testing process.

The purpose of testing the asphalt is to provide assurance that the products delivered and installed in the field meet the requirements of the project specifications. Asphalt is different from concrete because a structure may rely on the strength of the concrete to support the building. The strength of the asphalt will have a bearing on the performance of the asphalt mat, but generally, the failure of asphalt is not catastrophic or life-threatening. When the HMA does not meet project specifications, normally, the asphalt will not last as long as it was originally designed. Therefore, it is common on a project to a reduction in pay to the Contractor if the asphalt doesn't meet the specifications. This takes into account that the owner will most likely have to provide rehabilitation or replacement of the asphalt.

This book is going to focus on the testing of the final product, aggregate gradation, amount of asphalt cement, and compaction of the final asphalt mat. There are other tests performed on the final asphalt mat, such as smoothness and slope, to name a couple. This book will not focus on those tests and procedures. Testing of the material that makes up the HMA mix is not going to be discussed in this chapter of the book. Some testing has previously been discussed, such as for aggregates.

4.6.1 Sampling

Asphalt samples may be obtained from the batch plant or the laydown area. The samples that most represent field conditions are obtained from the laydown area. Samples obtained from the batch plant can be obtained from the stockpile, conveyor belts, or discharge from the conveyor belt. The sample procedures are the same as described in the soil section on sampling. The main difference is that the sample container must withstand material heated up to 350°F. Sampling conducted at the laydown site may include from the haul truck, windrow sampling, in-place pavement sampling, and core sampling. Sampling from the haul truck is difficult to perform without potentially segregating the material. Windrow sampling requires you to obtain the full sample from top to bottom in the section you remove. In-place sampling can be obtained by cutting out an area that was placed by the paver, sampling from the edge of the paved area, or my preferred method, laying a shovel flat on the area to be paved along the edge of the paver and placing all the sample laid on top of the shovel in the sample container.

4.6.2 Laboratory Quality Control

Samples obtained from the batch plant may include aggregate or HMA. The aggregate will be tested for the Atterberg Limits and gradation. This is especially true now with the nuclear asphalt ovens determining the asphalt content. The HMA may be tested for the Maximum Theoretical Specific Gravity (Rice Value), the asphalt content, and the aggregate properties such as Atterberg Limits and gradation.

4.6.2.1 Maximum Theoretical Specific Gravity (Rice Value)

The maximum theoretical Specific Gravity test is widely used to determine the percent compaction and, more importantly, the percent of air void. A known quantity of loosely compacted hot mix asphalt is placed in a flask. A vacuum is applied to the flask to remove all the air from the sample while the sample is shaken. The specific gravity is then calculated to determine the density of the HMA with zero voids.

4.6.2.2 Asphalt Content

The amount of asphalt cement needs to be determined to ensure the percentage of asphalt cement in the HMA is within the range specified in the specifications. Two main procedures are used to identify the amount of asphalt cement content. The first is the ignition oven, which is a destructive test, and the other test is a non-destructive test using a nuclear oven. A solvent extraction centrifuge used to be used as another procedure to determine asphalt content. This method is not used often because the solvent is a hazardous waste, and extra precautions are required.

4.6.2.2.1 Ignition Oven

Obtaining asphalt content from the ignition oven method requires taking a known quantity of HMA and placing the sample into a very hot oven exceeding 1000°F, cooking the asphaltic cement out of the sample. The remains of the sample are just aggregate and light ash that can easily be washed away. The aggregate can then be tested for gradation. This method is not acceptable for aggregates that are susceptible to high heat. The results will read a higher than accurate asphalt content considering the breakdown of the aggregate.

FIGURE 4.10 Ignition Oven

4.6.2.2.2 Nuclear Oven

The nuclear asphalt content gauge (nuclear oven) uses a radioactive source and backscatter detectors to identify the hydrogen ions found in the asphalt binder. The test procedure is relatively quick once the gauge is calibrated. Calibration requires obtaining numerous samples of known asphalt content to obtain a data history of different mixes. The nuclear asphalt content gauge will require to be licensed with either the EPA or the State Health Department. Since this gauge only identifies the amount of asphalt content and does not strip the asphalt binder from the aggregate, additional sampling of the aggregate must be conducted.

4.6.2.3 Aggregate

Asphalt pavement specifications normally require a gradation range the aggregate must meet as well as Atterberg Limits that are not to be exceeded. The aggregate can be sampled from the batch plant at approximately the same time the mix is being batched to get a reasonably representative sample. This method is required if the nuclear oven method is used to determine the asphalt content. A more representative sample of the aggregate is obtained from the ignition oven method. The aggregate coming from the ignition oven must be washed with detergent to wash the ash and clean the aggregate particle surface from the remains of the binder.

4.6.2.4 Bulk Specific Gravity

After all compaction efforts are completed, laboratory compacted cores or cored specimens obtained from the field are tested for bulk-specific gravity. These tests are used to calculate volumetric properties during the mix design process and determine the percentage of air voids from the laydown process.

4.6.3 Field Quality Control

Field Quality Control testing is provided to evaluate the product and the workmanship of the HMA pavement in accordance with the specifications. Field Quality Control also includes visual observations and field testing. For this book, we will only discuss field testing.

4.6.3.1 Temperature

Temperature control is critical to many HMA operations, including production, placement, and compaction. The temperature of the HMA is commonly checked in the haul trucks as well as in the hopper of the paver. Surface temperatures, either analog or infrared, are used to check the temperature of the asphalt pavement surface and are useful in determining when and which roller should be used.

4.6.3.2 Compaction

Once the HMA pavement has been laid and compacted, the compaction of the asphalt mat is tested. The compaction of the asphalt pavement is to achieve an optimum air void content and provide a smooth, uniform surface. The air void content is the single most important factor affecting the mixture's performance throughout the pavement's life. A high air void means the compaction is not within the specification range, which may cause future rutting and deflection in the pavement surface. An air void that does not meet the specifications on the low end means the pavement is over-compacted. This tends to lead to surface bleeding and/or unraveling of the aggregate from the mixture. The two most common methods for testing the compaction are with a nuclear density gauge or obtaining a core specimen. Both methods compare the density of the asphalt test to the maximum theoretical specific value (Rice Value) of the mix from the laboratory results.

The air voids are calculated by subtracting 100% from the percentage of the compaction results. For example, you have a "Rice Value" of 146.8 pcf. The nuclear density test or core test density determines the density of the test to be 138.2 pcf. The percent compaction is 138.2 pcf/146.8 pcf = 0.941 or 94.1% compaction. The air void is calculated from 100%, 100% - 94.1% = 5.9%. If the specification requires air voids to be between 4% and 6%, the asphalt pavement meets the air void requirements.

4.6.3.2.1 Cores

Core density is the most accurate means to obtain the density of the asphalt pavement as well as the thickness of the mat or each layer of the mat can be easily determined. Cores should not be obtained until the pavement has sufficient time to cool so the core will remain in tack while the coring procedure is conducted. The core specimen is obtained from a coring machine spinning a core bit core at a fast enough revolution per minute to cut through the asphalt smoothly. Water is pumped through the bit to wash the cuttings from the core. The core must be dried prior to conducting specific gravity testing. As you can see, this procedure takes time, and the Contractor will not get compaction test results for several days. The time required to obtain the percentage of air voids will not allow the Contractor to provide additional compaction efforts if needed. Therefore, a reduction in payment is generally applied for the installation of the HMA.

FIGURE 4.11 Core
Note two layers of asphalt

4.6.3.2.2 Nuclear Density Guage

Nuclear density tests are another means to obtain compaction test results, allowing the percentage of air voids to be calculated. The nuclear density gauge procedure is the same as how nuclear density tests are performed with soil, as described in Chapter 2. The one difference is that the nuclear gauge probe is not extended into the asphalt pavement but is placed in a backscatter mode, with the probe touching the asphalt surface. The nuclear isotopes penetrate the asphalt and obtain the readings from the nuclear returns. The compaction results are very quick, giving the results in the field, while the Contractor can try to obtain additional compaction in the mat. The test can be run at any temperature. The downside of this test is it relies on the surface being perfectly smooth. The installed pavement may have very minimal waves on the surface due to the pneumatic tire roller, which may influence the compaction results. This is why the core method is more accurate than the nuclear density gauge.

Problems – Chapter 4

1. The most suitable bituminous pavement produces what type of pavement?
   a. Stable.
   b. Durable.
   c. Flexible.
   d. Skid resistance.
   e. All of the above.

2. What ingredients make up asphalt?

3. The following properties are chosen for asphalt pavement. List all that apply.
   a. Cost of material
   b. Maximum particle size and grading.
   c. Toughness.
   d. Initial set of the asphalt.
   e. Specific gravity.

4. What does RAP stand for?

5. Which physical property listed below is not part of an optimum mixture?
   a. Resistance to permanent deformation.
   b. Durability.
   c. Skid resistance.
   d. Fatigue resistance.
   e. High temperature cracking.
   f. Workability.

6. The design of an HMA mix will consist of what following choices.
   a. Type and gradation of aggregate.
   b. Type and time of compaction equipment.
   c. Type and grade of asphalt binder.
   d. Amount of asphalt binder.

7. In the Superpave Method of asphalt design, how many revolutions are recommended in the design process for a State Highway?

8. Site preparation is as important or more important as the design of the mixed asphalt during pavement construction. True or False.

9. If the subgrade or base coarse distortion occurs during the proof roll test, the test is passed. True or False?

10. You can place an overlay layer of asphalt right after a section of road is milled. True or False?

11. In Appendix F, Section 321113 – Stabilized Subgrade, what is the minimum testing frequency for in-place soil density?

12. What is the minimum testing frequency for thickness in Appendix F, Section 321123—Aggregate Base Course?

13. In Appendix F, Section 321216 – Hot Mix Asphalt Pavement, what is the air and surface temperatures for mix placement on 2 inches to <3 inches for the top layer of pavement using PG 76-28 binder?

14. In Appendix F, Section 321216 – Hot Mix Asphalt Pavement, all HMAP shall be compacted to between 92.0 and 98.0 percent of Maximum Theoretical (Rice) Density. True or False?

15. In Appendix F, Section 321216 – Hot Mix Asphalt Pavement, if the Engineer finds the materials furnished, the work performed, or the finished product do not conform to the contract, a cost reduction will occur. True or False?

16. Match the following temperature ranges with the appropriate materials:

a. Aggregate drying drum	1. 250°F to 820°F
b. Mix discharge	2. 300°F to 350°F
c. Asphalt cement stored	3. 290°F to 325°F
d. RAP	4. 220°F to 280°F

17. Hot asphalt mix is usually transported to the laydown site by?
    a. End-dump truck.
    b. Bottom or belly dump truck.
    c. Live bottom truck.
    d. All of the above.

18. Pavers are the best means to place the HMA for pavements. True or False?

19. Paver's main purpose is to place the HMA to the desired? List all that apply.
    a. Asphalt content.
    b. Width.
    c. Cross slope.
    d. Grade.
    e. Thickness.
    f. All of the above.

20. Mark yes (Y) or no (N) as to why adequate compaction is so important in the performance of the HMA pavement.
    ____ Increases fatigue life.
    ____ Increases permanent deformation.
    ____ Reduces oxidation.
    ____ Increases moisture damage.
    ____ Increases strength.
    ____ Decreases stability.
    ____ Decreases low-temperature cracking.

21. The breakdown process is conducted right after the asphalt mat has been laid. True or False?

22. It is important to obtain a representative sample. True or False?

23. A field nuclear density test is more accurate on asphalt than obtaining density from a field core. True or False?

24. The asphalt mat has a "Rice Value" of 146.6 pcf. A core is obtained from the recently laid asphalt mat with a density of 138.0 pcf. The specifications require air voids between 4% and 6%. What is the % compaction of the core? What is the % air void of the core? Does the asphalt meet the specifications for compaction?

REFERENCES

1. Das, Braja M., Principles of Geotechnical Engineering, Second Edition, PWS-Kent Publishing Company, Boston, 1994.
2. Liu, Cheng, and Evett, Jack B., Soils and Foundations, PEARSON, Boston, 2019.
3. Derucher, Kenneth N., and Korfiatis, George P., Materials for Civil & Highway Engineers, Second Edition, Prentice Hall, Englewood Cliffs, N.J., 1988.
4. Day, Robert W., Soil Testing Manual, McGraw-Hill, Inc., New York, 2001.
5. Sowers, George F., Introductory Soil Mechanics and Foundations: Geotechnical Engineering, Fourth Edition, Macmillan Publishing Co., Inc., New York, 1979.
6. Lindeburg, Michael R., Civil Engineering Review Manual, Third Edition, Professional Publications, San Carlos, CA, 1984.
7. Lambe, T. William and Whitman, Robert V., Soil Mechanics, John Wiley & Sons, Inc., New York, 1969.
8. AASHTO, AASHTO Guide for Design of Pavement Structures, Volume 1, AASHTO, Washington, DC, 1993.
9. Kosmatha, Steven H. and Panarese, William C., Design and Control of Concrete Mixtures, Thirteenth Edition, Portland Cement Association, Skokie, Illinois, 1994.
10. ACI, ACI Certification, Technician Workbook, Concrete Field Testing Technician – Grade I, American Concrete Institute, Thirty-Seventh Edition, Farmington, Michigan, 2019.
11. Annual Book of ASTM Standards, Section 4, Construction, Volume 4.02, Concrete and Aggregate, ASTM International, West Conshohocken, PA, 2020.
12. Annual Book of ASTM Standards, Section 4, Construction, Volume 4.03, Road and Paving Materials, Vehicle Pavement Systems, ASTM International, West Conshohocken, PA, 2020.

13. Annual Book of ASTM Standards, Section 4, Construction, Volume 4.08, Soil and Rock (I), ASTM International, West Conshohocken, PA, 2020.
14. Annual Book of ASTM Standards, Section 4, Construction, Volume 4.09, Soil and Rock (II), ASTM International, West Conshohocken, PA, 2020.
15. Laboratory for the Certification of Asphalt Technicians (LabCAT), Level A – Laydown, Level B – Plant Materials Control, Rocky Mountain Asphalt Education Center.
16. The Asphalt Handbook, 7th Edition, Second Printing, Manual Series No. 4 (MS-4), The Asphalt Institute, College Park, MA, 2010.
17. Standard Specifications for Road and Bridge Construction, State Department of Highways, Division of Highways, State of Colorado, Denver, CO, 1986.
18. Bowles, Joseph E., Engineering Properties of Soils and Their Measurement, Second Edition, McGraw-Hill Book Company, New York, 1978.
19. Hot-Mix Asphalt Paving, Handbook 2000, US Army Corps of Engineers, AC 150/5370-14A, Appendix 1.
20. A Guide to Swelling Soil for Colorado Homebuyers and Homeowners, Second Edition, Colorado Geological Survey Special Publication 43, 2007.

APPENDIX A—GLOSSARY

Introduction

The following is a list of commonly used terms and definitions.

References

References used for the Glossary include:

- Soil Testing Manual
- AASHTO Guide for Design of Pavement Structures
- International Building Code
- American Society for Testing and Materials
    - ASTM D 8
    - ASTM C 125
    - ASTM D 653
    - ASTM D 4439

GLOSSARY

Absorption:	Defined as the mass of water in the aggregate divided by the dry mass of the aggregate. Absorption is used in soil mechanics to study of oversized particles or in the concrete mix design.
Adhesion:	Shearing resistance between two different materials. For example, for piles driven into clay deposits, there is adhesion between the surface of the pile and the surrounding clay.
Active Earth Pressure:	Horizontal pressure for a condition where the retaining wall has moved sufficiently to allow the backfill to shift its shear strength.
Aggregate:	A granular material used for pavement base, wall backfill, etc. It can be classified as coarse, fine, or open-graded.
Allowable Bearing Capacity:	The maximum allowable bearing pressure for the design of foundations.
Allowable Bearing Pressure:	The maximum pressure that can be imposed by a foundation onto soil or rock supporting the foundation. It is derived from experience and general usage and provides an adequate factor of safety against shear failure and excessive settlement.
Alluvium:	Detrital deposits resulting from the flow of water, including sediments deposited in river beds, canyons, floodplains, lakes, fans at the foot of slopes, and estuaries.
Anchor Bolts:	Bolts which fasten columns, girders or other members to concrete or masonry such as bolts used to anchor foundation plates or sill plates to the foundation.
Anisotropic Soil:	A soil mass having different properties in different directions at any given point referring primarily to stress-strain or permeability characteristics.
Apparent Opening Size:	For a geotextile, a property that indicates the approximate largest particle that would effectively pass through the geotextile.

Approval: A written engineering or geologic opinion by the responsible engineer, geologist of record, or responsible principal of the engineering company concerning the process and completion of the work, unless it specifically refers to the building official.

Approval Plans: The current plan, such as grading plans, bears the stamp of approval of the building official.

Approved testing agency: A facility whose testing operations are controlled and monitored by a registered civil engineer and which is equipped to perform and certify the tests as required by the local building code or building official.

Aquiclude: A relatively impervious rock or soil stratum that will not transmit groundwater fast enough to furnish an appreciable supply of water to a well or spring.

Aquifer: A relatively pervious rock or soil stratum that will transmit groundwater fast enough to furnish an appreciable supply of water to a well or spring.

Arching: The transfer of stress from an unconfined area to a less yielding or restrained structure. Arching is important in the design of pile or pier walls that have open gaps between the members.

Architect: A licensed architect who is registered to practice in the field of architecture. Architects often oversee the building process.

Artesian: Groundwater that is under pressure and is confined by impervious material. If the trapped pressurized water is released, such as by drilling a well, the water will rise above the groundwater table and may even rise above the ground surface.

As Graded. Generally called as built. The descriptor of the surface conditions at the completion of grading.

Asphalt: A dark brown to black cementitious material whose main ingredient is bitumen, that occurs in nature or is obtained from petroleum processing.

Asphalt Concrete (AC): A mixture of asphalt and aggregate that is compacted into a dense pavement surface.

At-Rest Earth Pressure:	Horizontal pressure for a condition where the retaining wall has not moved or compressed into the soil. This would also be applicable to a soil mass in its natural state.
Atterberg Limits:	Water contents correspond to different behavior conditions of plastic soil. See Liquid Limit, Plastic Limit, and Shrinkage Limit.
Average Degree of Consolidation:	The ratio, expressed as a percentage, of the settlement at any given time to the primary consolidation.
Backdrain:	Generally, a pipe and gravel or similar drainage system is placed behind earth-retaining structures such as buttresses, stabilization fills, and retaining walls.
Backfill:	Soil material placed behind or on top of an area that has been excavated. For example, a backfill is placed behind retaining walls and in utility trench excavations.
Base Coarse:	Sometimes called base. A layer of specified or selected material of planned thickness constructed on the subgrade or subbase for the purpose of providing support to the overlying concrete or asphalt concrete surface of roads and airfields.
Batter Pile:	A pile driven in at an angle inclined to the vertical to provide higher resistance to lateral loads.
Bearing Wall:	Any metal or wood stud wall that supports more than 100 pounds per linear foot of superimposed load. Any masonry or concrete wall that supports more than 200 pounds per linear foot of superimposed load or is more than one story.
Bedrock:	A more or less solid, relatively undisturbed rock in place either at the surface or beneath deposits of soil.
Bell:	The enlarged portion of the bottom of a drilled shaft foundation. A bell is used to increase the end-bearing resistance. Not all drilled shafts have bells.
Bell-and-Spigot Pipe:	A type of pipe that is expanded at one end (bell) to receive the unexpanded end (spigot) of the section. With the assistance of a proper sealing material, a tight joint is formed at the union of the two sections.

Bench: A relatively level step excavated into earth material on which fill is to be placed.

Bentonite: A soil or formational material that has a high concentration of the clay mineral montmorillonite. Bentonite is usually characterized by high swelling upon wetting. The term bentonite also refers to manufactured products that have a high concentration of montmorillonite, for example, bentonite pellets.

Berm: A raised bank or path of soil. For example, a berm is often constructed at the top of slopes to prevent water from flowing over the top of the slope.

Bid Bond: Security posted by a bidder to ensure performance in accordance with a bid.

Bid Documents: Drawings, details, and specifications for a particular project.

Binder, Soil: Typically clay-sized particles that can bind together or provide cohesion between soil particles. Organic matter and precipitation of cementing minerals can also bind together soil particles.

Bit: A device that is attached to the end of the drill stem and is used as a cutting tool to bore into soil and rock.

Bitumen: Dark brown to black cement-like residuum obtained from the distillation of suitable crude oils.

Bog: A peat-covered area with a high groundwater table. The surface is often covered with moss, and it tends to be nutrient-poor and acidic.

Boring: A method of investigating subsurface conditions by drilling a hole into the earth's materials. Usually, soil and rock samples are extracted from the boring. Field tests, such as the standard penetration test (SPT) and the vane shear test (VST), can also be performed in the boring.

Boring Log: A written record of the materials penetrated during the subsurface exploration.

Borrow: Earth material acquired from an off-site location for use in grading on a site.

Boulder:	A large detached rock fragment with an average dimension greater than 12 inches (300 mm).
Building Official:	The city engineer, director of the local building department, or a duly delegated representative.
Bulking:	The increase in the volume of soil or rock caused by its excavation. For example, rock or dense soil will increase in volume upon excavation or by being dumped into a truck for transportation.
Buttress Fill:	A fill mass, the configuration of which is designed by engineering calculations to stabilize a slope exhibiting adverse geologic features. A buttress is generally specified by minimum key width and depth and by maximum backcut angle. A buttress normally contains a back drainage system.
Caisson:	Sometimes large diameter piers. Another definition is a large structural chamber utilized to keep soil and water from entering a deep excavation or construction area. Caissons may be installed by being sunk in place or by excavating the bottom of the unit as it slowly sinks to the desired depth.
California Bearing Ratio:	The CBR can be determined for soil in the field or soil compacted in the laboratory. The CBR is frequently used for the design of roads and airfields.
Capillarity:	Also known as capillary action and capillary rise, the rise of water through soil due to the fluid property known as surface tension. Due to capillarity, pore water pressures are less than atmospheric because of the surface tension of pore water acting on the meniscus formed in void spaces between the soil particles. The height of capillary rise is inversely proportional to the pore size of the soil.
Casing:	A steel pipe that is temporarily inserted into a boring or drilled shaft to prevent the adjacent soil from caving.
Cation Exchange Capacity:	The capacity of clay-sized particles to exchange cations with the double layer.
Certificate of Occupancy:	A document stating that a building is approved for occupancy. The building authority issues the C.O.

Cesspool: A term often used interchangeably with seepage pit. A cesspool is defined as a pit that receives unsettled sewage.

Checking: A pattern of surface cracks running in irregular lines. When found in the top surface of asphalt it is the preliminary stage of alligatoring.

Civil Engineer: A professional engineer who is registered to practice in the field of civil works.

Civil Engineering: The application of the knowledge of the forces of nature, principles of mechanics, and properties of materials for the evaluation, design, and construction of civil works for the beneficial uses of humankind.

Clay Size Particles: Clay-size particles are finer than 0.002 mm. Most clay particles are flat or platelike in shape, and they have a large surface area.

Clearing, Brushing, Grubbing: The removal of vegetation by mechanical means.

Clogging: For a geotextile, a decrease in permeability is due to soil particles that either have lodged in the geotextile openings or have built up a restrictive layer on the surface of the geotextile.

Cluster Structure: A soil structure where soil grains consist of densely packed silt or clay-sized particles.

Coarse Aggregate: Gravel or crushed rock that is retained on the No. 4 (4.75 mm) sieve.

Coarse-Grained Soil: According to the Unified Soil Classification System, coarse-grained soils have more than 50 percent soil particles (by dry mass) retained on the No. 200 U.S. Standard sieve.

Cobble: A rock fragment, usually rounded or semi-rounded, with an average dimension between 3 and 12 inches (75 and 300 mm).

Coefficient of Compressibility: The change in void ratio divided by the corresponding change in vertical effective stress.

Coefficient of Curvature and Coefficient of Uniformity: These two parameters are used for the classification of coarse-grained soils (USCS) and nonplastic soils (ISBP). These tow parameters are used to distinguish a well graded soil from a poorly graded soil.

Cohesion: There are two types of cohesion: cohesion in terms of total stress and cohesion in terms of effective stress. For total cohesion, the soil particles are predominately held together by capillary tension. For effective stress cohesion, there must be actual bonding or attraction forces between the soil particles.

Cohesionless Soil: A soil, such as clean gravel or sand, that, when unconfined, falls apart in either a wet or dry state.

Cohesive Soil: A soil, such as silt or clay, that when unconfined, has considerable shear strength when dried and will not fall apart in a saturated state. Cohesive soil is also known as plastic soil or soil that has a plasticity index.

Collapsible Formations: A type of soil formation that tends to erode or decrease bearing capacity generally when water is added. Examples of collapsible formations include limestone and coal beds. Limestone can form underground caves and caverns, which can gradually enlarge, resulting in the collapse of the ground surface and the formation of sinkholes. Sites that are underlaid by coal or salt mines could also experience ground surface settlement when the underground mine collapses.

Collapsible Soil: Soil that is susceptible to large and sudden reduction in volume upon wetting. Collapsible soil usually has a low dry density and low moisture content. Such soil can withstand a large applied vertical stress with small compression but then experience much larger settlements after wetting, with no increase in vertical pressure. Collapsible soil can include fil compacted dry of optimum and natural collapsible soil, such as alluvium, colluvium, or loess.

Colluvium: Generally loose deposits usually found near the base of slopes and brought there chiefly by gravity through slow, continuous downhill creep.

Combination End-bearing and Friction Pile: A pile that derives its capacity from combined end-bearing resistance developed at the pile tip and frictional and/or adhesion resistance on the pile perimeter.

Community Well: A water well used for domestic water supply by more than one dwelling.

Compaction:	The densification of a fill by mechanical means.
Compaction Curve:	A curve showing the relationship between the dry density and the water content of a soil for a given compaction energy.
Compaction Test:	A laboratory compaction procedure whereby a soil at a known water content is compacted into a mold of specific dimensions. The procedure is repeated for various water contents to establish the compaction curve. The objective of the laboratory compaction test is to obtain the laboratory's maximum dry density and the optimum moisture content for the tested soil.
Composite Pavement:	A pavement structure composed of an asphalt concrete wearing surface and Portland cement concrete slab.
Compressibility:	A decrease in volume that occurs in the soil mass when it is subjected to an increase in loading. Some highly compressible soils are loose sands, organic clays, sensitive clays, highly plastic and soft clays, uncompacted fills, municipal landfills, and permafrost soils.
Concrete:	A mixture of aggregates and paste (Portland cement and water). The paste binds the aggregates together into a rocklike mass as the paste hardens because of the chemical reactions between the cement and the water.
Cone Penetration Test:	A field test used to identify and determine the in situ properties of soil deposits and soft rock.
Consolidation:	The decrease in volume due to loading. The consolidation of a saturated clay deposit is generally divided into three separate categories: initial, primary, and secondary.
Consolidation Test:	A laboratory test used to measure the consolidation properties of saturated cohesive soil. The specimen is laterally confined in a ring and is compressed between porous plates while a load is applied.
Construction Loan:	A loan provided by a lending institution specifically to construct or renovate a building.
Consultant:	Firm or individual retained by the building owner to design, assist in the bidding, assist in the contract, oversee construction, provide quality control, or assist in the project completion.

Contraction:	During the shearing of soil, the tendency of loose soil to decrease in volume.
Contractor:	A person or company under contract or otherwise retained by the client to perform demolition, grading, and other site improvements.
Controlled Strain Test:	A laboratory test in which the load is applied so as to control the rate of strain. The shear portion of triaxial compression tests is often performed by subjecting the soil specimen to a specific rate of axial strain.
Controlled Stress Test:	A laboratory test where the load is applied in increments. The consolidation test is often performed by subjecting the soil specimen to an incremental increase in load, with the soil specimen subjected to each load for a period of 24 hours.
Core Drilling:	Also known as diamond drilling, the process of cutting out cylindrical rock samples in the field.
Core Recovery (RQD):	The RQD is computed by summing the lengths of all pieces of the rock core equal to or longer than 4 inches (10 cm) and dividing by the total length of the core run. The RQD is multiplied by 100 to express it as a percentage.
Covenants:	Rules usually developed by a builder or developer regarding the physical appearance of buildings or the developed area. Typical covenants address building height, appropriate fencing, landscaping, and the exterior material that may be used.
Creep:	An imperceptibly slow and more or less continuous movement of slope forming soil or rock debris. In laboratory tests, drained creep occurs when a plastic soil experiences continued deformation under constant effective stress.
Critical Height:	The maximum height at which a vertical excavation will stand unsupported. The maximum height at which a sloped bank of soil or rock will stand unsupported.
Critical Slope:	The maximum slope at which a vertical excavation will stand unsupported at a given height. The maximum angle at which a sloped bank of soil or rock of a given height will stand unsupported.

Crown:	The highest point on the structure. For tunnels, the crown is the arched roof. For landslides, it is the area above the main scarp of the slide.
Curing:	In concrete application, the process in which the mortar and concrete harden. Favorable curing temperatures range from 50 to 70 degrees Fahrenheit. Design strength is achieved in 28 days.
Cut-Fill Transition:	The location on a grading site where one side consists of cut and the other side consists of fill.
Cutoff Wall:	The construction of tight sheeting or a barrier of impervious material extending downward to an essentially impervious lower boundary to intercept and block the path of groundwater seepage. Cutoff walls are often used in dam construction.
Dam:	A structure built to impound water or other fluid products such as tailing waste or wastewater effluent.
Damproofing:	A process used on concrete, masonry or stone surfaces to repel water, the main purpose of which is to prevent the coated surface from absorbing rain water while still permitting moisture vapor to escape from the structure.
Darby:	Also known as bullfloat. A flat tool used to smooth concrete flatwork immediately after screeding.
Dead Load:	Structural loads due to the weight of beams, columns, floors, roofs, and other fixed members. It does not include nonstructural items such as furniture, snow, wind, occupants, or inventory.
Debris:	All products of cleaning, grubbing, demolition, or contaminated soil material that are unsuitable for reuse as compacted fill and/or any other material so designated by the geotechnical engineer or building official.
Debris Flow:	An initial shear failure of a soil mass which then transforms itself into a fluid mass which can move rapidly over the ground surface.
Deep Foundation:	A foundation that derives its support by transferring loads to soil or rock at some depth below the structure.

Deflection: The amount of bending movement of any part of a structural member perpendicular to the axis of the member under an applied load.

Defloculating Agent: Used during the hydrometer test, a compound such as sodium hexametaphosphate that prevents clay size particles from forming into clumps.

Degree of Saturation: The volume of water in the void space divided by the total volume of voids expressed as a percentage. A completely dry soil has a degree of saturation of 0 percent, and a saturated soil has a degree of saturation of 100 percent.

Density: Mass per unit volume.

Deposition: The geologic process of laying down or accumulating natural material into beds, veins, or irregular masses. Deposition includes mechanical settling (such as sedimentation in lakes), precipitation (such as the evaporation of surface water to form halite), and the accumulation of dead plants (such as in peat bog).

Depth of Seasonal Moisture Change: Also known as the active zone. The layer of expansive soil is subjected to shrinkage during the dry season and to swelling during the wet season. This zone extends from the ground surface to the depth of significant moisture fluctuation.

Desiccation: The process of shrinkage of clays. The process involves a reduction in volume of the grain skeleton and subsequent cracking of the clay caused by the development of capillary stresses in the pore water as the soil dries.

Design Pressure: Specified pressure a product is designed to withstand.

Dewatering: The process used to remove water from a construction site, such as pumping from wells to lower the groundwater table during a foundation excavation.

Detritus: All material worn or broken down from rocks by mechanical means.

Deviator Stress: Difference between the major and minor principal stress in a triaxial test.

Differential Settlement: The difference in settlement between two foundation elements or between two points on a single foundation.

Digestion: The natural biological process which breaks down the organic matter into its components or basic parts.

Dilation: During the shearing of soil, the tendency of dense soil to increase in volume.

Direct Shear Test: A laboratory test used to obtain the effective shear strength properties of saturated soil. The test consists of applying a vertical pressure to the laterally confined soil specimen, submerging the soil specimen in distilled water, allowing the soil to consolidate, and then shearing the soil specimen by moving the top of the shear box relative to the fixe bottom. The soil specimen must be sheared slowly enough that excess pore water pressures do not develop.

Dispersed Structure: A soil structure where the clay-sized particles are oriented parallel to one another.

Distribution Box: A container that distributes or divides the flow of effluent from a septic tank into several seepage pits or leaching lines.

Domestic Well: A water well used for drinking and cooking purposes.

Double Layer: A grossly simplified interpretation of the positively charged water layer, together with the negatively charged surface of the particle itself in clay. Two reasons for the attraction of water to the clay particle are (1) the dipolar structure of water molecule causes it to be electrostatically attracted to the surface of the clay particle, and (2) the clay particles attract cations which contribute to the attraction of water by the hydration process. The absorbed water layer consists of water molecules that are tightly held to the clay particle faces, such as by the process of hydrogen bonding.

Downdrag: Also known as negative skin friction. Force induced on the deep foundation resulting from downward movement of adjacent soil relative to foundation element.

Drainage: The removal of surface water from the site.

Drawdown:	The lowering of the groundwater table that occurs in the vicinity of a well that is in the process of being pumped.
Earth Material:	Any rock, natural soil, fill, or any combination thereof.
Earth Pressure:	Usually used in reference to the lateral pressure imposed by soil mass against an earth supporting structure such as a retaining wall or basement wall. Three earth pressures are generally discussed: active, at rest, and passive.
Effective Stress:	The total stress minus the pore water pressure.
Effluent:	The liquid portion of sewage that flows out of a septic tank or has had the solids removed.
End-bearing Pile:	A pile whose support capacity is derived principally from the resistance of the foundation material on which the pile tip rests.
Engineering Geologist:	A geologist who is experienced and knowledgeable in the field of engineering geology.
Equivalent Fluid Pressure:	Horizontal pressure of soil, or soil and water in combination, which increases linearly with depth and is equivalent to that which would be produced by a soil of a given density. Equivalent fluid pressure is often used in the design of retaining walls.
Equivalent Single Axle Loads (ESAL):	Summation of equivalent 18,000 pound single axle loads used to combine mixed traffic to design traffic for the design period of pavement.
Erosion:	The wearing away of the ground surface as a result of the movement of wind, water, and/or ice.
Erosion Control Devices (temporary):	Devices that are removable and can rarely be salvaged for subsequent reuse. In most cases, they will last no longer than one rainy season. They include sandbags, gravel bags, plastic sheeting, silt fencing, straw bales and similar items.
Erosion Control System:	A combination of desilting facilities and erosion protection, including effective planting to protect adjacent private property, watercourses, public facilities, and receiving waters from any abnormal deposition of sediment or dust.

Excavation:	The mechanical removal of earth material.
Excess Pore Water Pressure:	The increment of pore water pressures greater than hydrostatic values, produced by consolidation stress in compressible materials or by shear strain.
Existing Grade:	The ground surface prior to grading.
Exit Gradient:	The hydraulic gradient near the toe of a dam or the bottom of an excavation through which groundwater seepage is exiting the ground surface.
Expansive Soil or Rock:	Soil or rock particles that are susceptible to increased volume change with the addition of water. Clay soils and shale or slate rock tend to be expansive.
Fabric of Soils:	Generally defined as the geometric arrangement of the soil particles. In contrast, the soil structure refers to both the geometric arrangement of soil particles and the interparticle forces which may act between them.
Fault:	A fracture in the earth's crust along which movement has occurred. A fault is considered active if the movement has occurred within the last 11,000 years.
Field Lateral:	Lengths of pipe joined together to form a line.
Fill:	A deposit of earth material placed by artificial means. An engineered (or structural) fill refers to a fill in which the geotechnical engineer has, during grading, made sufficient tests to enable the conclusion that the fill has been placed in substantial compliance with the recommendations of the geotechnical engineer and the governing agency requirements.
Fine-Grained Soil:	Per the Unified Soil Classification System, soil containing more than 50 percent (by dry mass) of particles finer than the No. 200 sieve
Fine Aggregate:	Often, sand (passes the No. 4 sieve and is retained on the No. 200 sieve.
Fines:	The silt and clay-sized particles in the soil. Soil particles that are finer than the No. 200 sieve.

Finished Grade:	The final grade of the site which conforms to the approved plan.
Flexible Pavement:	A pavement structure that maintains intimate contact with and distributes loads to the subgrade and depends on aggregate interlock, particle friction, and cohesion for stability.
Flocculated Structure:	Also known as card house structure. A soil structure where the clay size particles are oriented in edge-to-face arrangements.
Flocculation:	When in suspension in water, the process of fines attracting each other to form a larger particle or floc. In the hydrometer test, a dispersing agent is added to prevent the flocculation of fines.
Floor Plan:	The basic layout of the building or addition, which includes placement of walls, windows, and doors as well as dimensions.
Flow Line:	The path of travel traced by moving groundwater as it flows through a soil mass.
Flow Net:	A graphical representation used to study the flow of groundwater through soil.
Fold:	Bending or flexure of a layer or layers of rock. Usually, folds are created by the massive compression of rock layers.
Footing:	A structural member typically installed at a shallow depth that is used to transmit structural loads to the soil or rock strata. Common types of footings include combined footings, spread footing, pad footing, and strip footings.
Forms:	Structures usually made of wood or steel and used during the placement of concrete. Forms confine and support the fluid concrete as it hardens.
Foundation:	That part of the structure that supports the weight of the structure and transmits the load to underlying soil or rock. Foundations are considered deep or shallow.
Fracture:	A visible bread in a rock mass. Examples include joints, faults, and fissures.

Freeze: Also known as setup. An increase in the load capacity of a pile after it has been driven. Freeze is caused primarily by the dissipation of excess pore water pressures.

Friction Angle: In terms of effective shear stress, soil friction is usually considered to be due to the interlocking of the soil or rock grains and the resistance to sliding between the grains. A relative measure of a soil's frictional shear strength is the friction angle.

Friction Pile: A pile whose support capacity is derived principally from the resistance of the soil friction and/or adhesion mobilized along the side of the embedded pile.

Frost Heave: Refers to the development of ice layers or lenses within the soil that cause the ground surface to heave upward.

Geologist: An individual educated and trained in the field of geology.

Geophysical Techniques: Various methods of determining subsurface soil and rock conditions without performing subsurface exploration. A common geophysical technique is to induce a shock wave into the earth and then measure the seismic velocity of the wave's travel through the earth material. The seismic velocity has been correlated with the rip ability of the earth material.

Geosynthetic: A planar product manufactured from a polymeric material and typically placed in soil to form an integral part of a drainage, reinforcement, or stabilization system. Types include geotextiles, geogrids, geonets, and geomembranes.

Geotechnical Engineer: A licensed individual who performs an engineering evaluation of earth materials, including soil, rock, groundwater, and artificial materials, and their interaction with earth retention systems, structural foundations, and other civil engineering works.

Geotextile: A permeable geosynthetic composed solely of textiles.

Grade: The vertical location of the ground surface. The types of grades encountered are existing, finished, lowest adjacent, natural, and rough grade.

Grading:	Any operation consisting of excavation, filling, or combination thereof.
Grading Contractor:	A contractor licensed and regulated who specializes in grading work or is otherwise licensed to do grading work.
Grading Permit:	An official document or certificate issued by the building official authorizing grading activity as specified by approved plans and specifications.
Gravel Size Fragments:	Roc, fragments and soil particles that will pass the 3 inch (76 mm) sieve and be retained on a No. 4 sieve.
Groundwater Table:	Also known as the phreatic surface. The top surface of underground water, the location of which is often determined from piezometers, such as an open standpipe. A perched groundwater table refers to groundwater occurring in an upper zone separated from the main body of groundwater by underlying unsaturated rock or soil.
Grout:	A cement mortar mixture commonly used to fill joints and cavities of masonry or concrete.
Grouting:	The process of injecting grout into soil or rock formations to change their physical characteristics. Common examples include grouting to decrease the permeability of soil or rock strata or compaction grouting to densify loose soil or fill.
Gunite:	A construction material composed of cement, sand or crushed slag and water mixed and forced through a cement gun by pneumatic pressure, used in the construction of swimming pools.
Heave:	The upward movement of foundations or other structures caused by frost heave or expansive soil and rock.
Hillside Site:	A site which entails cut and/or fill grading of a slope that may be adversely affected by drainage and/or stability conditions within or outside the site, or which may cause an adverse effect on adjacent property.
Homogeneous Soil:	Soils that exhibit essentially the same physical properties at every point throughout the soil mass.

Honeycomb:	In reference to concrete construction; (1) Areas in a foundation wall where the aggregate is visible. Honeycombs can usually be remedied by applying a thin layer of grout or other cement product over the affected area. (2) Method by which concrete is poured and not puddled or vibrated, allowing the edges to have voids or holes after the forms are removed.
Honeycomb Structure:	A soil structure where loosely arranged bundles of soil particles, having a structure that resembles a honeycomb.
Horizon:	One of the layers of a soil profile that can be distinguished by its texture, color, and structure.
Hydraulic Conductivity:	Also known as coefficient of permeability. For laminar flow of water in soil, both terms are synonymous and indicate a measure of the soil's ability to allow water to flow through its soil pores.
Hydraulic Fill:	A fill placed by transporting soils through a pipe using large quantities of water. These fill are generally loose because they have little or no mechanical compaction during construction.
Hydrostatic Pore Water Pressure:	Pore water pressure or groundwater pressure is exerted under conditions of no flow where the magnitude of pore pressure increases linearly with the depth below the groundwater table.
Impervious Soil:	For sewage disposal systems, compact or clayey soil that has a very low percolation rate or does not permit the infiltration of water.
Inclinometer:	The horizontal movement preceding or during the movement of slopes can be investigated by successive surveys of the shape and position of flexible vertical casings installed in the ground. The surveys are performed by lowering an inclinometer probe into the flexible vertical casing.
Initial Consolidation:	The initial settlement of the structure caused by undrained shear deformation, or in some cases it contains plastic flow due to two or three-dimensional loading.
In situ:	Used in reference to the original in-place condition of the soil or rock.

Isotropic Soil:	A soil mass having essentially the same properties in all directions at any given point, referring primarily to stress-strain or permeability characteristics.
Invert:	The floor, bottom, or lowest portion of the internal cross-section of a closed conduit.
Jetting:	The use of water jet to facilitate the installation of a pile. It can also refer to the fluid placement of soil, such as jetting in the soil for a utility trench.
Key:	A designed compacted fill placed in a trench excavated in earth material beneath the toe of a proposed fill slope.
Keyway:	An excavated trench into competent earth material beneath the toe of a proposed fill slope.
Laboratory Maximum Dry Density:	The peak point of the compaction curve.
Laminar Flow:	Groundwater seepage in which the total head loss is proportional to the velocity.
Landslide:	Mass movement of soil or rock that involves shear displacement along one or several rupture surfaces, which are either visible or may be reasonably inferred.
Landslide Debris:	Material, generally porous and of low density, produced from instability of natural or artificial slopes.
Leaching:	The removal of soluble materials in soil or rock caused by percolating or moving groundwater.
Leaching Lines:	Tile or perforated pipes that permit the percolation (leaching) of effluent into the ground.
Lift:	During compaction operations, a layer of soil is dumped by the construction equipment and then subsequently compacted as structural fill.

Liquefaction:	The sudden, large decrease of shear strength of a cohesionless soil caused by collapse of the soil structure, produced by shock or earthquake induced shear strains, associated with a sudden but temporary increase of pore water pressures. Liquefaction causes the cohesionless soil to behave as a fluid.
Liquid Limit:	An Atterberg Limit property described as the water content corresponding to the behavior change between the liquid and plastic states of a soil. The liquid limit is arbitrarily defined as the water content at which a pat of soil, cut by a groove of standard dimensions, will flow together for a distance of 0.5 inches (12.7 mm) under the impact of 25 blows in a stand liquid limit device.
Live Load:	Structural load due to nonstructural members, such as furniture, occupants, inventory, wind, and snow.
Loess:	A wind deposited silt often has a high porosity and low density which is often susceptible to collapse of its soil structure upon wetting.
Lowest Adjacent Grade:	Adjacent to the structure, the lowest point of elevation of the finished surface of the ground, paving, or sidewalk.
Mineral:	An inorganic substance that has a definite chemical composition and distinctive physical properties. Most minerals are crystalline solids
Moisture Content:	Also known as water content. Moisture content and water content are synonymous. The definition of moisture content is the ratio of the mass of water in the soil divided by the dry mass of the soil, usually expressed as a percentage.
Necking:	A reduction in cross-sectional area of a drilled shaft as a result of the inward movement of the adjacent soils.
Negative Pore Water Pressure:	Pore water pressure that is less than atmospheric. An example is capillary rise, which can induce a negative pore water pressure in the soil. Another example is the undrained shearing of dense or highly overconsolidated soils, where the soil wants to dilate during shear, resulting in negative pore water pressure.
Non-Destructive:	A method of examining a component whereby no damage is done to the component itself.

Nonplastic Soil: A granular soil that cannot be rolled or molded at any water content. A nonplastic soil has a plasticity index equal to zero, or the plastic limit is greater than the liquid limit. Nonplastic soil is known as cohesionless soil.

Normally Consolidated: The condition that exists if a soil deposit has never been subjected to an effective stress greater than the existing overburden pressure and if the deposit is completely consolidated under the existing overburden pressure.

Nuclear Gauge: A device used to detect moisture and/or density by measuring slowed, deflected neutrons.

Open Graded Aggregate: Generally, a gravel that does not contain any soil particles fine than No. 4 sieve.

Open-Joint Tile: Pipes without tight joints.

Optimum Moisture Content: The moisture content, determine from a laboratory compaction test, at which the maximum dry density of the soil is obtained using a specific compaction effort.

Organic Soil: Soil that partly or predominately consists of organic matter.

Overburden: The soil that overlies bedrock. In other cases, it refers to all material overlying a point of interest in the ground, such as the overburden pressure exerted on a clay layer.

Overconsolidated: The condition that exists if a soil deposit has been subjected to an effective stress greater than the existing overburden pressure.

Overconsolidation Ratio (OCR): The ratio of the pre-consolidation vertical effective stress to the current vertical effective stress.

Oversize Particles: For fill compaction, the gravel and cobble size particles are retained on the ¾ inch or No 4 sieve.

Owner: Any person, agency, firm, or corporation having a legal or equitable interest in a given real property.

Particle Size Distribution: Also known as grain size distribution or gradation, the distribution of particle sizes in the soil is based on dry mass.

Passive Earth Pressure:	Horizontal pressure for a condition such as a retaining wall footing that has moved into and compressed the soil sufficiently to develop its maximum lateral resistance.
Pavement Performance:	The trend of serviceability with load applications.
Pavement Rehabilitation:	Work undertaken to extend the service life of an existing facility. This includes placement of additional surfacing material and/or other work necessary to return an existing roadway, including shoulders, to a condition of structural or functional adequacy. This could include the complete removal and replacement of the pavement structure.
Pavement Structure:	A combination of subbase, base course, and surface course placed on a subgrade to support the traffic load and distribute it to the roadbed.
Peak Shear Strength:	The maximum shear strength along a shear failure surface.
Peat:	A naturally occurring highly organic deposit derived primarily from plant materials.
Percolation:	The movement of effluent into or through the soil.
Percolation Test:	A field test that is used to determine the percolation rate of water based on specified test procedures.
Percussion Drilling:	A drilling process in which a borehole is advanced by using a series of impacts on the drill rods and attached bit.
Performance and Payment Bond:	Guaranty by a sure company that if a contractor fails to perform under a contract, the surety company will complete the work.
Performance Period:	Also known as design period. In the construction of asphalt pavements, This is the period of time an initially constructed or rehabilitated pavement structure will last (perform) before reaching its terminal serviceability.
Permafrost:	Perennially frozen soil, also defined as ground that remains below freezing temperatures for 2 or more years. The bottom of permafrost lies at depths ranging from a few feet to more than 1000 feet.

The Active layer is defined as the upper few inches to several feet of ground that is frozen in winter but thawed in summer.

Permanent Erosion Control Devices:	Improvements that remain throughout the life of the development. They include terrace drains, downdrains, slope landscaping, channels, storm drains, etc.
Permeability:	The ability of water or other fluid to flow through a soil by traveling through the void spaces. A high permeability indicates flow occurs rapidly. A measure of the soil's permeability is the hydraulic conductivity or coefficient of permeability.
Permit:	An official document or certificate issued by the building official authorizing the performance of a specified activity.
Pier:	A deep foundation system, similar to a cast-in-place pile, that consists of column-like reinforced concrete members. Piers are often of large enough diameter to enable down-hole inspection. Piers are also commonly referred to as drilled shafts, bored piles, or drilled caissons.
Piezometer:	A device installed to measure the pore water pressure (or pressure head) at a specific point within the soil mass.
Pile:	A deep foundation system consisting of relatively long, slender, column-like members that are often driven into the ground. Batter, combination end-bearing and friction, end-bearing, and friction are different types of piles.
Piping:	The movement of soil particles as a result of unbalanced seepage forces produced by percolating water, leading to the development of ground surface boils or underground erosion voids and channels.
Pit (Test Pit):	An excavation made for the purpose of observing subsurface conditions, performing field tests, and obtaining soil samples. A pit also refers to an excavation in the surface of the earth from which ore is extracted, such as an open pit mine.
Plan Submittal:	Submission of construction plans to the city or county in order to obtain a building permit.

Plans: Drawings prepared in a manner that shows the proposed construction, including dimensions. Plans are generally accompanied with specification for the construction to be approved and built.

Plastic Limit: An Atterberg Limit property is described as the water content corresponding to the behavior change between a soil's plastic and semi-solid states. The plastic limit is arbitrarily defined as the water content at which the soil will just begin to crumble when rolled into a thread approximately 1/8 inch (3.2 mm) in diameter.

Plasticity: Term applied to silt and clay to indicate the soil's ability to be rolled and molded without breaking apart. A measure of a soil's plasticity is the plasticity index.

Plasticity Index: The liquid limit minus the plastic limit, expressed as a whole number.

Plastic Soil: A soil that exhibits plasticity. Plastic soil is also known as cohesive soil.

Plat: A map of a geographical area as recorded by the county depicting the boundary of the property. Plats need to be prepared by a licensed land surveyor.

Plot Plan: A bird's eye view showing how a building sits on the building lot, typically showing setbacks (how far the building must sit from the property line or other objects), easements, rights of way, and drainage.

Ponding: A condition where water stands in an area for prolonged periods due to poor drainage.

Pore Water Pressure: The water pressure that exists in the soil void spaces. Pore water pressure is usually described as excess, hydrostatic, or negative.

Porosity: The ratio of the volume of voids to the total volume of the soil or rock expressed as a percentage.

Porous Soil: For sewage disposal systems, a soil that has a high permeability that readily permits percolation of effluent into and through the soil.

Portland Cement: A mixture of certain minerals that, when mixed with water form a gray-colored paste and cure into a very hard mass.

Post:	A vertical member of wood, steel, concrete, or other material that transfers weight from the top of the post to whatever the post is resting on.
Post & Beam Construction:	The most common type of wall framing, using posts which carry horizontal beams on which joists are supported. It allows for fewer bearing partitions, and less material.
Pozzolan:	For concrete mix design, a siliceous or siliceous and aluminous material that will chemically react with calcium hydroxide within the cement paste to form compounds having cementitious properties.
Precast Concrete:	Concrete building components that are formed and cured at a factory and then transported to a work site for erection.
Precise Grading Permit:	A permit that is issued on the basis of approved plans that show the precise structure location, finish elevations, and all on-site improvements.
Preconstruction Pressure:	The greatest vertical effective stress to which a soil, such as a clay layer, has been subjected.
Prepared Roadbed:	In-place roadbed soils compacted or stabilized according to provisions for applicable specifications.
Pressure:	Can also be referred to as stress. The load is divided by the area over which it acts.
Primary Consolidation:	The compression of clays under load that occurs as excess pore water pressures slowly dissipate with time.
Progressive Failure:	Formation and development of localized stresses that lead to fracturing of the soil, which spreads and eventually forms a continuous rupture surface and a failure condition. Stiff fissured clay slopes are especially susceptible to progressive failure.
Pumping:	The ejection of foundation material, either wet or dry, through joints or cracks or along edges of rigid slabs resulting from vertical movements of the slab under a load.
Punching Shear Failure:	Shear failure where the foundation pushes into the soil due to the compression of soil directly below the footing as well as vertical shearing around the footing perimeter.

Quick Clay: A clay that has a sensitivity greater than 16. Upon remolding, such clays can exhibit fluid conditions.

Quick Condition: Also known as quicksand. A condition in which groundwater is flowing upward with a sufficient hydraulic gradient to produce a zero effective stress condition in the sand deposit.

Rebar: Reinforcing bar, normally made of steel, used to increase the tensile strength of concrete.

Refusal: During subsurface exploration, an inability to excavate any deeper with the boring equipment. Refusal could also be due to many different factors, such as hard rock, boulders, or a layer of cobbles.

Reinforced Concrete: A combination of steel and concrete using the best properties of each. The steel consists of rebar and is placed in the forms prior to pouring the concrete.

Relative Compaction: The degree of compaction (expressed as a percentage) defined as the field dry density divided by the laboratory maximum dry density.

Relative Density: Term applied to a sand deposit to indicate its relative density state, defined as the ratio of (1) the difference between the void ratio in the loosest state and the in situ void ratio or (2) the difference between the void rations in the loosest and in the densest states.

Residual Soil: Soil is derived by the in-place weathering of the underlying material.

Retaining Wall: A wall designed to resist the lateral displacement of soil or other materials.

Rigid Pavement: A pavement structure that distributes loads to the subgrade, having as one course a Portland cement concrete slab of relatively high bending resistance.

Ripping: The characteristic of rock or dense and rocky soils that can be excavated without blasting. Ripping is accomplished by using equipment such as dozer-rippers, ripper-scarified, or impact ripper.

Riprap: Rocks that are generally less than 2 tons (1800 kg) in mass that are placed on the ground surface, on slopes, or at the toe of slopes, or on top of structures to prevent erosion by wave action or strong currents.

Roadbed:	The graded portion of a road between top and side slopes, prepared as a foundation for the pavement structure.
Roadbed Material:	The material below the subgrade in cuts and embankments and in embankment foundations, extending to such depth as affects the support of the pavement structure.
Rock:	A relatively solid mass that has permanent and strong bonds between the minerals. Rock can be classified as being sedimentary, igneous, or metamorphic.
Rock Mechanics:	The application of the knowledge of the mechanical behavior of rock to engineering problems dealing with rock. Rock mechanics overlaps with engineering geology.
Rotary Drilling:	A drilling process in which a borehole is advanced by rotation of a drill bit under constant pressure without impact.
Rough Grade:	The stage at which the grade approximately conforms to the approved plan.
Rubber Tired Roller:	A roller with rubber tires commonly used for compacting subgrade or aggregate base.
Rubble:	Rough stones of irregular shape and sizes that are naturally or artificially broken from larger masses of rock. Rubble is often created during quarrying, stone cutting, and blasting.
Running Soil:	In tunneling or trench excavations, a granular material that tends to flow or "run" into the excavation.
Sand Boil:	The ejection of sand at the ground surface, usually forming a cone shape, caused by underground piping.
Sand Equivalent (SE):	A measure of the amount of silt or clay contamination in fine aggregate.
Sand Size Particles:	Soil particles that will pass the No. 4 (4.75 mm) sieve and be retained on the No. 200 (0.075 mm) sieve.

Screeding: The wood or metal straightedge used to strike off or level newly place concrete when casting concrete. Screeds can be the leveling device used or the form work used to level or establish the level of the concrete. Screeds can be hand used or mechanical.

Secondary Consolidation: The final component of settlement, the part that occurs after essentially all the excess pore water pressures have dissipated.

Seep: A small area where water oozes from the soil or rock.

Seepage: The infiltration or percolation of water through soil and rock.

Seepage Pit: A hole in the ground that is loosely walled and receives sewage liquids from a septic tank. The seepage pit permits the sewage liquids to filter into the ground.

Seepage Analysis: An analysis to determine the quantity of groundwater flowing through a soil deposit. For example, by using a flow net, the quantity of groundwater flowing through or underneath an earth dam can be determined.

Seepage Force: The frictional drag of water flowing through the soil voids.

Seepage Velocity: The velocity of flow of water in the soil. Superficial velocity is the velocity of flow into or out of the soil.

Select Material: A suitable native material obtained from a specified source, such as a particular roadway cut or borrow area, of a suitable material having specified characteristics to be used for a specific purpose.

Serviceability: The ability at time of observation of a pavement to serve traffic (autos and trucks) which use the facility.

Settlement: The permanent downward vertical movement experienced by structures as the underlying soil consolidates, compresses, or collapses due to structural load or secondary influences. Settlement can be described as differential or total.

Sewage Spring: Underground sewage water that reaches the surface of the ground or flows out of the ground surface.

Shallow Foundation:	A foundation that derives its support by transferring load directly to soil or rock at a shallow depth.
Shear Failure:	A failure in a soil or rock mass caused by shearing strain along one or more slip surfaces.
Shear Key:	Similar to a buttress, however, it is generally constructed by excavating a slot within a natural slope in order to stabilize the upper portion of the slope without grading encroachment into the lower portion of the slope. A shear key is also often used to increase the factor of safety of an ancient landslide.
Shear Plane:	A plane along which failure of soil or rock occurs by shearing.
Shear Strength:	The maximum shear stress that a soil or rock can sustain.
Shear Stress:	Stress that acts parallel to the surface element.
Shotcrete:	Mortar or concrete pumped through a hose and projected at high velocity onto a surface. Shotcrete can be applied by a wet or dry mix method.
Shrinkage Factor:	When the loose fill material is worked into a compacted state, the ratio of the volume of compacted material to the volume of borrow material.
Shrinkage Limit:	An Atterberg Limit property is described as the water content corresponding to the behavior change between the semisolid and solid states of soil. The shrinkage limit is also defined as the water content at which any further reduction in water content will not result in a decrease in the volume of the soil mass.
Sidewalls:	The sides of the seepage pit, where most of the percolation takes place.
Sieve:	Laboratory equipment consisting of a pan with a screen at the bottom. Sieves are used to separate particles of soil samples into their various sizes.
Single Axle Load:	The total load transmitted by all wheels of a single axle extending the full width of the vehicle.

Silt Size Particles:	That portion of soil that is finer than the No. 200 (0.075 mm) sieve and coarser than 0.002 mm.
Single Grained Structure:	A soil structure where an arrangement composed of individual soil particles. This is a common structure of sands.
Site:	The particular lot or parcel of land where grading or other development is performed.
Skeleton Structure:	A soil structure where coarser soil grains form a skeleton with the void spaces partly filled by a relatively loose arrangement of soil fines.
Slab On Grade:	A type of construction in which footings are needed but little or no foundation wall is poured.
Slaking:	The crumbling and disintegration of earth materials when exposed to air or moisture. Slaking can also refer to the breaking up of dried clay when submerged in water, due either to compression of entrapped air by inwardly migrating water or to the progressive swelling and sloughing off of the outer layer.
Slickensides:	Surfaces within a soil mass that have been smothered and striated by shear movements on these surfaces.
Slope:	An inclined ground surface. For graded slopes, the steepness is generally specified as a ratio of horizontal : vertical (for example, 2 : 1 slope). Common types of slopes include natural (unaltered) slopes, cut slopes, false slopes (temporary slopes generated during fill compaction operations), and fill slopes.
Slope Wash:	Soil and/or rock material that has been transported down a slope by mass wasting assisted by runoff water not confined by channels.
Slough:	Loose, non-compacted fill material generated during grading operations. Slough can also refer to a shallow slope failure, such as sloughing of the slope face.
Sludge:	The solid part of sewage that settles out and digests in a septic tank.
Slump:	In the placement of concrete, a measure of consistency of freshly mixed concrete as measured by the slump test. In geotechnical engineering, a slump could also refer to a slope failure.

Slump Test: A field test to measure the consistency of the concrete mix or its stiffness. The slump is measured in inches from the top of the slump cone to where the concrete has sagged (slumped) when the cone is removed. The maximum reading is the height of the cone, 12 inches. If the slump test results are high, one likely cause would be too much water.

Slurry Seal: In the construction of asphalt pavements, a fluid mixture of bituminous emulsion, fine aggregate, mineral filler, and water. A slurry seal is applied to the top surface of an asphalt pavement to seal its surface and prolong its wearing life.

Soil: Sediments or other accumulations of mineral particles produced by the physical and chemical disintegration of rocks. Inorganic soil does not contain organic matter, while organic soil contains organic matter.

Soil Absorption System: Tile line or a seepage pit that permits the sewage liquids to filter into the ground.

Soil Matrix: For fill compaction, that portion of the soil that is finer than the ¾ inch or No. 4 (4.75 mm) sieve.

Soil Sampler: A device used to obtain soil samples during subsurface exploration. Based on the inside clearance ratio and the area ratio, soil samples can be either disturbed or undisturbed.

Soil Stabilization: The treatment of soil to improve its properties. There are many methods of soil stabilization such as adding gravel, cement, or lime to the soil. The soil could also be stabilized by using geotextiles, by drainage, or through the use of compaction.

Soil Structure: Generally refers to both the geometric arrangement of the soil particles and the interparticle forces that may act between them. Common soil structures are cluster, dispersed, flocculated, honeycomb, single-grained, and skeleton.

Soils Mechanics: The application of the laws and principles of mechanics and hydraulics to engineering problems dealing with soil as an engineering material.

Soils Engineer: See Geotechnical Engineer.

Soils Engineering:	See Geotechnical Engineering.
Spalling:	The chipping or flaking of concrete, bricks, or other masonry where improper drainage or venting and freeze/thaw cycling exists.
Span:	The horizontal distance between supporting structures such as beams, trusses, or columns.
Specific Gravity:	Specific gravity is a measure of an object compared to water. Specific gravity is defined as the ratio of the density of the soil particles divided by the density of water. The specific gravity of soil or oversized particles can be determined in the laboratory.
Specification:	A precise statement in the form of specific requirements. The requirements could be applicable to a material, product, system, or engineering service.
Stabilization Fill:	Similar to a buttress fill, the configuration of which is typically related to slope height and is specified by the standards of practice for enhancing the stability of locally adverse conditions. A stabilization fill is normally specified by minimum key width and depth and by maximum backcut angle. A stabilization fill usually has a back drainage system.
Staking:	During grading, the surveyor places wood stakes that indicate the elevation of the existing ground surface and the cut or fill to obtain the final proposed elevation per the grading plans.
Standard Penetration Test (SPT):	A field test that consists of driving a thick-walled sampler (inside diameter = 1.5 inches and outside diameter of 2.0 inches) into the soil by using a 140 pound hammer falling 30 inches. The number of blows to drive the sampler 18 inches is recorded. The N value (penetration resistance) is defined as the number of blows required to drive the sampler from a depth interval of 6 to 18 inches.
Starters or Activators:	Substances or compounds that assist in carrying out digestion in a septic tank.
Steel Trowel:	Tool used for non-porous smooth finishes of concrete. It is a flat steel tool used to spread and smooth plaster, mortar, or concrete. For a smooth finish, use a trowel when the concrete begins to stiffen.

Strike and Dip: A planar structure, such as a shear surface, fault, bed, etc. The strike is the compass direction of a level line drawn on the planar structure. The dip angle is measured between the planar structure and a horizontal surface.

Strain: The change in shape of soil when it is acted upon by stress. In geotechnical engineering, strain is positive when it results in compression of the soil.

Structure: That which is built or constructed, an edifice or building of any kind, or any piece of work artificially built up or composed of parts joined together in some definite manner.

Structural Number: An index number derived from an analysis of traffic, roadbed soil conditions, and environment, which may be converted to the thickness of flexible pavement layers through the use of suitable layer coefficients related to the type of material being used in each layer of the pavement structure.

Subbase: The layer or layers of specified or selected material of designed thickness placed on a subgrade to support a base course (or in the case of ridged pavements, the Portland cement concrete slab).

Sub Contractor: A contractor who specializes in a particular trade generally hired and working under the direction of the general contractor.

Subdrain: A pipe and gravel or similar drainage system placed in the alignment of where the drainage beneath the surface is proposed to flow, or canyons or former drainage channels. After placement of the subdrain, structural fill is placed on top of the subdrain.

Subgrade: For roads and airfields, the underlying soil or rock that supports the pavement section (subgrade, base, and wearing surface). The subgrade is also referred to as the basement soil or foundation soil.

Subgrade Modulus: Also known as the modulus of subgrade reaction. Value often obtained from field plate load tests and used in the design of pavements and airfields.

Subsidence: Settlement of the ground surface over a very large area, such as caused by the extraction of oil from the ground or the pumping of groundwater from wells.

Substructure: The foundation.

Subsoil Profile: Developed from subsurface exploration, a cross-section of the ground that shows the soil and rock layers. A summary of field and laboratory tests could also be added to the subsoil profile.

Sulfate (SO_4): A chemical compound occurring in some soils which, at above certain levels of concentration, has a corrosive effect on ordinary Portland cement concrete and some metals.

Sump: A small pit excavated in the ground or through the basement floor to serve as a collection basin for surface runoff or groundwater. A sump pump is used to periodically drain the sump pit when it fills with water.

Superstructure: The portion of the structure located above the foundation (includes beams, columns, floors, and other structural and architectural members).

Surface Course: Also known as wearing course. One or more layers of a pavement structure are designed to accommodate the traffic load, the top layer of which resists skidding, traffic abrasion, and the disintegrating effects of climate.

Swell: Increase in soil volume, typically referring to volumetric expansion of clay due to an increase in water content.

Tack Coat: In the construction of asphalt pavements, a bituminous material is applied to an existing surface to provide a bond between different layers of the asphalt concrete.

Tailings: In terms of grading, nonengineered fill accumulates on or adjacent to equipment haul roads.

Tandem Axle Load: The total load transmitted to the road by two consecutive axles extending across the full width of the vehicle.

Tensile Test: For a geosynthetic, a laboratory test in which the geosynthetic is stretched in one direction to determine the force-elongation characteristics, breaking force, and the breaking elongation.

Terrace:	A relatively level step constructed in the face of a graded slope surface for drainage control and maintenance purposes.
Texture of Soil:	The degree of fineness of the soil, such as smooth, gritty, or sharp, when the soil is rubbed between the fingers.
Thixotropy:	The property of a remolded clay that enables it to stiffen (gain shear strength) in a relatively short time.
Tight Pipe:	A solid pipe that has tight joints.
Tile Line:	Plain end tile line laid with loose joints in a trench which has been partly filled with gravel or crushed stone. The tile is also surrounded and covered with loose gravel or crushed stone and permits sewage effluent to filter into the ground.
Till:	Material created directly by glaciers, without transportation or sorting by water. Till often consists of a wide range in particle sizes, including boulders, gravel, sand, and clay.
Topsoil:	The fertile upper zone of soil which contains organic matter and is usually darker in color and loose.
Torsional Ring Shear Test:	A laboratory test where a relatively thin soil specimen of circular or annular cross section is consolidated and then sheared at a slow rate in order to obtain the drained residual friction angle.
Total Settlement:	The absolute vertical movement of the foundation.
Total Stress:	The effective stress plus the pore water pressure. The vertical total stress for uniform soil and a level ground surface can be calculated by multiplying the total unit weight of the soil by the depth below ground surface.
Triaxial Test:	A laboratory test in which a cylindrical specimen of soil or rock encased in an impervious membrane is subjected to a confining pressure and then loaded axially to failure.
Ultimate Bearing Capacity:	The bearing pressure that causes failure of the soil or rock supporting the foundation.

Underconsolidation:	The condition that exists if a soil deposit is not fully consolidated under the existing overburden pressure and excess pore water pressures exist within the soil. Undersolidation occurs in areas where cohesive soil is being deposited very rapidly and not enough time has elapsed for the soil to consolidate under its own weight.
Underpinning:	Piles or other types of foundations built to provide new support for an existing foundation. Underpinning is often used as a remedial measure.
Unit Weight:	Weight per unit volume.
Vane Shear Test (VST):	An in-situ field test that consists of inserting a four-bladed vane into the borehole and then pushing the vane into the clay deposit located at the bottom of the borehole. Once it is inserted into the clay, the maximum torque required to rotate the vane and shear the clay is measured. Based on the dimensions of the vane and the maximum torque, the undrained shear strength of the clay can be calculated.
Vibrodensification:	The densification or compaction of cohensionless soils by imparting vibrations into the soil mass so as to rearrange soil particles, resulting in less voids in the overall mass.
Void Ratio:	The volume of voids divided by the volume of soil solids.
Water-Cement Ratio:	For concrete mix design, the ratio of the mass of water (exclusive of that part absorbed by the aggregates) to the mass of cement.
Waterproofing:	The process where a building component is made totally resistant to the passage of water and/or water vapor.
Welded Wire Fabric:	A two-way reinforcement system for rigid slabs fabricated from cold-drawn steel wire, having parallel longitudinal wires welded at regular intervals to parallel transverse wires.
Well Point:	During the pumping of groundwater, the perforated end section of a well pipe where the groundwater is drawn into the pipe.
Wetland:	Land which has a groundwater table at or near the ground surface, or land that is periodically under water and supports various types of vegetation that are adapted to a wet environment.

Windrow: A string of large rock buried within engineered fill in accordance with guidelines set forth by the geotechnical engineer or governing agency requirements. Also, the accumulation of fill, generally provided in a line or row, is produced by a grader during the grading procedure prior to final placement and compaction.

Workability of Concrete: The ability to manipulate a freshly mixed quantity of concrete with a minimum loss of homogeneity.

APPENDIX B—SYMBOLS

Symbol	Definition
α	Correction factor for specific gravity
C_c	Coefficient of curvature
C_u	Coefficient of uniformity
D	Soil particle diameter
D_{10}	Particle size for a percent finer (the number represents percent passing)
D_r	Relative density
e	Void ratio of soil
F	Percent finer (percent passing)
G_b	Bulk specific gravity
G_s	Specific gravity of solids
LI	Liquidity index
LL	Liquid limit
M (W)	Mass (Weight) of compacted soil
M_c (W_c)	Mass (Weight) of the empty container
M_{dc} (W_{dc})	Mass (Weight) of the container plus dry soil
M_R (W_R)	Dry mass (weight) of soil retained on No. 200 sieve
M_s (W_s)	Dry mass (weight) of soil solids
M_w (W_w)	Mass (Weight) of water in soil
M_{wc} (W_{wc})	Mass (Weight) of container plus wet soil
PI	Plasticity index

PL	Plastic limit
R_{DS}	Cumulative amount of dry soil retained on a given sieve
RC	Relative compaction
S	Degree of saturation of soil
T	Time (minutes)
T	Temperature (°F)
V	Total volume of soil
V_g	Volume of gas in soil
V_s	Volume of soil solids
V_v	Volume of voids in soil
V_w	Volume of water in soil
ω	Water content
ω_{opt}	Optimum moisture content
γ_d	Dry unit weight of soil
γ_w	Unit weight of water
ρ_d	Dry density of compacted soil
ρ_s	Density of soil solids
ρ_t	Total density of soil
ρ_w	Density of water
ϕ	Friction angle

APPENDIX C— CONVERSION FACTORS

LENGTH

1 ft	=	12.0 in
1 yd	=	3.0 ft
1 mi	=	5280.0 ft
1 in	=	2.54 cm
1 m	=	3.28084 ft
1 mi	=	1.609344 km

AREA square yard (yd^2)

1 ft^2	=	144.0 in^2
1 yd^2	=	9.0 ft^2
1 ac	=	43,560 ft^2
1 m^2	=	10.76391 ft^2
1 ha	=	107,639.1 ft^2
1 ha	=	10,000.0 m^2
1 ha	=	2.471054 ac

ABBREVIATIONS

inch (in)
foot (ft)
yard (yd)
mile (mi)
centimeter (cm)
kilometer (km)
square inch (in^2)
square yard (ft^2)

acre (ac)
square meter (m^2)
hectare (ha)
cubic inch (in^3)
cubic foot (ft^3)
cubic yard (yd^3)
cubic meter (m^3)
cubic centimeter (cm^3)
liter (L)
fluid oz – us (oz)

VOLUME

1 ft^3	=	1,728.0 in^3
1 yd^3	=	27.0 ft^3
1 ft^3	=	28.316847 L
1 ft^3	=	957.506492 oz
1 ft^3	=	7.480519 gal
1 pt	=	16.0 oz
1 qt	=	2.0 pt
1 gal	=	4.0 qt
1 qt	=	3.943137 cup
1 in^3	=	16.387064 cm^3
1 yd^3	=	764.554858 L
1 m^3	=	35.314667 ft^3
1 L	=	33.814023 oz
1 L	=	1.056688 qt

FORCE

1 kip	=	1,000.0 lbf
1 lbf	=	4.448222 N

WEIGHT (MASS)

1 lb	=	16.0 oz
1 ton	=	2,000.0 lb
1 ton	=	907.18474 kg
1 kg	=	2.204623 lb
1 lb	=	453.59237 gm
1 oz	=	28.349523 gm

STRESS

1 atm	=	2,116.216837 psf
1 atm	=	14.69595 psi
1 atm	=	4036.782505 water column inches
1 atm	=	1,033.227555 water column cm
1 atm	=	101,325.01 Pa or N/m^2
1 atm	=	29.921258 inches mercury (°C)
1 atm	=	1.01325 bar

gallon – us liquid (gal)
quart – us liquid (qt)
pint – us liquid (pt)
cup (cup)
pound-force (lbf)
newton (N)
kilogram (kg)
gram (gm)
pound (lb) or (#)
ton (ton)
atmosphere (atm)
pound per square foot (psf)
pound per square in (psi)
pascal (Pa)
newton per square meter (N/m^2)
pound per cubic foot (pcf)
kilogram per cubic meter (kg/m^3)
mile per hour (mph)
feet per second (fps)
meter per second (mps)
minutes (min)
seconds (s)

UNIT WEIGHT (DENSITY)

1 pcf	=	16.018593 kg/m3

VELOCITY

Mph	=	1.466667 fps
Mps	=	3.28084 fps

TIME

1 year	=	365 days
1 day	=	24 hours
1 hour	=	60 min
1 min	=	60 s

1 in mercury (0°C)	=	70.726198 psf
1 in mercury (0°C)	=	0.491154 psi
1 in mercury (0°C)	=	13.5951 water column inches
1 in mercury (0°C)	=	34.531554 water column cm
1 in mercury (0°C)	=	3,386.388667 Pa or N/m^2
1 in water (4°C)	=	5.20233 psf
1 in water (4°C)	=	0.036127 psi
1 in water (4°C)	=	249.088908 Pa or N/m^2
1 in water (4°C)	=	0.073556 inches mercury (0°C)
1 ft water (4°C)	=	62.42796 psf
1 in water (4°C)	=	0.433524 psi
1 in water (4°C)	=	2,989.06786 Pa or N/m^2
1 in water (4°C)	=	0.882672 inches mercury (0°C)
1 psi	=	144.0 psf
1 psf	=	47.880259 Pa or N/m^2

TEMPERATURE

°F	=	°C(9/5) + 32
°C	=	(°F-32) x (5/9

COMMON SI PREFEXES		
Mega	M	1,000,000
Kilo	K	1,000
Centi	C	0.01
Milli	M	0.001
Micro	µ	0.0000001

Appendix D
Soil Specifications

NOTE: The specifications in this book are strictly to be used as an example as a working copy of specification that may be found on real life projects. Only portions of the specifications are shown in this book that the author felt were relevant to the subject matter discussed. The specifications included are not intended to be complete or a full representation of the project specifications.

REFERENCE: Aims Community College
North Campus Refresh
Greeley, Colorado
Project Manual – Volume 2 or 2
July 29, 2016
Prepared by DLR Group

SECTION 312000 - EARTH MOVING

PART 1 GENERAL

1.01 SUMMARY

A. Work Included:

1. Provide excavation, rough grading, filling, backfilling, compaction, and disposal of spoil materials as shown on the Drawings or specified herein, or as required to complete the Work.

B. Related Sections:

1. Section 014523, Testing and Inspection Services

2. Section 015713, Temporary Erosion and Sediment Control

3. Section 015726, Watering for Dust Control

1.02 REFERENCES

A. Reference Standards: Comply with the requirements of the reference standards noted herein, except where more stringent requirements are listed herein or otherwise required by the Contract Documents.

1.03 DEFINITIONS

A. Structural Fill: The term “structural fill”, as used herein, includes soil materials used for fill beneath and within 5'-0" of buildings, including fill used under fill-supported slabs and the upper 3'-0" of fill under pavements.

B. Overlot Fill: The term “overlot fill”, as used herein, includes soil materials used for general site filling outside 5'-0" beyond building lines and below the upper 3'-0" of fill under pavements, but does not include materials used for exterior wall or trench backfilling.

C. Exterior Backfill: The term “exterior backfill”, as used herein, includes soil materials used for backfilling the exterior sides of grade beam, foundation wall, and basement wall excavations, or for backfilling site retaining walls.

D. Interior Backfill: The term “interior backfill”, as used herein, includes soil materials used for backfilling the interior sides of grade beam or foundation wall excavations.

E. Trench Backfill: The term “trench backfill”, as used herein, includes soil materials used for backfilling utility or other trenches outside the building line.

F. Compacted Backfill: The term “compacted fill”, as used herein, includes soil material used for embankment and general site backfilling. All fill on this project shall be considered compacted fill.

G. Over Excavation: The term "over excavation", as used herein, includes the excavation and removal of materials beyond subgrade elevations or dimensions shown or specified, as directed or approved by the ENGINEER and the testing and inspections agency to correct unsatisfactory conditions.

H. Unauthorized Excavation: The term "unauthorized excavation", as used herein, includes the inadvertent or unintentional excavation and removal of materials beyond subgrade elevations or dimensions shown or specified. Unauthorized excavation, including disposition of over excavated materials and other work resulting from slides, cave-ins, swelling, upheaval, or remedial work shall be the CONTRACTOR's expense.

I. Unclassified Excavation: The term "unclassified excavation", as used herein, includes excavation of all material of whatever character required for the Work, including surface boulders. All excavation on this project is unclassified excavation.

J. Topsoil: The term "topsoil", as used herein, includes the top 6" of soil in its undisturbed state. All topsoil in construction areas shall be stripped and stockpiled to be used for final grading over disturbed areas.

1.04 QUALITY ASSURANCE

A. Regulatory Requirements: Comply with applicable local, State, and Federal rules, regulations, and ordinances concerning shoring, bracing, or sloping of excavations and safety of workers.

1. Shoring Design: Provide the services of a professional ENGINEER to design all shoring, bracing, and underpinning required to protect excavations and adjacent structures or improvements.

B. Erosion Control: Submit for the OWNER's approval a temporary erosion control plan to be implemented during the early phases of the site excavation and site development and maintained throughout the construction period in order to protect adjacent properties, streets and storm sewers from erosion and sediment runoff during the construction process. Do not commence excavation and grading until the OWNER has approved the temporary erosion control plan.

C. Testing Agency:

1. All quality control testing and inspecting required herein will be performed by an independent testing and inspections agency, employed in accordance with the provisions of Section 014523, Testing and Inspection Services.

2. Notify the testing and inspections agency not less than 48 hours in advance of all work requiring testing or inspection services.

1.05 SUBMITTALS

A. Samples: Submit representative samples of all materials proposed for use in filling or backfilling operations to the testing and inspections agency for analysis and determination of compliance with the requirements specified herein.

1.06 JOBSITE CONDITIONS

A. Existing Contours and Elevations: Contours of existing ground elevations at the site, and elevations of finish grade cuts, fills, and excavations for the Work are shown on Drawings. Contours for existing ground lines are based on information provided by others, and are believed to be correct, but may not be absolutely so. Existing contours and elevations should therefore be considered approximate, and should be verified at the site prior to submission of bids.

B. Verification of Existing Conditions: Visit the site prior to submission of bids. Verify existing conditions, elevations, and contours. In the event of discrepancies between existing conditions and the requirements of the Contract Documents, contact the ENGINEER for clarification.

C. Existing Bench Marks: Carefully preserve and maintain existing bench marks, monuments, property line pipes and pins, and other reference points. If disturbed or destroyed, restore or replace at no additional cost to the OWNER.

1.07 GEOTECHNICAL REPORT

A. The Contractor shall follow the specifications and recommendations of the geotechnical report prepared for this project titled Geotechnical Exploration Report Aims Community College, North Campus Refresh, Greeley, Colorado, EEC Project No. 1162022. The geotechnical report was prepared by Earth Engineering Consultants, 4396 Greenfield Drive, Windsor, CO 80550.

PART 2 PRODUCTS

2.01 MATERIALS

A. General: All fill material, regardless of intended use category, must be clean and free from organic matter, roots, brush or other vegetation, trash, debris or other detrimental substances, and rocks or unbroken lumps larger than 4", and must be tested and approved by the testing and inspection agency prior to placement.

B. Overlot Fill Materials:

1. Existing soils obtained from on-site excavations, including granular or aggregate base course from removed pavements, broken and pulverized claystone or claystone-sandstone bedrock.

2. Bedrock must be processed and broken or pulverized so that the maximum particle or fragment size does not exceed 4".

3. If sufficient materials meeting the above requirements are not available from on-

site sources, provide additional material obtained from off-site sources and approved by the testing and inspections agency, at no additional cost to the OWNER.

C. Structural Fill:

1. Unless otherwise recommenced by the Project Soils Report, provide select, non-expansive fill with liquid limit of 25 or less, plasticity index (P.I.) of 10 or less, with no more than 20% passing a No. 200 sieve and maximum particle size not to exceed 3".

2. Natural clayey sand and sandy clay soils obtained from on-site excavations, and granular or aggregate base material from existing removed pavements, which meet the above requirements and which are approved by the testing and inspection agency may be used for structural fill.

3. If sufficient materials meeting the above requirements are not available from on-site sources, provide additional material obtained from off-site sources and approved by the testing and inspections agency, at no additional cost to the OWNER.

D. Topsoil

1. Where construction limits are outside of frequently traveled public rights-of-way, topsoil shall be removed and stockpiled in a manner that will clearly separate topsoil from additional material removed from trenches. Stockpiles of topsoil shall be managed so as to avoid driving of equipment on the piles, or loss, or contamination of the material from other causes.

2. When trench backfill is placed, the surface shall be finished to allow for topsoil replacement in a reasonably uniform layer.

3. Should trench settlement occur on private easements, the space shall be filled with topsoil material equal to that of the surrounding surface and the area reseeded.

4. If sufficient topsoil is not available from stockpiles of existing material taken from on-site, provide additional topsoil as required to complete the Work at no additional cost to the OWNER.

5. Imported topsoil shall comply with the requirements of the specifications for landscape work.

2.02 SOURCE QUALITY CONTROL

A. Fill and Backfill Materials: Test existing on-site soils and borrow material proposed for use in filling and backfilling operations as follows:

1. Moisture Content: ASTM D2216

2. Maximum Index Density: ASTM D4253

3. Moisture Density Relations: ASTM D698 (Standard Proctor)

4. Plasticity Index: ASTM D4318

5. Particle Size Analysis: ASTM D422

PART 3 EXECUTION

3.01 EXAMINATION

A. Verification of Conditions: Examine areas and conditions under which the work of this Section will be performed. Do not proceed with the work until unsatisfactory conditions have been corrected. Commencement of work implies acceptance of all areas and conditions.

3.02 PREPARATION

A. Existing Utilities:

1. General: Locate existing underground utilities in areas of the Work. If utilities that are to remain in place, provide protection during earthwork operations. Should uncharted or incorrectly charted piping or other utilities be encountered during excavation, consult OWNER or ENGINEER immediately for direction. Cooperate with utility companies in keeping respective services and facilities in operation. Repair damaged utilities to the satisfaction of the appropriate utility company.

2. Active Utilities: Do not interrupt existing utilities serving facilities occupied and used by the OWNER or by adjacent properties, except when permitted in writing by the ENGINEER, and then only after acceptable temporary utility services have been provided. Remove or relocate utilities only as indicated or specified.

3. Inactive Utilities: Report inactive or abandoned utilities encountered in excavating or grading operations, and remove, plug, or cap as required. In the absence of specific requirements, plug or cap such utility lines at least 5'-0" outside new building walls, or as required by local rules or regulations. If septic systems are encountered, abandon, remove, and fill in accordance with local requirements and specifications.

4 Removal: Demolish and completely remove from the Project site all existing underground utilities indicated to be removed. Coordinate with utility companies for discontinuance of services if lines are active.

B. Protection of Persons and Property:

1. Provide all necessary measures to protect workmen and passersby. Barricade open excavations occurring as part of the Work as required by municipal or other

authorities having jurisdiction.

2. Protect adjacent streets, roadways, and properties throughout the entire operation. Protect newly graded areas from destruction by weather or runoff. Protect structures, utilities, sidewalks, pavements, and other improvements from damage caused by settlement, lateral movement, undermining, washout, and other hazards created by earthwork operations.

C. Layout and Staking: Provide the services of a registered land surveyor to perform all layout and staking of the Work.

D. Explosives: Use of explosives will not be permitted.

E. Dewatering:

1. Prevent surface and subsurface or ground water from flowing into excavations or from flooding the Project site and surrounding area. Do not allow water to accumulate in excavations.

2. Remove water to prevent softening of the excavation bottoms or other soil changes detrimental to stability of subgrades and foundations. Provide and maintain pumps, well points, sumps, suction and discharge lines, and other dewatering system components necessary to convey water away from excavations.

3. Convey water removed from excavations and rain water to collection or runoff areas. Establish and maintain temporary drainage ditches and diversions outside excavation limits for each structure. Do not use trench excavations as temporary drainage ditches.

4. Locate and retain soil materials away from edge of excavations.

F. Dust Control: Provide dust control as required to alleviate dust nuisance to the public, to adjacent properties, and to other work underway at the project site or at off-site borrow areas. Reference Section 015726, Watering for Dust Control.

G. Erosion And Sediment Control: Provide and maintain temporary pollution controls in accordance with Section 015713, Temporary Erosion And Sediment Control.

H. Access: Maintain vehicular access to the site and adjacent areas at all times.

I. Cleanup: Remove excess materials not required for grading or filling purposes, including unacceptable excavated material, accumulated debris, and rubbish from site. Burning of waste material is prohibited.

3.03 EXCAVATING

A. General:

1. Remove material of every nature or description encountered in obtaining required

lines and grades. Excavate and/or place and compact fill to provide for grade elevation(s) required by Drawings.

2. Pitch grading around excavation to prevent water from running into excavated areas.

3. Pre-rip hardpan and soft bedrock with single-tooth ripper or other suitable equipment to facilitate excavation with conventional earth-moving equipment.

B. Stability:

1. Slope sides of excavations in compliance with OSHA requirements an local codes or ordinances. Shore and brace where sloping is not possible because of space restrictions or stability of material excavated.

2. Continuously monitor cut slopes for distress. Take all necessary precautions to safeguard workers, structures, and utilities.

3. Provide all necessary shoring, sheeting, or bracing of sides of excavations required to prevent caving, erosion, and gullying. Provide underpinning of existing structures or other improvements adjacent to excavations and which are subject to damage therefrom.

C. Unanticipated Conditions: Notify the ENGINEER immediately upon finding evidence of previous structures, filled materials which penetrate below designated excavation levels, groundwater or water-bearing strata, or other conditions which are not shown or which cannot be reasonably assumed from existing surveys and geotechnical reports. Secure the ENGINEER's instruction before proceeding with further work in such areas.

D. Unsatisfactory Soils: Remove or otherwise correct unsanitary, sour, or otherwise unsatisfactory soil. Remove contaminated or unsuitable material from under building and paved areas.

E. Additional Excavation: When excavation has reached required subgrade elevations, notify the testing and inspection agency, who will make an observation of conditions. If unsuitable bearing materials are encountered at required subgrade elevations, carry excavations deeper and replace excavated material as directed by the testing and inspection agency and approved by the ENGINEER, at mutually agreed unit or lump sum prices.

3.04 FILLING AND BACKFILLING

A. Ground Surface Preparation: Scarify existing subgrade or ground surface to depth of 6" prior to compacting and placing fill. Moisture condition to between 1% below and 3% above optimum moisture content, and recompact to at least 95% or maximum standard Proctor density (ASTM D698).

B. Cut Areas: Proof roll and compact cut areas to produce a compaction equal to that of filled areas. If soft or yielding material is encountered in cuts as a result of trapped water, and cannot be satisfactorily stabilized by moisture control and recompaction, excavate unsuitable material to depth required by the testing and inspection agency and approved by the ENGINEER. Fill resulting excavation with suitable material and compact as specified herein for corrective fills.

C. Benching: When fill meets a natural slope of 20 degrees (5:1) or steeper, cut benches approximately 10ft wide in existing slope to serve as keys to connect existing grade with newly placed fill. Slope tops of benches 2% to outslope.

D. Overlot Fill Placement: Place fill materials in layers not more than 8" in loose depth for material compacted by heavy compaction equipment, and not more than 4" in loose depth for material compacted by hand operated tampers. Before compaction, moisten or aerate each layer as necessary to provide moisture content between 1% below and 3% above optimum.

E. Corrective Fills:

1. Under footings, foundations, or retaining walls, fill over excavation or unauthorized excavation by extending indicated bottom elevation of wall or base to excavation bottom, without altering required top elevation. Or, at the option of the CONTRACTOR and with the ENGINEER's approval, fill to indicated subgrades with compacted structural fill as specified in this Section or with lean concrete.

2. Elsewhere, backfill and compact over excavations or unauthorized excavations as specified for authorized excavations of same classification, unless otherwise directed.

F. Backfilling:

1. Place backfill material in maximum 6" layers up to required subgrade elevations. Before compaction, moisten or aerate each layer as necessary to provide moisture content between 2% below and 2% above optimum for sand soils an sandstone, an between optimum and 3% above optimum for clays.

2. Cold Weather Work: Do not place fill material that contains frost, snow, or ice, or place new or additional fill on material that contains frost, snow, or ice.

3.05 COMPACTING

A. Fill and Backfill Areas:

1. Compact each layer to required percentage of maximum density for each area classification. Place fill and backfill materials evenly to required elevations. Accomplish compaction by use of power rollers, sheep foot rollers, machine tampers, or other mechanical equipment as may be required to obtain density specified.

2. Take care to prevent wedging action of backfill against structures by carrying material uniformly around structure to approximately same elevation in each lift. Take care when backfilling against walls, keeping fill elevations equal on each

side of wall where applicable.

3. Thoroughly compact by means of mechanical tampers portions of earth fill between projections which cannot be properly compacted by means of rolling equipment.

B. Moisture Control:

1. Where subgrade, fill, backfill, or any layer of soil material must be moisture conditioned before compaction, uniformly apply water to surface of material in a manner and quantity to prevent free water from appearing on surface during or subsequent to compaction operations.

2. Remove and replace or scarify and air dry material that is too wet to permit compaction to specified density. Material that has been removed because it is too wet to permit compaction may be stockpiled or spread and allowed to dry.

C. Density Requirements:

1. Compact subgrade, fill, and backfill to not less than following percentages of maximum standard Proctor density, as determined by ASTM D698:
   a. Structures, Including Slabs-on-Grade....................................95% min.
   b. Walkways, Terraces, and Patios ....95% min.
   c. Roadway, Driveway, and Parking Area Subgrades.................95% min.
   d. Backfill, Top 3 ft Below Exterior Grade ....95% min.
   e. Backfill, Below Top3 ft. ....90% min.
   f. Lawn or Unpaved Areas..90% min.

2. For soils which do not exhibit a well defined moisture density relationship, compact to not less than equivalent percentages of maximum density, as determined by ASTM D4253.

3.06 ROUGH GRADING

A. Uniformity: Make finish grades not otherwise indicated to uniform levels on slopes between points where elevations are given or between such points and existing finish grade. Round abrupt changes in slopes. Should figures or spot elevations for finished grades conflict with finish contours shown, figures or spot elevations shall govern.

B. Drainage: Maintain site in a freely draining condition at all times. Provide drainage for ground water or springs that may be encountered.

C. Rounding: Finish swales and gutters to drain readily. Provide rounding at top and bottom of banks and at other breaks in grade.

D. Levels: Do all cutting, filling, backfilling and grading necessary to bring areas within contract limit lines to the following approximate subgrade levels:

1. Floors: To bottom of concrete slabs-on-grade, within tolerance of ±3/8" when tested with a 10-foot straightedge.

2. Paved Areas: Refer to Civil Drawings for thickness of pavements; grade to within tolerance of ±1/2".

3. Unpaved Areas: To 6" below grade shown to allow for topsoil, planting, and lawns, and to bottom of exterior slab thickness noted on concrete flatwork (walks, terraces, patios, curbs, and gutters), within tolerances of ±3/4".

E. Rough Grading Finish: Degree of rough grading finish expected is that ordinarily obtained from either blade or scraper operation. Surfaces shall be compacted and free irregular surface changes. Limit grading to areas shown. Grade to provide positive drainage of water away from building.

F. Pavement Subgrade: Just prior to commencement of paving operations, at such time as will cause no delay to paving installer, place final fill, regrade, and compact to establish blue top levels. Place subgrade material in 8" loose layers. Before compaction, moisten or aerate each layer as necessary to provide moisture content between 1% below and 3% above optimum.

G. Redistribution of Topsoil: Spread topsoil over entire surface of areas to receive planting and lawns as necessary to establish finish grades indicted on drawings, and to minimum depth of 6". When necessary, loosen and reshape compacted subgrade to preserve minimum 6" depth below finish grade.

3.07 FIELD QUALITY CONTROL

A. Field Density Tests: Perform field density tests in accordance with ASTM D1556, ASTM D2167, or ASTM D2922. Perform field moisture tests in accordance with ASTM D3017.

B. Test Intervals: As determined by Section 01 45 23, Testing and Inspection Services.

C. Failings: If, based on the testing and inspection agency reports and inspections, compacted subgrade or fills are found to be below specified density, provide additional compaction and testing at the CONTRACTOR's expense.

3.08 ADJUSTMENT AND PROTECTION

A. Protection of Graded Areas: Protect newly graded areas from traffic and erosion. Keep free of trash and debris.

B. Reconditioning Compacted Areas: When completed compacted areas are disturbed by subsequent construction operations or adverse weather, scarify surface, reshape, and compact to required density prior to further construction.

C. Drying: Protect existing and newly placed fills and soil materials from excessive drying prior to placement of floor slabs or other construction.

END OF SECTION 312000

SECTION 312333 - TRENCHING AND BACKFILLING

PART 1 GENERAL

1.01 SCOPE

A. This section covers excavation and trenching, including but not limited to dewatering, preparation of subgrades, pipe bedding, backfilling, compacting, groundwater barriers, materials testing, and finish grading for underground pipelines and appurtenances.

1.02 RELATED SECTIONS

A. Section 014523, Testing and Inspection Services.

B. Section 312313, Structural Backfill.

1.03 REFERENCES

A. AASHTO T99, *The Moisture–Density Relations of Soils Using a 2.5-kg (5.5-lb) Rammer and a 305-mm (12-in.) Drop (Method A)*, latest revision.

B. ASTM D448 (AASHTO M43), *Standard Classification for Sizes of Aggregate for Road and Bridge Construction*, latest revision.

C. ASTM D698, *Standard Test Methods for Laboratory Compaction Characteristics of Soil Using Standard Effort (12,400 ft-lbf/ft3 (600 kN-m/m3))*, latest revision.

D. ASTM D2487, *Standard Practice for Classification of Soils for Engineering Purposes (Unified Soil Classification System)*, latest revision.

E. ASTM D2922, *Standard Test Methods for Density of Soil and Soil-Aggregate in Place by Nuclear Methods (Shallow Depth)*, latest revision.

F. ASTM D3017, *Standard Test Method for Water Content of Soil and Rock in Place by Nuclear Methods (Shallow Depth)*, latest revision.

G. ASTM D4253, *Standard Test Methods for Maximum Index Density and Unit Weights of Soils Using a Vibratory Table*, latest revision.

H. ASTM D4254, *Standard Test Methods for Minimum Index Density and Unity Weight of Soils and Calculation of Relative Density*, latest revision.

1.04 CONSTRUCTION STAKING

A. Construction staking shall be performed under the direct supervision of a Professional Land Surveyor licensed in the State of Colorado.

B. Adequate staking shall be provided to establish acceptable horizontal and vertical control.

C. Offsets shall be staked so that the OWNER may check vertical and horizontal alignment.

D. All survey notes and construction staking notes shall be entered into bound, hard cover field books, kept at the construction site for the duration of the project, and shall be made available to the OWNER upon request.

E. All survey data, which is developed by the Contractor or the Design Engineer in performing surveys required by the work, shall be available to the OWNER for examination and reproduction throughout the construction and warranty periods.

F. The Design Engineer and OWNER shall be informed of all field changes to the OWNER accepted Construction Drawings. Approval for the changes shall be required from both the Design Engineer and OWNER prior to the changes being made in the field.

1.05 JOB CONDITIONS

A. Drainage and groundwater.

1. Keep excavations and trenches free of water during construction. Divert surface runoff and utilize sumps, gravel blankets, well points, drain lines or other means of dewatering, as necessary.

   a. Dewater the excavation or trench until the structure, pipe, or other, to be installed therein, is completed to the extent that no damage from hydrostatic pressure, floatation, or other cause will result.

   b. Water shall be removed from the trench to the extent necessary in order to provide a firm subgrade and dry conditions for pipeline installation.

2. The pipeline being constructed shall not be used for dewatering.

3. The piping used to dewater the trench shall not be left in the trench when backfilled.

4. Groundwater barriers shall be installed in high groundwater conditions or if deemed necessary by the OWNER.

5. Prior to beginning dewatering operations, the Contractor shall obtain all necessary permits and appropriate authorization to start dewatering. If groundwater will be discharged or drained into an irrigation ditch, pond, stream or waterway, a Colorado Department of Public Health and Environment (CDPHE) Dewatering Permit will be required.

   a. Permit applications may take up to thirty (30) days to be reviewed by CDPHE.

   b. The Contractor is required to complete and process the Discharge Monitoring Report (DMR) that is typically a part of the Dewatering Permit.

c. Upon completion of the work, the Contractor shall be responsible for completing

d. CDPHE Discharge Termination Notice.

6. Copies of all CDPHE permits shall be submitted to the OWNER prior to the start of dewatering operations. Copies of the permits shall also be kept onsite for the duration of the work.

B. Blasting is not permitted within the jurisdiction of the OWNER unless otherwise authorized by the OWNER.

C. Sequencing

1. Backfill shall be completed, at the end of each day, to the extent that no damage from hydrostatic pressure, floatation, or other cause will result.

2. Where excavation is a hazard to automotive or pedestrian traffic, the amount of open trench and the time duration of that opening shall be minimized. The OWNER shall direct the amount of open trench that is acceptable for the condition encountered.

3. During construction, maintain access to private residence and businesses.

D. Underground Obstructions

1. It is the Contractor/Owner's responsibility to call for utility locates. Call UNCC at 1- 800-922-1987 or dial 811 for locates.

2. The Contractor shall expose and verify the size, location, and elevation of underground utilities and other obstructions, sufficiently in advance of construction to permit changes to be made to the Construction Drawings in the event there is a conflict with the proposed and existing utilities. In the event there is a conflict, the Contractor shall notify the Design Engineer, OWNER, and affected utility company immediately.

3. Protect and support utilities, appurtenances, structures, etc., by shoring, bracing or other means necessary.

E. Weather

1. Do not install pipe or place pipe bedding on frozen soil in the trench bottom.

2. Do not place frozen materials, snow or ice in backfill, fill, or embankments.

3. Do not deposit, tamp, roll or otherwise mechanically compact backfill in water.

PART 2 PRODUCTS

2.01 GENERAL

A. All material shall be free from frozen matter, stumps, roots, brush, other organic matter, cinders, corrosive material, debris, broken asphalt and concrete, and any other objectionable material that is not suitable in the opinion of the OWNER.

B. If job excavated material is not sufficient or suitable, suitable material shall be imported approved by the OWNER.

2.02 STABILIZATION MATERIAL

A. If the existing soil in the trench bottom is judged to be unsuitable by the OWNER or the Design Engineer, at a minimum, the top six-inches (6") of the trench subgrade shall be removed and replaced with stabilization material.

B. If deemed necessary by the OWNER, more than six-inches (6") of material from the trench bottom may require removal and replacement with a stabilization material.

C. Stabilization material is crusher-run rock, conforming to CDOT #357 (ASTM D448, AASHTO M43).

TABLE 2.2-C: Stabilization Material – CDOT #357

Size (inch)	Percent (%) Passing
2 ½"	100
2"	95-100
1"	35-70
½"	10-30
#4	0-5

Or approved equivalent.

2.03 GEOTEXTILE (FILTER) FABRIC

A. Geotextile fabric shall be used in conjunction with stabilization material unless approved otherwise by the OWNER.

B. Geotextile fabric shall conform to Colorado Department of Transportation, Division of Highways, State of Colorado *"Standard Specifications for Road and Bridge Construction", Section 712.08, Class A Table 712-2*, latest edition.

C. Acceptable geotextile fabric manufacturers are:

1. TenCate – Mirafi 500X

2. Webtec, Inc. Geosythetics – TerraTex GS

3. Or approved equivalent.

2.04 BEDDING ZONE MATERIALS

A. The bedding zone shall extend six-inches (6") below the invert of the pipe to six-inches (6") above top of pipe.

B. Bedding material is angular crushed rock, 3/8-inch minus squeegee and gravel and shall meet the following requirements when tested with laboratory sieves for course aggregates in accordance with ASTM C136.

Sieve Designation	Percent by Weight Passing Lab Sieve
3/8"	100
#200	0-5

C. Groundwater Barrier

1. Unified Soil Classifications

a. GC – clayey gravels, clayey sandy gravels

b. SC – clayey sands, clayey gravelly sands

c. CL – inorganic clays of low to medium plasticity, gravelly clays, sandy clays, silty clays, clean clays.

d. Material shall not be lumpy or hard but shall be finely divided, suitable, and free from stones.

2. Or Flow Fill in conformance with Section 312313, Structural Backfill.

2.05 INSULATION BOARD

A. Insulation board shall be installed above the bedding zone when the depth of cover over the water line is less than five (5) feet.

B. Insulation board shall be a minimum one-inch (1") thick. Acceptable insulation board manufacturers are:

1. Dow Chemical Company - Styrofoam™

2. Owens-Corning

3. Or approved equivalent.

2.06 TRENCH BACKFILL MATERIAL

A. Generally, soil materials for trench backfill shall be as recommended in the approved soils report.

A. Trench backfill material shall be placed from a point six-inches (6") above the pipe to six-inches (6") below the ground surface, or bottom of topsoil layer, or bottom of the pavement subgrade, whichever is applicable.

B. Class 2 Structural Backfill shall be composed of suitable materials developed on the project. To be suitable for use under this classification, backfill shall be free of frozen material, organic material, trash, glass, broken concrete, and other corrosive or deleterious material.

1. If the material contains rock fragments or other materials that, in the opinion of the Engineer, will be injurious to the structure, the native material shall not be used for backfilling, and the Contractor shall be required to furnish "Class 1 Structural Backfill" material in accordance with Section 31 23 13, Structural Backfill.

C. Reference Section 312313, Structural Backfill for flow fill construction specification requirements.

PART 3 EXECUTION

3.01 PREPARATION

A. Topsoil shall be stripped from areas which are to be disturbed by construction and stockpiled.

B. Topsoil shall be segregated from non-organic trench material and debris.

3.02 TRENCHING

A. Excavate trenches by open cut methods, except where a boring is indicated on the Construction Drawings, required by jurisdictional agencies, or desired by the Contractor and approved by the OWNER.

B. Do not use mechanical equipment in locations where its operation would cause damage to trees, buildings, culverts, utilities, structures or other property above or below ground. In all such locations, hand-excavating methods shall be used.

C. Use mechanical equipment designed and operated so the rough excavated trench bottom elevation can be controlled with uniform trench width and vertical sidewalls from an elevation one (1) foot above the top of installed pipe to the bottom of the trench. The trench alignment shall be sufficiently accurate to permit pipe to be aligned properly between the pipe and sidewalls of the trench. Do not undercut the trench sidewall to obtain clearance.

D. Contractor shall follow the most current regulations concerning excavations set forth by OSHA: 29 CFR Part 1926.

E. Excavation in Rock

1. Over-excavate a minimum of six-inches (6") below the bottom of the trench.

2. Backfill with compacted bedding material.

F. Preparation of Trench Bottom

1. Grade trench bottom uniformly to provide clearance for each section of pipe.

2. Remove loose materials, water and foreign objects.

3. Provide firm subgrade suitable for placement of bedding material.

4. Wherever unstable material is encountered in the bottom of the trench, over-excavate such material to a depth suitable for constructing a stable subgrade or as determined by the OWNER.

5. Backfill over-excavation with stabilization material and compact. A geotextile fabric layer shall be placed between the stabilization material and the bedding material.

G. Stockpiling Excavated Materials

1. Pile suitable material for backfilling in an orderly manner a sufficient distance from trench banks to avoid overloading and to prevent slide or cave-ins.

2. Do not stockpile excavated material against existing structures or appurtenances.

3. The Contractor shall follow the most current OSHA regulations concerning excavations.

H. Trench Widths

1. Trench width shall be maintained to within three-inches (3") of that specified on the Standard Drawings unless otherwise specified by the OWNER.

3.03 PIPE BEDDING

A. Placement and Compaction

1. Distribute, grade, and compact bedding material to provide uniform and continuous support beneath the pipe at all points between bells and pipe joints.

2. Bell holes shall be dug deep enough to provide a minimum of two-inches (2") of clearance between the bell and bedding material. The pipe shall not be supported by the pipe bell.

3. Deposit bedding material and compact uniformly and simultaneously on each side of the pipe to prevent lateral displacement.

4. Compact granular bedding material by vibrating, slicing with a shovel, or bent tee-bar.

B. Place pipe bedding in accordance with the Standard Drawings.

3.04 GROUND WATER BARRIERS

A. Ground water barriers shall be constructed in such a manner to impede the passage of water through the bedding material and shall be installed when high groundwater conditions exist or as directed by the OWNER.

B. Ground water barriers shall be keyed at least one (1) foot into the trench wall and bottom, and spaced ten (10) feet upstream of each manhole for gravity sanitary sewers or every 400 feet on water lines and sanitary sewer force mains.

C. At a minimum, ground water barriers shall extend one (1) foot above the bedding material.

D. Refer to Standard Drawings for additional installation requirements.

3.05 INSULATION BOARD

A. Install insulation board in accordance with the Standard Drawings.

B. Insulation board shall be installed above the bedding zone wherever the depth of cover over the water main is less than five (5) feet.

1. Insulation board installation shall consist of two (2) overlapping boards, one-inch (1") minimum thickness per board, with off-set joints.

2. Insulation board shall be placed across the full trench width.

3.06 BACKFILLING AND COMPACTION

A. Backfill trench promptly after completion of pipe bedding, but only after the OWNER has inspected the work.

B. Backfill shall consist of approved materials uniformly distributed in layers brought up equally. Each layer of backfill shall not exceed six inches before compacting to the required density and before successive layers are placed.

1. Field compaction densities for backfill soils for improved areas shall be as indicated in the soil report but not less than 95% of maximum density or within + two percent (2%) of optimum moisture content as determined by AASHTO T99.

2. Field compaction densities for backfill soils for unimproved areas shall be not less than 90 percent of maximum density within + two (2%) percent of optimum moisture content as determined by AASHTO T99. Do not compact topsoil.

C. Use backfilling and compaction methods and equipment appropriate for the backfill material. Do not use equipment or methods that will transmit damaging shocks to the pipe.

D. Do not perform compaction by jetting or water settling.

E. Rock and bedrock encountered in the excavation shall not be used in backfill.

F. For areas not receiving surface improvements after construction, return the final grading to the depth of stripping over all areas disturbed by construction operations and replace topsoil.

3.07 MATERIALS AND QUALITY CONTROL TESTING

A. The Contractor is responsible for quality control testing and the testing shall be performed by an independent testing agency employed by the Contractor.

B. For backfill compaction and moisture requirements and the required materials testing, frequency of tests, and standard testing methods, reference Section 01 45 23, Testing and Inspection Services.

C. The following requirements shall also apply:

1. Groundwater Barriers

a. Compaction – 95%

b. Moisture – ±2%

2. Bedding Material

a. Compaction – 80% of relative density

D. Services

1. Water services shall have a minimum of one (1) moisture/density test per service.

2. Sanitary sewer services shall have a minimum of two (2) moisture/density tests per service or at the OWNER's discretion.

3. Moisture/density tests in the vicinity of vaults, valve boxes and manholes shall be performed at a minimum of one (1) foot away from the edge of vault/manhole sections or valve boxes.

a. Tests shall be performed in random directions from the vault, manhole, or valve box, on separate lifts.

b. A minimum of one (1) test shall be performed, on opposite sides of the vault, manhole or valve box, for every two (2) feet of backfill material.

4. The Contractor shall keep copies of all quality control test results in a notebook at the job site for the duration of the project. Test results shall be made available to the OWNER at all times.

3.08 COMPACTION TEST FAILURE

A. If the required compaction and moisture is not obtained, it shall be the responsibility of the Contractor to recompact or rework the material to the required state of compaction and moisture.

B. In cases where there is a failure to achieve the required compaction or moisture, the OWNER may require that the backfill be removed and recompacted or replaced entirely with suitable materials.

C. Water line and sanitary sewer line/manhole testing may be required after recompaction if the testing had been performed prior to recompaction.

1. Water line testing shall be performed between valves on both sides of the recompacted area.

2. Sanitary sewer line testing shall be performed between manholes on both sides of the recompacted area.

3. Sanitary sewer manhole testing shall be performed if recompaction occurs in the vicinity of the manhole.

END OF SECTION 312333

SECTION 313700 - RIPRAP

PART 1 GENERAL

1.01 SUMMARY

A. Section Includes:

1. Stone Riprap

2. Filter Material

B. Related Sections:

1. Section 334100, Reinforced Concrete Pipe

1.02 SUBMITTALS

A. Product Data: Submit service records or tests of riprap indicating suitability for application indicated and that proposed rock meets specified requirements. Submit certified test reports from cloth filter manufacturer or from a qualified testing laboratory showing compliance with all specified requirements.

1.03 DELIVERY, STORAGE AND HANDLING

A. Keep filter cloth wrapped during shipment, storage and handling. Shade filter cloth from direct sunlight and do not expose to extreme heat or cold.

1.04 PROJECT/SITE CONDITIONS

A. Protection of Persons and Property: Protect structures, utilities, sidewalks, pavements, and other facilities from damage caused by placing of riprap.

B. Cleaning: Maintain adjacent streets free of dirt accumulation arising out of work of this section. Use suitable means of cleaning equipment, streets or both and to meet requirements of authorities having jurisdiction.

PART 2 PRODUCTS

2.01 STONE MATERIALS

A. Provide riprap of hard rock, angular in shape; resistant to weathering and to water action; free from overburden, spoil, shale and organic material; and meeting gradation requirements as specified. Neither breadth nor thickness of a single stone shall be less than one-third its length. Rounded stone or boulders will not be accepted unless authorized by the ENGINEER in writing. Thin slab stone or shale and stone with shale seams is not acceptable. Provide rock with a minimum specific gravity of 2.50 (CDOT) as computed by multiplying the specific gravity (Bulk-Saturated-Surface-Dry Basis, AASHTO T85) times 62.4 pounds per cubic foot.

B. Provide rock from sources selected well in advance of the time when rock will be required in the work. The acceptability of the rock will be determined by service records or by suitable tests. If testing is required, suitable sample of stone shall be taken in the presence of the ENGINEER at least 25 days in advance of the time when the placing of riprap is expected to begin. The approval of some rock fragments from a particular quarry site shall not be construed as constituting the approval of all rock fragments taken from that quarry.

C. In the absence of service records, resistance to disintegration from the type of exposure to which the stone will be subjected will be determined by any or all of the following tests:

1. When the riprap must withstand abrasive action from material transported by the stream, the abrasion test in the Los Angeles machine shall also be used. When the abrasion in the Los Angeles machine ASTM C535 is used, the stone shall have a percentage loss of not more than 40 after 500 revolutions.

2. When the freezing and thawing test (AASHTO T103) for ledge rock procedure A) is used as a guide to resistance to weathering, the stone shall have a loss not exceeding 10% after 12 cycles of freezing and thawing.

D. Provide riprap meeting the gradation requirements shown on the drawings. Concrete masonry and concrete pavement may not be used for riprap.

2.02 BEDDING MATERIALS

A. Granular Bedding: Provide porous, free draining material consisting of sand, gravel, cinders, slag, crushed stone or other approved free draining material. Provide material meeting the gradation shown on the drawings.

B. Filter Cloth: Provide filter cloth fabricated from wither a pervious sheet woven of polyvinylidene or polypropylene monofilament yarn, or a non-woven fabric made from hetrofilaments of polypropylene and polyethylene. The yarn shall consist by eight of a least 85% of the base plastic and shall contain stabilizers and inhibitors added to the base plastic to make the filaments resistant to deterioration due to ultraviolet or heat exposure. Provide cloth free of defects or flaws which significantly affect its physical or filtering properties.

C. The cloth sheets may be sewn together with thread consisting of the same materials as the cloth, either at the point of manufacture or at the job site, at the CONTRACTOR's option, to form sections of the required size. All edges of the woven cloth shall be selvedged to prevent unraveling.

D. Provide filter cloth meeting the following requirements:

1. Minimum Thickness (ASTM D1910): 30 to 40 mils.

2. Minimum Weight (ASTM D1910): 3.5 to 6.5 ounces per square yard for polypropylene and 10.8 ounces per square yard to polyvinylidene.

3. Percent Elongation before breaking: 60 to 100%

4. Equivalent size of openings (sieve): No. 70 to 100.

E. Provide securing pins of galvanized wire or other approved material. Wire pins shall be made of wire 0.091" or larger in diameter and "U" shaped having legs 6" long and a 1" crown.

PART 3 EXECUTION

3.01 FILTER CLOTH

A. Prepare the surface to receive the cloth to a relatively smooth condition free of obstructions depressions and debris. Do not lay the cloth in a stretched condition, but lay the cloth loosely with the long dimension perpendicular to the channel centerline. Overlap the panels a minimum of 3" with securing pins inserted through both layers at not greater than 3 foot intervals along a line through the approximate mid-point of the overlap. Place the cloth so that the upstream panel will overlap the downstream panel. Install additional pins as necessary to prevent any slippage of the filter cloth, regardless of location. No more than two horizontal overlaps shall be permitted. Stagger vertical overlaps a minimum of five feet. Lengths and widths of individual sheets are at the CONTRACTOR's option. Grommets are not required.

B. Filter cloths can be placed by using steel rods as weights fastened along the edges. Additional intermediate weights can be used to assist sinking the cloth into place.

C. Filter cloth damaged or displaced before or during installation or during placement of overlying layers of stone shall be replaced or repaired to the satisfaction of the ENGINEER at the CONTRACTOR's expense.

3.02 GRANULAR BEDDING

A. Place the granular bedding in prepared area after placement of the filter cloth to the full specified depth of the area in one operation, using methods which will not cause segregation of particle sizes within the bedding. The surface of the finish layers should be reasonably even and free from mounds and windrows. Place additional layers of granular bedding, when required, in the same manner, using methods which will not cause mixture of the material in the different layers.

3.03 RIPRAP

A. Place stone for riprap on the prepared slope or area in a manner which will produce a reasonably well graded mass of stone with the minimum practicable percentage of voids. The entire mass of stone shall be placed so as to be in conformance with the lines, grades, and thickness' shown on the drawings. Place riprap to its full course thickness in one operation and in such a manner as to avoid displacing the underlying material. Placing of riprap in layers, or by dumping into chutes, or by similar methods likely to cause segregation will not be permitted.

B. Do not drop the stone on the filter cloth from a height greater than two feet.

C. Well distribute the larger stones and ensure the entire mass of stone conforms to the gradation specified. Place and distribute all material going into riprap protection so that there will be no larger accumulations of either the larger or smaller sizes of stone.

D. It is the intent of these specifications to produce a fairly compact riprap protection in which all sizes of material are placed in their proper proportions. Hand placing or rearranging of individual stones by mechanical equipment may be required to the extent necessary to secure the results specified.

END OF SECTION 313700

Appendix E
Concrete Specifications

NOTE: The specifications in this book are strictly to be used as an example as a work ng copy of specification that may be found on real life projects. Only portions of the specifications are shown in this book that the author felt were relevant to the subject matter discussed. The specifications included are not intended to be complete or a full representation of the project specifications.

REFERENCE: Aims Community College
North Campus Refresh
Greeley, Colorado
Project Manual – Volume 2 or 2
July 29, 2016
Prepared by DLR Group

SECTION 031005 - CIVIL CONCRETE FORMWORK

PART 1 GENERAL

1.01 SUMMARY

A. Work Included: This Section includes the design, construction, maintenance, and removal of formwork for cast-in-place concrete, as shown on the Drawings or specified herein, or as required to complete the Work.

B. Related Work:

1. Section 013300, Submittal Procedures.

2. Section 032005, Civil Concrete Reinforcement.

3. Section 033005, Civil Cast-in-Place Concrete.

1.02 REFERENCES

A. Reference Standards: Comply with the requirements of the reference standards noted herein, except where more stringent requirements are listed herein or otherwise required by the Contract Documents.

1.03 DEFINITIONS

A. Exposed Finishes: Exposed finishes include those which will be rubbed, dash-coated, painted, or left unfinished in finished work.

1.04 SYSTEM DESCRIPTION

A. Design Responsibility: Design of concrete formwork, including all necessary shoring and bracing, shall be the responsibility of the CONTRACTOR, who shall provide all special engineering design or analysis of formwork requirements.

B. Design Requirements:

1. Comply with the applicable design, erection, maintenance, and removal provisions of ACI 347.

2. Design, erect, support, brace, and maintain formwork so that it will safely support vertical and lateral loads that might be applied, until such loads can be supported by the in-place concrete structure.

3. Design forms and falsework to include assumed values of live load, dead load, weight of moving equipment operated on the formwork, concrete mix, height of concrete drop, vibrator frequency, ambient temperature, foundation pressures,

stresses, lateral stability, and other factors affecting the safety of the structure during construction.

4. Provide camber in formwork as required for anticipated deflections due to weight and pressures of fresh concrete and construction loads.

1.05 DELIVERY, STORAGE, AND HANDLING

A. Delivery: Deliver manufacture forming materials to the jobsite in manufacturer's original packaging or bundles.

B. Storage and Protection: Protect all formwork materials, including lumber, plywood, metal, fiberglass or fiber forms, and form lines from exposure to adverse weather and keep clean and free of mud, rust, and other deleterious substances.

PART 2 PRODUCTS

2.01 FORM MATERIALS

A. Wood Forms:

1. Forms for Unexposed Surfaces: #2 common or better lumber or moisture-resistant plywood.

2. Forms for Exposed Surfaces: New or like-new, APA B-B Class I Plyform plywood, Exterior type, minimum 5/8" thick. Surface shall be smooth, free of ripples, scratches, indentations, unsound surface knots, prominent grain, warps, or breaks. Provide overlay type (HDO or MDO) where required for very smooth finish.

3. Examine Drawings for areas to receive special forming treatment for architectural effect.

B. Fiberglass Forms: Fiberglass reinforced plastic round column forms as indicated, in new or like-new condition. Provide standard shapes or sizes as indicated or required, including all fillers, stiffeners, and other accessories.

C. Tubular Fiber Forms: Spirally constructed tubular forms consisting of 6" wide laminated fiber strips, using non-water sensitive adhesive. Interior shall be lined with a plastic coating; exterior shall be uniformly wax-impregnated for weather and moisture protection. Provide diameters or sizes as indicated on the Drawings; wall thickness shall be as published by the manufacturer.

D. Void Forms: Moisture-resistant treated paper faces, biodegradable, having a uniform cellular interior construction, and capable of sustaining all vertical and horizontal loads applied until such loads can be supported by the concrete structure. Provide pre-manufactured, non-field cut, sealed void forms with curved, radial, vertical edges adjacent to drilled piers at circular edges of drilled piers at the intersection with grade beams or structural slabs.

2.02 MISCELLANEOUS MATERIALS

A. Preformed Construction Joints: Galvanized steel tongue and groove keyed type, minimum 24 gauge, overall width less than slab thickness, pre-punched for stake pins and dowels. Provide joint forms with removable caps where forms will be left in place as control joints. Acceptable products include Burke "Keyed Kold", Vulcan Metal Products "Super Screed", or equivalent.

B. Isolation (Expansion) Joint Forming Material: Preformed bituminous, rubber, or cork fuller material conforming to ASTM D1751 or D1752, precut in width required by slab thickness.

C. Form Ties: Regular adjustable length "snap-ties" of type which do not leave an opening through concrete or leave metal exposed within 1" of surface of concrete. Ties shall be of type which permit neat and solid patching of tie hole. Wire ties will not be permitted.

D. Chamfer Strips: 3/4" x 3/4" triangular wood or plastic strips.

E. Anchor Slots: Standard 1" x 22 gauge galvanized steel "dovetail" type, size and length as required. Furnish with removable polyfoam filler.

F. Waterstops: Extruded, high-quality virgin polyvinyl chloride (PVC), of size and type indicated on the Drawings or required for particular application.

G. Miscellaneous Accessories: Provide miscellaneous reglets, hangers, ceiling inserts, threaded inserts, and similar anchoring devices which will be embedded in finished concrete.

H. Form Releasing Agent: Commercially formulated, chemically neutral releasing agent of type recommended for form material and approved by ENGINEER. Releasing agent shall not bond with, stain, or adversely affect concrete surfaces requiring bond or adhesion, not impede the curing of the concrete or adhesion of subsequently applied finish materials. Products manufactured with "waste" oils will not be accepted.

PART 3 EXECUTION

3.01 EXAMINATION

A. Verification of Conditions: Examine areas and conditions under which the work of this Section will be performed. Do not proceed with the work until unsatisfactory conditions have been corrected. Commencement of work implies acceptance of all areas and conditions.

B. Inspect formwork for size, shape, profile, condition of interior surfaces and joints, and attachment of accessories and embedded items prior to each placement of concrete.

C. Structural design and fabrication of formwork and shoring are the sole responsibility of the CONTRACTOR, who shall provide all necessary inspections thereof prior to concrete placements.

3.02 FORMWORK CONSTRUCTION

A. General Requirements - All Form Materials:

1. Construct formwork so that concrete members and structures will be of correct size, shape, alignment, elevation, and position.

2. Make formwork sufficiently tight to prevent leaking of cement paste during concrete placement. Solidly butt joints, and provide backup material as required to prevent leakage of paste and formation of "fins".

3. Fit forms placed in successive units for continuous surfaces to accurate alignment, free from irregularities, and within acceptable tolerances.

4. Carry all vertical and lateral loads to the ground by the formwork system, including shoring and bracing, and in-place construction that has attained adequate strength to support the added loads.

5. Provide trussed supports when adequate foundations for shores cannot be secured.

6. Provide shores and struts with positive means for adjustment capable of taking up formwork settlement during concrete placement. Use jacks, wedges, shims, or a combination thereof.

7. Verify location and dimensions of slab depressions, slopes, blockouts, floor drains, and similar engineering or mechanical features.

B. Wood Forms:

1. Use lumber and form plywood in longest practical lengths. Make surfaces straight and plumb, with joints tight and in proper alignment.

2. Fabricate forms for easy removal without hammering or prying against concrete surfaces. Provide crush plates where stripping may damage concrete surfaces.

3. Form exposed corners or edges with 3/4" x 3/4" continuous chamfers. Miter chamfers at intersections or changes in direction to provide uniform, straight edges.

4. Provide extra studs, wales, or blocking to prevent bowing of formwork during concrete placement.

C. Fiberglass Forms:

1. Place forms in exact alignment.

2. Provide all reveals and form accessories required to maintain details shown.

D. Tubular Column Forms:

1. Provide tubular forms in continuous lengths for full height of columns; joints in tubular forms will not be permitted.

2. Erect forms manually or with block and tackle or lifting equipment. Place forms over pre-tied and erected column reinforcement, taking care not to damage inside face of forms.

3. Provide bracing collars and diagonal bracing at intervals not to exceed 8'-0" vertically. Columns forms shall be straight, plumb, and firmly braced prior to concrete placement.

E. Void Forms:

1. Provide void spaces of full size and extent shown on the Drawings.

2. Protect fiberboard void forms from weather at all times prior to and during installation. Do not allow forms, once in place, to remain exposed prior to placement of concrete.

3. Construct void forms to proper dimensions to fit various conditions shown on the Drawings or encountered in the field. Accurately cut forms to shape of piers, columns, or other abutting work.

4. Place void forms in largest pieces practical, and adequately secure in place with joints sealed to prevent leakage of concrete into the void space.

5. Place forms in a neat, workmanlike manner, on ground surfaces that are clean, dry, firmly compacted, and properly shaped. Staple or otherwise fasten forms together in accordance with manufacturer's directions.

6. Continuously overlay fiberboard void forms with 1/8" water-resistant Masonite or plywood.

7. Lay out and install void forms only in areas where concrete may be placed immediately. In order to protect forms from exposure to weather. Provide runways or walkways to protect void forms from point loading or damage prior to or during concrete placement.

8. Provide evidence satisfactory to the ENGINEER and the Structural ENGINEER that proper void spaces have been provided.

F. Temporary Openings: Provide temporary openings for cleaning and inspection of forms.

G. Accommodations for Other Trades: Verify size and location of openings, recesses, chases, or sleeves required to accommodate the work of other trades.

H. Embedded Items: Accurately place and secure in place all required anchor plates, weld plates, or other items to be built into or embedded in the concrete.

I. Earth Forms: Not permitted

3.03 JOINTS

A. Expansion Joints: Install pre-formed compressible expansion joint fillers at locations indicated or required. Do not continue reinforcement across expansion joints.

B. Construction Joints in Formed Elements:

1. General: Locate and install construction joints as indicated on the Drawings, or if not shown on Drawings, locate so as not to impair strength or appearance of the structure. Submit construction joint locations not shown on the Drawings to the ENGINEER for acceptance.

2. Keyways: Provide keyways as detailed on the Drawings. Where not detailed, provide 1-1/2" minimum depth continuous keyways in all construction joints in walls and slabs and between walls and footings.

3. Place construction joints in the center one-third of spans of beams and girders unless indicated otherwise. Continue reinforcement across construction joints. Submit construction joint locations not shown on the Drawings for the ENGINEER's acceptance.

3.04 PREPARATION FOR CONCRETE PLACEMENT

A. Cleaning: Remove all dirt, trash, debris, or other foreign matter from inside of forms. Close up and seal temporary openings used for cleaning and inspection.

B. Form Coating: Coat contact surfaces of forms with form release agent prior to placing reinforcement. Mix an apply release agent in strict accordance with manufacturer's directions. Do not allow excess release agent to accumulate in forms, or to come in contact with reinforcement or embedded items.

3.05 FORM REMOVAL

A. General Requirements:

1. Do not disturb or remove forms until concrete has hardened sufficiently to permit form removal with complete safety.

2. Unless otherwise directed by the ENGINEER, side forms of beams, columns, and walls may be removed no sooner than 48 hours after concrete placement, provided concrete is sufficiently hard not to be damaged by form removal

operations, and that curing and protection of exposed concrete surfaces is maintained after removal of forms.

B. Protection of Concrete Surfaces: Exercise care in removing forms from finished concrete so that surfaces are not marred, gouged, or otherwise damaged, and that corners remain true, sharp, and unbroken.

C. Concrete Patching: Patching of tie holes, honeycombs, voids or other defects is specified in Section 03 30 05 Civil Cast-In-Place Concrete.

D. Re-Use of Forms: Clean and repair surfaces of forms that are to be reused in the Work. Split, frayed, delaminated, or otherwise damaged form surfaces will not be allowed in future work. Re-align and tighten all joints.

3.06 CLEANING

A. Construction Cleaning: Remove excess materials, tools, and debris as the work progresses; keep the jobsite and work area clean at all times.

END OF SECTION 031005

SECTION 032005 - CIVIL CONCRETE REINFORCEMENT

PART 1 GENERAL

1.01 SUMMARY

A. Work Included: This Section includes the furnishing, fabrication, and installation of reinforcement for concrete, including all necessary accessories, as shown on the Drawings or specified herein, or as required to complete the Work.

B. Related Work:

1. Section 013300, Submittal Procedures.

2. Section 031005, Civil Concrete Formwork.

3. Section 033005, Civil Cast-in-Place Concrete.

1.02 REFERENCE

A. Reference Standards: Comply with the requirements of the reference standards noted herein, except where more stringent requirements are listed herein or otherwise required by the Contract Documents.

B. ACI 318, "Building Code Requirements for Reinforced Concrete", latest editions in effect at the Contract Date.

1.03 SUBMITTALS

A. Submittal Procedures: Refer to Section 013300.

B. Shop Drawings:

1. Submit shop drawings for fabrication, bending and placement of concrete reinforcement. Comply with ACI Detailing Manual (AP 66). Provide 1/4" scale elevations of all walls and grade beams with reinforcing shown.

2. Show all openings in walls and structural slabs. Include special reinforcement required at openings through concrete structure.

3. Size and placement of openings shall be reviewed by and stamped by the mechanical, electrical, plumbing, fire protection, elevator, or other subcontractors whose work relates to the openings prior to submitting for ENGINEER's review.

C. Certificates: When requested by the ENGINEER, submit certified evidence that reinforcing materials and accessories comply with the requirements of this Section.

1.04 QUALITY ASSURANCE

A. Qualification of Workmen: Use only workmen who are experienced in the fabrication and placement of concrete reinforcement.

B. Coordination of Drawings and Specifications: Carefully examine Drawings for references to concrete reinforcement. In the event of conflict between the Drawings and Specifications, the more stringent requirements will apply.

1.05 DELIVERY, STORAGE, AND HANDLING

A. Tagging and Identification: Tag all reinforcing steel for identification in the field. Take necessary precautions to maintain identification after bundles of reinforcing steel are broken.

B. Storage and Protection: Protect reinforcing materials and accessories before, during, and after installation. Store in a manner to prevent excessive rusting or fouling with dirt, grease, or other materials which would inhibit bonding of concrete to reinforcement.

PART 2 PRODUCTS

2.01 MATERIALS

A. Manufacturer: Unless otherwise approved in writing by the ENGINEER, concrete reinforcement shall be fabricated form domestic-manufactured steel.

B. Deformed Steel Reinforcing Bars (ReBar):

1. Unless otherwise indicated, provide bars complying with ASTM A615, Grade 60, bar diameters as indicated on the Drawings.

2. Provide reinforcing bars complying with ASTM A706 where welded connections are used.

C. Welded Wire Fabric (WWF): ASTM A185, fabricated from smooth wire conforming to ASTM A82; wire and mesh size as indicated on the Drawings. Furnish in flat sheets only. (May also be referred to on Drawings as welded wire mesh (WWM).

D. Wire Ties: ASTM A82, 16 gauge.

E. Support or Spacing Accessories:

1. Comply with ACI SP-66 and CRSI Manual of Standard Practice.

2. Provide stainless steel, galvanized, or plastic tipped accessories where resting on forms where the underside will be exposed to view or where plaster, stucco, dash coat, or paint finish.

2.02 FABRICATION

A. General: Fabricate reinforcing bars to conform to required shapes and dimensions, with fabrication tolerances conforming to ACI 318 and CRSI Manual of Standard Practice. In case of fabrication errors, do not re-bend or straighten bars in a manner which will injure or weaken the material.

B. Cause for Rejection: Bars with reduced cross-section due to excessive rusting, bars that have been fabricated in excess of ACI or CRSI dimensional tolerances, and bars with bends or kinks not shown or required will be rejected.

PART 3 EXECUTION

3.01 EXAMINATION

A. Verification of Conditions: Examine areas and conditions under which the work of this Section will be performed. Do not proceed with the work until unsatisfactory conditions have been corrected. Commencement of work implies acceptance of all areas and conditions.

3.02 INSTALLATION

A. General:

1. Clean reinforcement to remove excessive rust, mill scale, dirt, grease, or other coatings that would reduce or inhibit bonding to concrete.

2. Provide bolsters, chairs, or other approved means of support of slab reinforcement as required to maintain correct position as shown on the Drawings. Lifting of reinforcement by hooks or other means during concrete placement will not be permitted.

3. Position, support, and secure reinforcement against displacement by formwork or construction operations, including concrete placement. Space chairs an other supports in accordance with ACI SP-66.

4. Do not place reinforcing bars more than 2" beyond last leg of any continuous bar support.

5. Do not use reinforcement supports as bases for runways for concrete conveying equipment and similar construction loads.

6. Arrange, space and securely tie reinforcing bars and supports together using specified tie wire. Set tie wires so that twisted ends are directed away from exposed concrete surfaces.

7. Field bends will not be permitted without approval of ENGINEER.

8. Install welded wire fabric in lengths as long as practical. Lap adjoining pieces at least one and one-half full meshes.

B. Splices:

1. Do not splice bars except at locations indicated on the Drawings, except as otherwise specifically approved by the ENGINEER.

2. Necessary splices not shown on the Drawings or otherwise noted shall be in accordance with ACI Class "B" lap splices for bar sizes up to #11, but in no event shall be less than 40 bar diameters.

3. Splices in bars larger than #11 shall be made with approved thermal or mechanical coupling devices. Welding shall comply with AWS D1. 4-79.

C. Slab Openings: Unless otherwise indicated, provide the following additional reinforcement at slab openings:

1. At slab openings of 12" or less, spread main reinforcing around opening, leaving not less than the required coverage of concrete.

2. At openings greater than 12", provide diagonal bars as indicated on the Structural Drawings.

D. Conduit in Slab: Run conduit or piping under top layer of reinforcing. Provide a minimum of 1-1/2" clear between conduits and pipes or between conduits or pipes and parallel reinforcing. Do not "bundle" conduit or pipes. Coordinate with Section 03 30 05 Civil Cast-in-Place Concrete for thickening of slabs, if required, at conduits or piping.

E. Construction Joints:

1. Provide construction joints at locations indicated or approved by the ENGINEER. Run reinforcing or dowels continuously through construction joints, allowing for 30-diameter splices in columns and 40-diameter splices in beams or slabs.

2. Form construction joints in slabs with galvanized metal or wood keyed forms. Make joints during concrete placement.

3. If keyed metal forms are to be left in place to create control joints, place with top edge of form ½" below tip of slab and provide removable wood backup to provide screed. Lap edges of keyed metal forms at least 12", and firmly stake or secure to prevent offsetting during concrete placement.

3.03 FIELD QUALITY CONTROL

A. Notification: Notify the ENGINEER and the testing and inspection agency after reinforcement is in place, but not less than 48 hours prior to scheduled placement of concrete, to allow sufficient time for observation and inspection of in-place reinforcement, and correction of rejected work.

B. Reinforcing Steel Inspection: Inspect concrete reinforcement fabrication and placement for bar size, splices, form clearance, supports, and attachment of accessories and embedded items prior to each concrete placement. Do not place concrete until deficiencies are corrected and reinspected

END OF SECTION 032005

SECTION 033000 - CAST-IN-PLACE CONCRETE

PART 1 - GENERAL

1.1 RELATED DOCUMENTS

A. Drawings and general provisions of the Contract, including General and Supplementary Conditions and Division 01 Specification Sections, apply to this Section.

1.2 SUMMARY

A. This Section specifies cast-in place concrete, including formwork, reinforcement, concrete materials, mixture design, placement procedures, and finishes, for the following:

1. Footings.
2. Foundation walls.
3. Building frame members.
4. Building walls.

1.3 DEFINITIONS

A. Cementitious Materials: Portland cement alone or in combination with fly ash; subject to compliance with requirements.

1.4 SUBMITTALS

A. Product Data: For each type of product indicated.

B. Design Mixtures: For each concrete mixture. Submit alternate design mixtures each with its own identification number when characteristics of materials, Project conditions, weather, test results, or other circumstances warrant adjustments.

C. Indicate the minimum following information:

1. Mix Identification Number.
2. Mix-use type.
3. Required 28-day compressive strength.
4. Cement content.
5. Coarse aggregate type and quantity.
6. Fine aggregate type and quantity.
7. Total aggregate gradation.
8. Water quantity.
9. Admixture types and quantity.
10. Slump measurement.
11. Air content.

12. 28-day shrinkage rate.
13. 28-day concrete strength test.
14. Alkali Silicate Reactivity (ASR).
15. Chloride-ion content.
16. Amounts of mixing water to be withheld for later addition at Project site.

D. Steel Reinforcement Shop Drawings: Placing drawings that detail fabrication, bending, and placement. Include bar sizes, lengths, material, grade, bar schedules, stirrup spacing, bent bar diagrams, bar arrangement, splices and laps, mechanical connections, tie spacing, hoop spacing, and supports for concrete reinforcement.

E. Welding certificates.

F. Material Test Reports: For the following, from a qualified testing agency, indicating compliance with requirements:

1. Aggregates. Include service record data indicating absence of deleterious expansion of concrete due to alkali aggregate reactivity.

G. Material Certificates: For each of the following, signed by manufacturers:

1. Cementitious materials.
2. Admixtures.
3. Form materials and form-release agents.
4. Steel reinforcement and accessories.
5. Waterstops.
6. Curing compounds.
7. Bonding agents.
8. Adhesives.
9. Semirigid joint filler.
10. Joint-filler strips.
11. Repair materials.

H. Minutes of preinstallation conference.

1.5 QUALITY ASSURANCE

A. Manufacturer Qualifications: A firm experienced in manufacturing ready-mixed concrete products and that complies with ASTM C 94/C 94M requirements for production facilities and equipment.

1. Manufacturer certified according to NRMCA's "Certification of Ready Mixed Concrete Production Facilities."

B. Testing Agency Qualifications: An independent agency, acceptable to authorities having jurisdiction, qualified according to ASTM C 1077 and ASTM E 329 for testing indicated, as documented according to ASTM E 548.

1. Personnel conducting field tests shall be qualified as ACI Concrete Field Testing Technician, Grade 1, according to ACI CP-01 or an equivalent certification program.
2. Personnel performing laboratory tests shall be ACI-certified Concrete Strength Testing Technician and Concrete Laboratory Testing Technician - Grade I. Testing Agency laboratory supervisor shall be an ACI-certified Concrete Laboratory Testing Technician - Grade II.

C. Source Limitations: Obtain each type or class of cementitious material of the same brand from the same manufacturer's plant, obtain aggregate from one source, and obtain admixtures through one source from a single manufacturer.

D. Welding: Qualify procedures and personnel according to AWS D1.4, "Structural Welding Code--Reinforcing Steel."

E. ACI Publications: Comply with the following unless modified by requirements in the Contract Documents:

1. ACI 301, "Specification for Structural Concrete," Sections 1 through 5."
2. ACI 117, "Specifications for Tolerances for Concrete Construction and Materials."

F. Concrete Testing Service: Engage a qualified independent testing agency to perform material evaluation tests and to design concrete mixtures.

G. Preinstallation Conference: Conduct conference at Project site to comply with requirements in Division 01 Sections.

1. Before submitting design mixtures, review concrete design mixture and examine procedures for ensuring quality of concrete materials. Require representatives of each entity directly concerned with cast-in-place concrete to attend, including the following:

   a. Contractor's superintendent.
   b. Independent testing agency responsible for concrete design mixtures.
   c. Ready-mix concrete manufacturer.
   d. Concrete subcontractor.
   e. Owner's independent testing and inspection agency.
   f. Architect.

2. Review special inspection and testing and inspecting agency procedures for field quality control, concrete finishes and finishing, cold- and hot-weather concreting procedures, curing procedures, construction contraction and isolation joints, and joint-filler strips, semirigid joint fillers, forms and form removal limitations, shoring and reshoring procedures, anchor rod and anchorage device installation tolerances, steel reinforcement installation, concrete repair procedures, and concrete protection.

1.6 DELIVERY, STORAGE, AND HANDLING

A. Steel Reinforcement: Deliver, store, and handle steel reinforcement to prevent bending and damage.

B. Waterstops: Store waterstops under cover to protect from moisture, sunlight, dirt, oil, and other contaminants.

PART 2 - PRODUCTS

2.1 MANUFACTURERS

A. In other Part 2 articles where titles below introduce lists, the following requirements apply to product selection:

1. Products: Subject to compliance with requirements, provide one of the products specified.
2. Manufacturers: Subject to compliance with requirements, provide products by one of the manufacturers specified.

2.2 FORM-FACING MATERIALS

A. Smooth-Formed Finished Concrete: Form-facing panels that will provide continuous, true, and smooth concrete surfaces. Furnish in largest practicable sizes to minimize number of joints.

1. Plywood, metal, or other approved panel materials.
2. Exterior-grade plywood panels, suitable for concrete forms, complying with DOC PS 1, and as follows:

a. Medium-density overlay, Class 1 or better; mill-release agent treated and edge sealed.

B. Rough-Formed Finished Concrete: Plywood, lumber, metal, or another approved material. Provide lumber dressed on at least two edges and one side for tight fit.

C. Forms for Pedestals and Supports: Metal, glass-fiber-reinforced plastic, paper, or fiber tubes that will produce surfaces with gradual or abrupt irregularities not exceeding specified formwork surface class. Provide units with sufficient wall thickness to resist plastic concrete loads without detrimental deformation.

D. Chamfer Strips: Wood, metal, PVC, or rubber strips, 3/4 by 3/4 inch (19 by 19 mm), minimum.

E. Rustication Strips: Wood, metal, PVC, or rubber strips, kerfed for ease of form removal.

F. Form-Release Agent: Commercially formulated form-release agent that will not bond with, stain, or adversely affect concrete surfaces and will not impair subsequent treatments of concrete surfaces.

1. Formulate form-release agent with rust inhibitor for steel form-facing materials.

G. Form Ties: Factory-fabricated, removable or snap-off metal or glass-fiber-reinforced plastic form ties designed to resist lateral pressure of fresh concrete on forms and to prevent spalling of concrete on removal.

1. Furnish units that will leave no corrodible metal closer than 1 inch (25 mm) to the plane of exposed concrete surface.
2. Furnish ties that, when removed, will leave holes no larger than 1 inch (25 mm) in diameter in concrete surface.
3. Furnish ties with integral water-barrier plates to walls indicated to receive dampproofing or waterproofing.

2.3 STEEL REINFORCEMENT

A. Reinforcing Bars: ASTM A 615/A 615M, Grade 60 (Grade 420), deformed.

B. Low-Alloy-Steel Reinforcing Bars: ASTM A 706/A 706M, deformed.

C. Plain-Steel Wire: ASTM A 82, as drawn.

D. Plain-Steel Welded Wire Reinforcement: ASTM A 185, plain, fabricated from as-drawn steel wire into flat sheets.

2.4 REINFORCEMENT ACCESSORIES

A. Joint Dowel Bars: ASTM A 615/A 615M, Grade 60 (Grade 420), plain-steel bars, cut bars true to length with ends square and free of burrs.

B. Bar Supports: Bolsters, chairs, spacers, and other devices for spacing, supporting, and fastening reinforcing bars and welded wire reinforcement in place. Manufacture bar supports from steel wire, plastic, or precast concrete according to CRSI's "Manual of Standard Practice," of greater compressive strength than concrete.

2.5 CONCRETE MATERIALS

A. Cementitious Material: Use the following cementitious materials, of the same type, brand, and source, throughout Project:

1. Portland Cement: ASTM C 150, Type I/II, gray Supplement with the following at contractor's option for concrete other than slabs and flatwork:

   a. Fly Ash: ASTM C 618, Class C.

B. Normal-Weight Aggregates: ASTM C 33, Class 4S coarse aggregate or better, graded. Provide aggregates from a single source with documented service record data of at least 10 years' satisfactory service in similar applications and service conditions using similar aggregates and cementitious materials.

1. Maximum Coarse-Aggregate Size: As indicated.
2. Fine Aggregate: Free of materials with deleterious reactivity to alkali in cement.

C. Lightweight Aggregate: ASTM C 330.

D. Water: ASTM C 94/C 94M and potable.

2.6 ADMIXTURES

A. Air-Entraining Admixture: ASTM C 260.

B. Chemical Admixtures: Provide admixtures certified by manufacturer to be compatible with other admixtures and that will not contribute water-soluble chloride ions exceeding those permitted in hardened concrete. Do not use calcium chloride or admixtures containing calcium chloride.

1. Water-Reducing Admixture: ASTM C 494/C 494M, Type A.
2. Retarding Admixture: ASTM C 494/C 494M, Type B.
3. Water-Reducing and Retarding Admixture: ASTM C 494/C 494M, Type D.
4. High-Range, Water-Reducing Admixture: ASTM C 494/C 494M, Type F.
5. High-Range, Water-Reducing and Retarding Admixture: ASTM C 494/C 494M, Type G.
6. Plasticizing and Retarding Admixture: ASTM C 1017/C 1017M, Type II.

2.7 WATERSTOPS

A. Self-Expanding Butyl Strip Waterstops: Manufactured rectangular or trapezoidal strip, butyl rubber with sodium bentonite or other hydrophilic polymers, for adhesive bonding to concrete, 3/4 by 1 inch (19 by 25 mm).

1. Products: Subject to compliance with requirements, provide one of the following:

a. Colloid Environmental Technologies Company; Volclay Waterstop-RX.
b. Concrete Sealants Inc.; Conseal CS-231.
c. Greenstreak; Swellstop.
d. Henry Company, Sealants Division; Hydro-Flex.
e. JP Specialties, Inc.; Earthshield Type 20.
f. Progress Unlimited, Inc.; Superstop.
g. TCMiraDRI; Mirastop.

2.8 CURING MATERIALS

A. Evaporation Retarder: Waterborne, monomolecular film forming, manufactured for application to fresh concrete.

1. Products: Subject to compliance with requirements, provide one of the following:

a. Axim Concrete Technologies; Cimfilm.
b. Burke by Edoco; BurkeFilm.
c. ChemMasters; Spray-Film.
d. Conspec Marketing & Manufacturing Co., Inc., a Dayton Superior Company Aquafilm.
e. Dayton Superior Corporation; Sure Film.
f. Euclid Chemical Company (The); Eucobar.
g. Kaufman Products, Inc.; Vapor Aid.
h. Lambert Corporation; Lambco Skin.
i. L&M Construction Chemicals, Inc.; E-Con.
j. MBT Protection and Repair, Div. of ChemRex; Confilm.
k. Meadows, W. R., Inc.; Sealtight Evapre.
l. Metalcrete Industries; Waterhold.
m. Nox-Crete Products Group, Kinsman Corporation; Monofilm.
n. Sika Corporation, Inc.; SikaFilm.
o. Symons Corporation, a Dayton Superior Company; Finishing Aid.
p. Unitex; Pro-Film.
q. US Mix Products Company; US Spec Monofilm ER.
r. Vexcon Chemicals, Inc.; Certi-Vex EnvioAssist.

B. Absorptive Cover: AASHTO M 182, Class 2, burlap cloth made from jute or kenaf, weighing approximately 9 oz./sq. yd. (305 g/sq. m) when dry.

C. Moisture-Retaining Cover: ASTM C 171, polyethylene film or white burlap-polyethylene sheet.

D. Water: Potable.

E. Clear, Waterborne, Membrane-Forming Curing Compound: ASTM C 309, Type 1, Class B, dissipating.

1. Products: Subject to compliance with requirements, provide one of the following:

a. Anti-Hydro International, Inc.; AH Curing Compound #2 DR WB.
b. Burke by Edoco; Aqua Resin Cure.
c. ChemMasters; Safe-Cure Clear.
d. Conspec Marketing & Manufacturing Co., Inc., a Dayton Superior Company; W.B. Resin Cure.
e. Dayton Superior Corporation; Day Chem Rez Cure (J-11-W).
f. Euclid Chemical Company (The); Kurez DR VOX.
g. Kaufman Products, Inc.; Thinfilm 420.
h. Lambert Corporation; Aqua Kure-Clear.
i. L&M Construction Chemicals, Inc.; L&M Cure R.
j. Meadows, W. R., Inc.; 1100 Clear.
k. Nox-Crete Products Group, Kinsman Corporation; Resin Cure E.
l. Symons Corporation, a Dayton Superior Company; Resi-Chem Clear Cure.
m. Tamms Industries, Inc.; Horncure WB 30.
n. Unitex; Hydro Cure 309.
o. US Mix Products Company; US Spec Maxcure Resin Clear.
p. Vexcon Chemicals, Inc.; Certi-Vex Enviocure 100.

2.9 RELATED MATERIALS

A. Expansion- and Isolation-Joint-Filler Strips: ASTM D 1752, cork or self-expanding cork.

B. Bonding Agent: ASTM C 1059, Type II, non-redispersible, acrylic emulsion or styrene butadiene.

C. Epoxy Bonding Adhesive: ASTM C 881, two-component epoxy resin, capable of humid curing and bonding to damp surfaces, of class suitable for application temperature and of grade to suit requirements, and as follows:

1. Types IV and V, load bearing, for bonding hardened or freshly mixed concrete to hardened concrete.

D. Reglets: Fabricate reglets of not less than 0.0217-inch- (0.55-mm-) thick, galvanized steel sheet. Temporarily fill or cover face opening of reglet to prevent intrusion of concrete or debris.

E. Dovetail Anchor Slots: Hot-dip galvanized steel sheet, not less than 0.0336 inch (0.85 mm) thick, with bent tab anchors. Temporarily fill or cover face opening of slots to prevent intrusion of concrete or debris.

2.10 REPAIR MATERIALS

A. Repair Underlayment: Cement-based, polymer-modified, self-leveling product that can be applied in thicknesses from 1/8 inch (3.2 mm) and that can be feathered at edges to match adjacent floor elevations.

1. Cement Binder: ASTM C 150, portland cement or hydraulic or blended hydraulic cement as defined in ASTM C 219.
2. Primer: Product of underlayment manufacturer recommended for substrate, conditions, and application.
3. Aggregate: Well-graded, washed gravel, 1/8 to 1/4 inch (3.2 to 6 mm) or coarse sand as recommended by underlayment manufacturer.
4. Compressive Strength: Not less than 4100 psi (29 MPa)at 28 days when tested according to ASTM C 109/C 109M.

B. Repair Overlayment: Cement-based, polymer-modified, self-leveling product that can be applied in thicknesses from 1/8 inch (3.2 mm) and that can be feathered at edges to match adjacent floor elevations.

1. Cement Binder: ASTM C 150, portland cement or hydraulic or blended hydraulic cement as defined in ASTM C 219.
2. Primer: Product of topping manufacturer recommended for substrate, conditions, and application.
3. Aggregate: Well-graded, washed gravel, 1/8 to 1/4 inch (3.2 to 6 mm) or coarse sand as recommended by topping manufacturer.
4. Compressive Strength: Not less than 5000 psi (34.5 MPa) at 28 days when tested according to ASTM C 109/C 109M.

2.11 CONCRETE MIXTURES, GENERAL

A. Prepare design mixtures for each type and strength of concrete, proportioned on the basis of laboratory trial mixture or field test data, or both, according to ACI 301.

1. Use a qualified independent testing agency for preparing and reporting proposed mixture designs based on laboratory trial mixtures.

B. Cementitious Materials: Limit percentage, by weight, of cementitious materials other than portland cement in concrete as follows:

1. Fly Ash: 15 percent.

C. Limit water-soluble, chloride-ion content in hardened concrete to 0.3 percent by weight of cement.

D. Admixtures: Use admixtures according to manufacturer's written instructions.

1. Use water-reducing, high-range water-reducing or plasticizing admixture in concrete, as required, for placement and workability.
2. Use water-reducing and retarding admixture when required by high temperatures, low humidity, or other adverse placement conditions.
3. Use water-reducing admixture in pumped concrete, concrete for heavy-use industrial slabs and parking structure slabs, concrete required to be watertight, and concrete with a water-cementitious materials ratio below 0.50.

2.12 CONCRETE MIXTURES FOR BUILDING ELEMENTS

A. Proportion normal-weight and lightweight concrete mixtures to equal or exceed the minimum 28 day concrete compressive strength and other specified criteria indicated on the drawings.

2.13 FABRICATING REINFORCEMENT

A. Fabricate steel reinforcement according to CRSI's "Manual of Standard Practice."

2.14 CONCRETE MIXING

A. Ready-Mixed Concrete: Measure, batch, mix, and deliver concrete according to ASTM C 94/C 94M and ASTM C 1116, and furnish batch ticket information.

1. When air temperature is between 85 and 90 deg F (30 and 32 deg C), reduce mixing and delivery time from 1-1/2 hours to 75 minutes; when air temperature is above 90 deg F (32 deg C), reduce mixing and delivery time to 60 minutes.

PART 3 - EXECUTION

3.1 FORMWORK

A. Design, erect, shore, brace, and maintain formwork, according to ACI 301, to support vertical, lateral, static, and dynamic loads, and construction loads that might be applied, until structure can support such loads.

B. Construct formwork so concrete members and structures are of size, shape, alignment, elevation, and position indicated, within tolerance limits of ACI 117.

C. Limit concrete surface irregularities, designated by ACI 347R as abrupt or gradual, as follows:

1. Class A, 1/8 inch (3.2 mm) for smooth-formed finished surfaces.
2. Class B, 1/4 inch (6 mm) for rough-formed finished surfaces.

D. Construct forms tight enough to prevent loss of concrete mortar.

E. Fabricate forms for easy removal without hammering or prying against concrete surfaces. Provide crush or wrecking plates where stripping may damage cast concrete surfaces. Provide top forms for inclined surfaces steeper than 1.5 horizontal to 1 vertical.

1. Install keyways, reglets, recesses, and the like, for easy removal.
2. Do not use rust-stained steel form-facing material.

F. Set edge forms, bulkheads, and intermediate screed strips for slabs to achieve required elevations and slopes in finished concrete surfaces. Provide and secure units to support screed strips; use strike-off templates or compacting-type screeds.

G. Provide temporary openings for cleanouts and inspection ports where interior area of formwork is inaccessible. Close openings with panels tightly fitted to forms and securely braced to prevent loss of concrete mortar. Locate temporary openings in forms at inconspicuous locations.

H. Chamfer exterior corners and edges of permanently exposed concrete.

I. Form openings, chases, offsets, sinkages, keyways, reglets, blocking, screeds, and bulkheads required in the Work. Determine sizes and locations from trades providing such items.

J. Clean forms and adjacent surfaces to receive concrete. Remove chips, wood, sawdust, dirt, and other debris just before placing concrete.

K. Retighten forms and bracing before placing concrete, as required, to prevent mortar leaks and maintain proper alignment.

L. Coat contact surfaces of forms with form-release agent, according to manufacturer's written instructions, before placing reinforcement.

3.2 EMBEDDED ITEMS

A. Place and secure anchorage devices and other embedded items required for adjoining work that is attached to or supported by cast-in-place concrete. Use setting drawings, templates, diagrams, instructions, and directions furnished with items to be embedded.

1. Install anchor rods, accurately located, to elevations required and complying with tolerances in Section 7.5 of AISC's "Code of Standard Practice for Steel Buildings and Bridges."
2. Install reglets to receive waterproofing and to receive through-wall flashings in outer face of concrete frame at exterior walls, where flashing is shown at lintels, shelf angles, and other conditions.
3. Install dovetail anchor slots in concrete structures as indicated.

3.3 REMOVING AND REUSING FORMS

A. General: Formwork for sides of beams, walls, columns, and similar parts of the Work that does not support weight of concrete may be removed after cumulatively curing at not less than 50 deg F (10 deg C) for 24 hours after placing concrete, if concrete is hard enough to not be damaged by form-removal operations and curing and protection operations are maintained.

1. Leave formwork for structural elements that supports weight of concrete in place until concrete has achieved at least 75 percent of its 28-day design compressive strength.
2. Remove forms only if shores have been arranged to permit removal of forms without loosening or disturbing shores.

B. Clean and repair surfaces of forms to be reused in the Work. Split, frayed, delaminated, or otherwise damaged form-facing material will not be acceptable for exposed surfaces. Apply new form-release agent.

C. When forms are reused, clean surfaces, remove fins and laitance, and tighten to close joints. Align and secure joints to avoid offsets. Do not use patched forms for exposed concrete surfaces unless approved by Architect.

3.4 STEEL REINFORCEMENT

A. General: Comply with CRSI's "Manual of Standard Practice" for placing reinforcement.

1. Do not cut or puncture vapor retarder. Repair damage and reseal vapor retarder before placing concrete.

B. Clean reinforcement of loose rust and mill scale, earth, ice, and other foreign materials that would reduce bond to concrete.

C. Accurately position, support, and secure reinforcement against displacement. Locate and support reinforcement with bar supports to maintain minimum concrete cover. Do not tack weld crossing reinforcing bars.

1. Weld reinforcing bars according to AWS D1.4, where indicated.

D. Set wire ties with ends directed into concrete, not toward exposed concrete surfaces.

E. Install welded wire reinforcement in longest practicable lengths on bar supports spaced to minimize sagging. Lap edges and ends of adjoining sheets at least one mesh spacing. Offset laps of adjoining sheet widths to prevent continuous laps in either direction. Lace overlaps with wire.

3.5 JOINTS

A. General: Construct joints true to line with faces perpendicular to surface plane of concrete.

B. Construction Joints: Install so strength and appearance of concrete are not impaired, at locations indicated or as approved by Architect.

1. Place joints perpendicular to main reinforcement. Continue reinforcement across construction joints, unless otherwise indicated. Do not continue reinforcement through sides of strip placements of floors and slabs.
2. Locate joints for beams, slabs, joists, and girders in the middle third of spans. Offset joints in girders a minimum distance of twice the beam width from a beam-girder intersection.
3. Locate horizontal joints in walls and columns at underside of floors, slabs, beams, and girders and at the top of footings or floor slabs.
4. Space vertical joints in walls as indicated. Locate joints beside piers integral with walls, near corners, and in concealed locations where possible.
5. Use a bonding agent at locations where fresh concrete is placed against hardened or partially hardened concrete surfaces.
6. Use epoxy-bonding adhesive at locations where fresh concrete is placed against hardened or partially hardened concrete surfaces.

C. Doweled Joints: Install dowel bars and support assemblies at joints where indicated. Lubricate or asphalt coat one-half of dowel length to prevent concrete bonding to one side of joint.

3.6 WATERSTOPS

A. Self-Expanding Strip Waterstops: Install in construction joints and at other locations indicated, according to manufacturer's written instructions, adhesive bonding, mechanically fastening, and firmly pressing into place. Install in longest lengths practicable.

3.7 CONCRETE PLACEMENT

A. Before placing concrete, verify that installation of formwork, reinforcement, and embedded items is complete and that required inspections have been performed.

B. Before test sampling and placing concrete, water may be added at Project site, subject to limitations of ACI 301.

1. Do not add water to concrete after adding high-range water-reducing admixtures to mixture.

C. Deposit concrete continuously in one layer or in horizontal layers of such thickness that no new concrete will be placed on concrete that has hardened enough to cause seams or planes of weakness. If a section cannot be placed continuously, provide construction joints as indicated. Deposit concrete to avoid segregation.

1. Deposit concrete in horizontal layers of depth to not exceed formwork design pressures and in a manner to avoid inclined construction joints.
2. Consolidate placed concrete with mechanical vibrating equipment according to ACI 301.
3. Do not use vibrators to transport concrete inside forms. Insert and withdraw vibrators vertically at uniformly spaced locations to rapidly penetrate placed layer and at least 6 inches (150 mm) into preceding layer. Do not insert vibrators into lower layers of concrete that have begun to lose plasticity. At each insertion, limit duration of vibration to time necessary to consolidate concrete and complete embedment of reinforcement and other embedded items without causing mixture constituents to segregate.

D. Cold-Weather Placement: Comply with ACI 306.1 and as follows. Protect concrete work from physical damage or reduced strength that could be caused by frost, freezing actions, or low temperatures.

1. When average high and low temperature is expected to fall below 40 deg F (4.4 deg C) for three successive days, maintain delivered concrete mixture temperature within the temperature range required by ACI 301.
2. Do not use frozen materials or materials containing ice or snow. Do not place concrete on frozen subgrade or on subgrade containing frozen materials.
3. Do not use calcium chloride, salt, or other materials containing antifreeze agents or chemical accelerators unless otherwise specified and approved in mixture designs.

E. Hot-Weather Placement: Comply with ACI 301 and as follows:

1. Maintain concrete temperature below 90 deg F (32 deg C) at time of placement. Chilled mixing water or chopped ice may be used to control temperature, provided water equivalent of ice is calculated to total amount of mixing water. Using liquid nitrogen to cool concrete is Contractor's option.
2. Fog-spray forms, steel reinforcement, and subgrade just before placing concrete. Keep subgrade uniformly moist without standing water, soft spots, or dry areas.

3.8 FINISHING FORMED SURFACES

A. Rough-Formed Finish: As-cast concrete texture imparted by form-facing material with tie holes and defects repaired and patched. Remove fins and other projections that exceed specified limits on formed-surface irregularities.

1. Apply to concrete surfaces not exposed to public view such as mechanical rooms and storage rooms where cast-in-place concrete walls occur.

B. Related Unformed Surfaces: At tops of walls, horizontal offsets, and similar unformed surfaces adjacent to formed surfaces, strike off smooth and finish with a texture matching adjacent formed surfaces. Continue final surface treatment of formed surfaces uniformly across adjacent unformed surfaces, unless otherwise indicated.

3.9 MISCELLANEOUS CONCRETE ITEMS

A. Filling In: Fill in holes and openings left in concrete structures, unless otherwise indicated, after work of other trades is in place. Mix, place, and cure concrete, as specified, to blend with in-place construction. Provide other miscellaneous concrete filling indicated or required to complete the Work.

B. Equipment Bases and Foundations: Provide machine and equipment bases and foundations as shown on Drawings. Set anchor bolts for machines and equipment at correct elevations, complying with diagrams or templates from manufacturer furnishing machines and equipment.

C. Steel Pan Stairs: Provide concrete fill for steel pan stair treads, landings, and associated items. Cast-in inserts and accessories as shown on Drawings. Screed, tamp, and trowel-finish concrete surfaces.

3.10 CONCRETE PROTECTING AND CURING

A. General: Protect freshly placed concrete from premature drying and excessive cold or hot temperatures. Comply with ACI 306.1 for cold-weather protection and ACI 301 for hot-weather protection during curing.

B. Evaporation Retarder: Apply evaporation retarder to unformed concrete surfaces if hot, dry, or windy conditions cause moisture loss approaching 0.2 lb/sq. ft. x h (1 kg/sq. m x h) before and during finishing operations. Apply according to manufacturer's written instructions after placing, screeding, and bull floating or darbying concrete, but before float finishing.

C. Formed Surfaces: Cure formed concrete surfaces, including underside of beams, supported slabs, and other similar surfaces. If forms remain during curing period, moist cure after loosening forms. If removing forms before end of curing period, continue curing for the remainder of the curing period.

D. Unformed Surfaces: Begin curing immediately after finishing concrete. Cure unformed surfaces, including floors and slabs, concrete floor toppings, and other surfaces.

E. Cure concrete according to ACI 308.1, by one or a combination of the following methods:

1. Moisture Curing: Keep surfaces continuously moist for not less than seven days with the following materials:

a. Water.
b. Continuous water-fog spray.
c. Absorptive cover, water saturated, and kept continuously wet. Cover concrete surfaces and edges with 12-inch (300-mm) lap over adjacent absorptive covers.

2. Moisture-Retaining-Cover Curing: Cover concrete surfaces with moisture-retaining cover for curing concrete, placed in widest practicable width, with sides and ends lapped at least 12 inches (300 mm), and sealed by waterproof tape or adhesive. Cure for not less than seven days. Immediately repair any holes or tears during curing period using cover material and waterproof tape.

   a. Moisture cure or use moisture-retaining covers to cure concrete surfaces to receive floor coverings.
   b. Moisture cure or use moisture-retaining covers to cure concrete surfaces to receive penetrating liquid floor treatments.
   c. Cure concrete surfaces to receive floor coverings with either a moisture-retaining cover or a curing compound that the manufacturer certifies will not interfere with bonding of floor covering used on Project.

3. Curing Compound: Apply uniformly in continuous operation by power spray or roller according to manufacturer's written instructions. Recoat areas subjected to heavy rainfall within three hours after initial application. Maintain continuity of coating and repair damage during curing period.

   a. After curing period has elapsed, remove curing compound without damaging concrete surfaces by method recommended by curing compound manufacturer unless manufacturer certifies curing compound will not interfere with bonding of floor covering used on Project.

4. Curing and Sealing Compound: Apply uniformly to floors and slabs indicated in a continuous operation by power spray or roller according to manufacturer's written instructions. Recoat areas subjected to heavy rainfall within three hours after initial application. Repeat process 24 hours later and apply a second coat. Maintain continuity of coating and repair damage during curing period.

3.11 JOINT FILLING

A. Prepare, clean, and install joint filler according to manufacturer's written instructions.

   1. Defer joint filling until concrete has aged at least one month. Do not fill joints until construction traffic has permanently ceased.

B. Remove dirt, debris, saw cuttings, curing compounds, and sealers from joints; leave contact faces of joint clean and dry.

C. Install semirigid joint filler full depth in saw-cut joints and at least 2 inches (50 mm) deep in formed joints. Overfill joint and trim joint filler flush with top of joint after hardening.

3.12 CONCRETE SURFACE REPAIRS

A. Defective Concrete: Repair and patch defective areas when approved by Architect. Remove and replace concrete that cannot be repaired and patched to Architect's approval.

B. Patching Mortar: Mix dry-pack patching mortar, consisting of one part portland cement to two and one-half parts fine aggregate passing a No. 16 (1.18-mm) sieve, using only enough water for handling and placing.

C. Repairing Formed Surfaces: Surface defects include color and texture irregularities, cracks, spalls, air bubbles, honeycombs, rock pockets, fins and other projections on the surface, and stains and other discolorations that cannot be removed by cleaning.

1. Immediately after form removal, cut out honeycombs, rock pockets, and voids more than 1/2 inch (13 mm) in any dimension in solid concrete, but not less than 1 inch (25 mm) in depth. Make edges of cuts perpendicular to concrete surface. Clean, dampen with water, and brush-coat holes and voids with bonding agent. Fill and compact with patching mortar before bonding agent has dried. Fill form-tie voids with patching mortar or cone plugs secured in place with bonding agent.
2. Repair defects on surfaces exposed to view by blending white portland cement and standard portland cement so that, when dry, patching mortar will match surrounding color. Patch a test area at inconspicuous locations to verify mixture and color match before proceeding with patching. Compact mortar in place and strike off slightly higher than surrounding surface.
3. Repair defects on concealed formed surfaces that affect concrete's durability and structural performance as determined by Architect.

D. Repairing Unformed Surfaces: Test unformed surfaces, such as floors and slabs, for finish and verify surface tolerances specified for each surface. Correct low and high areas. Test surfaces sloped to drain for trueness of slope and smoothness; use a sloped template.

1. Repair finished surfaces containing defects. Surface defects include spalls, popouts, honeycombs, rock pockets, crazing and cracks in excess of 0.01 inch (0.25 mm) wide or that penetrate to reinforcement or completely through unreinforced sections regardless of width, and other objectionable conditions.
2. After concrete has cured at least 14 days, correct high areas by grinding.
3. Correct localized low areas during or immediately after completing surface finishing operations by cutting out low areas and replacing with patching mortar. Finish repaired areas to blend into adjacent concrete.
4. Correct other low areas scheduled to receive floor coverings with a repair underlayment. Prepare, mix, and apply repair underlayment and primer according to manufacturer's written instructions to produce a smooth, uniform, plane, and level surface. Feather edges to match adjacent floor elevations.
5. Correct other low areas scheduled to remain exposed with a repair topping. Cut out low areas to ensure a minimum repair topping depth of 1/4 inch (6 mm) to match adjacent floor elevations. Prepare, mix, and apply repair topping and primer according to manufacturer's written instructions to produce a smooth, uniform, plane, and level surface.
6. Repair defective areas, except random cracks and single holes 1 inch (25 mm) or less in diameter, by cutting out and replacing with fresh concrete. Remove defective areas with clean, square cuts and expose steel reinforcement with at least a 3/4-inch (19-mm) clearance all around. Dampen concrete surfaces in contact with patching concrete and apply bonding agent. Mix patching concrete of same materials and mixture as original concrete except without coarse aggregate. Place, compact, and finish to blend with adjacent finished concrete. Cure in same manner as adjacent concrete.

7. Repair random cracks and single holes 1 inch (25 mm) or less in diameter with patching mortar. Groove top of cracks and cut out holes to sound concrete and clean off dust, dirt and loose particles. Dampen cleaned concrete surfaces and apply bonding agent. Place patching mortar before bonding agent has dried. Compact patching mortar and finish to match adjacent concrete. Keep patched area continuously moist for at least 72 hours.

E. Perform structural repairs of concrete, subject to Architect's approval, using epoxy adhesive and patching mortar.

F. Repair materials and installation not specified above may be used, subject to Architect's approval.

3.13 FIELD QUALITY CONTROL

A. Testing and Inspecting: Owner will engage a special inspector and qualified testing and inspecting agency to perform field tests and inspections and prepare test reports.

B. Inspections:

1. Steel reinforcement placement.
2. Steel reinforcement welding.
3. Headed bolts and studs.
4. Verification of use of required design mixture.
5. Concrete placement, including conveying and depositing.
6. Curing procedures and maintenance of curing temperature.
7. Verification of concrete strength before removal of shores and forms from beams and slabs.

C. Concrete Tests: Testing of composite samples of fresh concrete obtained according to ASTM C 172 shall be performed according to the following requirements:

1. Testing Frequency: Obtain one composite sample for each day's pour of each concrete mixture exceeding 5 cu. yd. (4 cu. m), but less than 25 cu. yd. (19 cu. m), plus one set for each additional 50 cu. yd. (38 cu. m) or fraction thereof.

    a. When frequency of testing will provide fewer than five compressive-strength tests for each concrete mixture, testing shall be conducted from at least five randomly selected batches or from each batch if fewer than five are used.

2. Slump: ASTM C 143/C 143M; one test at point of placement for each composite sample, but not less than one test for each day's pour of each concrete mixture. Perform additional tests when concrete consistency appears to change.
3. Air Content: ASTM C 231, pressure method at point of placement, for normal-weight concrete; one test for each composite sample, but not less than one test for each day's pour of each concrete mixture.
4. Concrete Temperature: ASTM C 1064/C 1064M; one test hourly when air temperature is 40 deg F (4.4 deg C) and below and when 80 deg F (27 deg C) and above, and one test for each composite sample.

5. Unit Weight: ASTM C 567, fresh unit weight of structural lightweight concrete; one test for each composite sample, but not less than one test for each day's pour of each concrete mixture.
6. Compression Test Specimens: ASTM C 31/C 31M.

   a. Cast and laboratory cure two sets of two standard cylinder specimens for each composite sample.
   b. Cast and field cure two sets of two standard cylinder specimens for each composite sample to verify strength is adequate where tilt-up concrete is tilted into place.

7. Compressive-Strength Tests: ASTM C 39/C 39M; test one set of two laboratory-cured specimens at 7 days and one set of two specimens at 28 days.

   a. Test one set of two field-cured specimens at 7 days and one set of two specimens at 28 days.
   b. A compressive-strength test shall be the average compressive strength from a set of two specimens obtained from same composite sample and tested at age indicated.

8. When strength of field-cured cylinders is less than 85 percent of companion laboratory-cured cylinders, Contractor shall evaluate operations and provide corrective procedures for protecting and curing in-place concrete.
9. Strength of each concrete mixture will be satisfactory if every average of any three consecutive compressive-strength tests equals or exceeds specified compressive strength and no compressive-strength test value falls below specified compressive strength by more than 500 psi (3.4 MPa).
10. Test results shall be reported in writing to Architect, concrete manufacturer, and Contractor within 48 hours of testing. Reports of compressive-strength tests shall contain. Project identification name and number, date of concrete placement, name of concrete testing and inspecting agency, location of concrete batch in Work, design compressive strength at 28 days, concrete mixture proportions and materials, compressive breaking strength, and type of break for both 7- and 28-day tests.
11. Nondestructive Testing: Impact hammer, sonoscope, or other nondestructive device may be permitted by Architect but will not be used as sole basis for approval or rejection of concrete.
12. Additional Tests: Testing and inspecting agency shall make additional tests of concrete when test results indicate that slump, air entrainment, compressive strengths, or other requirements have not been met, as directed by Architect. Testing and inspecting agency may conduct tests to determine adequacy of concrete by cored cylinders complying with ASTM C 42/C 42M or by other methods as directed by Architect.
13. Additional testing and inspecting, at Contractor's expense, will be performed to determine compliance of replaced or additional work with specified requirements.
14. Correct deficiencies in the Work that test reports and inspections indicate does not comply with the Contract Documents.

END OF SECTION 033000

SECTION 033005 - CIVIL CAST-IN-PLACE CONCRETE

PART 1 GENERAL

1.01 SUMMARY

A. Section Includes:

1. Cast-in-place concrete.

B. Related Sections:

1. Section 013300, Submittal Procedures.
2. Section 031005, Civil Concrete Formwork.
3. Section 032005, Civil Concrete Reinforcement.

1.02 REFERENCE

A. Reference Standards: Comply with the requirements of the reference standards noted herein, except where more stringent requirements are listed herein or otherwise required by the Contract Documents.

B. Applicable provisions of ACI 301, "Specifications for Structural Concrete for Buildings", and ACI 318, "Building Code Requirements for Reinforced Concrete", latest editions in effect at the Contract Date.

1.03 SUBMITTALS

A. Submittal Procedures: Refer to Section 013300, Submittal Procedures.

B. Product Data: Submit manufacturer's published product data for admixtures, chemicals, and other manufactured products used in the mixing, placing, curing, and treatment of concrete specified herein. Manufacturer's product data shall include written certification by an independent testing laboratory indicating compliance of products with the requirements of this Section.

C. Certification of Compatibility: Submit written verification or certification of compatibility of curing products with other finish materials and surface treatments, including adhesives and setting materials used for application of finish flooring materials.

D. Concrete Mix Designs:

1. Provide current mix designs for each type and quality of concrete specified herein or used in the Work. Prepare mix designs using the aggregates and Portland cement brands proposed for use in the Work. Submit proposed mix designs not less than two (2) weeks prior to first concrete placement.

2. Include the following information in mix design reports:
    a. Mix identification designation (unique for each mix submitted).
    b. Statement of intended use for mix.
    c. Identification of materials used in design mix by brand name or source, including fly ash or other pozzolans and admixtures.
    d. Unit weight, cement type, water/cementitious materials ratio, air content, slump, and 28-day compressive strength based on standard cylinder tests.
    e. Required average strength qualification data (based on previous field experience of trial mixtures) per ACI 301, Paragraphs 3.9.1, 3.9.2, and 3.9.3.
    f. Field test data submitted for qualification of average strength under ACI 301, Paragraphs 3.9.1, 3.9.2, and 3.9.3 shall include copies of the independent testing agency's reports from which the data was compiled.
    g. For lightweight concrete, submit splitting tensile strength qualification data acceptable to the ENGINEER. Such data may be in the form of tests conducted in accordance with ASTM C496 within the last year on concrete made with similar materials to those proposed and produced to meet a specified strength within 1000 psi of that required.
    h. Submit separate qualification data for each production facility that will supply concrete to the project.

3. Submit separate mix designs for each class or category of concrete, including each combination of compressive strength and type and/or quantity of mix material including admixtures, each change in slump limits, and each change in entrained air content.

4. Submit mix designs for mix types of classes of concrete not specified herein but proposed by the CONTRACTOR due to his methods of construction.

5. For lightweight concrete placed by pumping, submit separate mix designs or each 100 feet of vertical or horizontal distance from the pump to the point of discharge. This requirement may be waived if evidence acceptable to the ENGINEER is submitted demonstrating, by previous successful experience, that the proposed mix will meet all requirements of these specifications, when sampled at the point of discharge, over the full range of distances required.

6. Submit one copy of a representative concrete batch trip ticket containing information as specified herein.

7. Do not commence any concrete work until all mix designs are approved.

E. Test Reports: Submit copies of all test and inspection reports to the OWNER and the OWNER's Consulting Structural ENGINEER.

1.04 QUALITY ASSURANCE

A. Regulatory Requirements: Comply with applicable provisions of the state and local building and safety codes.

B. Coordination:

1. Coordinate the work of this Section with other work related in any way to this Section. Such coordination includes, but is not limited to:
   a. Placement of anchor bolts, weld plated, and other embedded items furnished under other Sections.
   b. Location and dimensions of blockouts and recesses to receive the work of other Sections.
   c. Compatibility of curing products with subsequent finishes of floor treatments, including adhesives or setting material used in the installation of finish flooring products.

2. Additional compensation will not be allowed for damages or corrective action required as a result of failure to coordinate the work of this Section with related work.

C. Pre-installation Conference:

1. Schedule and conduct a pre-installation conference for the purpose of reviewing all quality control procedures and construction requirements for all equipment, materials, and systems used in the performance of the work of this Section and related Sections.

2. Schedule conference not less than 10 days prior to the start of any work relating to cast-in-place concrete, formwork construction, and reinforcement fabrication.

3. Invite the ENGINEER, the OWNER, the testing and inspections agency, the CONTRACTOR's project manager and superintendent, representatives of subcontractors and material suppliers who's work relates in any way to concrete work, and any others deemed necessary by the CONTRACTOR.

4. Distribute written minutes of the pre-installation conference to all parties in attendance.

D. Quality Control:

1. All quality control testing and inspections required herein will be performed by an independent testing and inspections agency.

2. Cooperate with testing and inspection agency. Give minimum 48 hours notice for work requiring testing or inspection services. Submit samples for testing in a timely manner so as not to delay work.

3. Provide adequate facilities for testing, including suitable protected storage for freshly poured concrete test cylinders.

4. See Section 013300, Submittal Procedures for submittal requirements.

E. Record of Work: Keep a record listing the time and date of placement of all concrete for the structure. Such record shall be kept until the completion of the project and shall be available to the ENGINEER for examination at any time.

F. Batch Tickets: Collect and retain all batch tickets. Batch tickets shall contain information specified in ASTM C94, Paragraph entitled "Batch Ticket Information".

1.05 PROJECT AND SITE CONDITIONS

A. Cold Weather Concreting:

1. Comply with the performance requirements of ACI 306R, excluding Chapter 6 thereof and excluding the use of accelerating additives.

2. Heat formwork, reinforcing, and accessories to not less than 40°F [4°C] prior to placing concrete. Temperature of concrete at time of mixing and placement shall conform to ACI 306R, Table 1.4.1, but in no instance shall be less than 50°F [10°C].

3. Provide heated enclosures during the placement and finishing of concrete work; maintain enclosures until concrete has fully cured and attained the specified compressive strength.

4. Submit proposed system of heating to the ENGINEER for review prior to commencing cold weather placement.

B. Hot Weather Concreting:

1. Comply with the performance requirements of ACI 305R, except as modified herein.

2. Moisten forms and reinforcing before placing concrete; keep exposed surfaces continually moist.

3. Reduce temperature of water used in mixing concrete by chilling or adding crushed ice. Temperature of concrete at time of placement should not exceed 90°F.

4. Take precautions during concrete placement to prevent rapid drying or flash setting of concrete. Provide a continuous fog spray of water immediately after screeding, and provide sun shading and wind screens as required to protect non-formed surfaces.

5. Use of admixtures to retard concrete setting time or reduce water requirements during hot weather placement will be subject to design mix requirements and ENGINEERs review.

C. Rain: Do not place concrete during rain unless freshly placed concrete can be adequately protected until it has sufficiently hardened so that it will not be damaged. In the event that rain begins to fall during concrete placement, take necessary measures to assure that the strength of the structure will not be impaired and that surface finishes will be protected.

1.06 SEQUENCING AND SCHEDULING

A. Conditions for Placement: Do not commence mixing and placement of concrete until all the following conditions have been met:

1. Concrete mix designs have been approved by the ENGINEER.

2. Pre-installation conference has been held, with appropriate meeting minutes or written instructions distributed to all concerned parties.

3. Formwork and reinforcing for area to be concreted have been inspected and approved by the testing and inspections agency.

B. Coordination: Coordinate placement of concrete with other trades whose work will be built into or relates to concrete work in any way. In particular, assume full responsibility for the coordination and placement of the following:

1. Sleeves, plates, anchors, hangers, and other embedments furnished under other Sections of these Specifications. Secure ENGINEER's approval for installation of sleeves, conduits, of other openings or penetrations through structural members.

2. Provide adequate structural framing and reinforcing for mechanical or other equipment, and at all openings through beams, slabs, joists, walls, or decks, and for distribution of equipment loads to structural members.

3. Refer to Drawings for all concrete finishes, location and dimensions of slab drops or depressions, block-outs, reglets, chamfers, reveals, rustication, and other special architectural treatments.

PART 2 PRODUCTS

2.01 CONCRETE MATERIALS

A. Cementitious Materials:

1. Portland Cement:
   a. Provide a standard brand of portland cement complying with ASTM C150, Type II for concrete in direct contact with soil and Type I for all other concrete, unless otherwise indicated.
   b. Use one brand of cement throughout the Work.

2. Fly Ash:
   a. Comply with ASTM C618, Class C or Class F.

b. Where fly ash is proposed and approved in mix designs, provide fly ash from single source (i.e., single source of original coal and single combustion source).

c. Fly ash shall not exceed 20% of cement content by weight.

d. Do not use fly ash in mixes containing Type V cement.

B. Fine Aggregate: Washed, natural sand having strong, hard, durable particles, conforming to the physical and grading requirements of ASTM C33, and containing no more than 2% by weight of deleterious matter such as clay lumps, mica, shale, or schist.

C. Coarse Aggregate: Clean, hard, fine-grained, sound crushed rock or washed gravel, or a combination of both, conforming to the physical and grading requirements of ASTM C33. Limits of deleterious substances such as clay lumps, friable particles, chert, flat, elongated, or laminated pieces shall not exceed the allowable for severe weather regions, as defined in ASTM C33. Maximum size of coarse aggregate shall not exceed 1-1/4", except where smaller size is specified in mix designs for pumped concrete.

D. Lightweight Aggregate: Natural volcanic pumice or expanded clay, shale or slate produced in rotary kilns and crushed, conforming to the physical and grading requirements of ASTM C330, with minimum Fsp of 5.5 times the square root of the specified 28-day compressive strength and a stain index of 20 or less, as measured by ASTM C641. Obtain lightweight aggregate from an established source acceptable to the ENGINEER.

E. Water: Clean and potable, free from deleterious amounts of oil, acid, alkali, salt, and organic materials.

F. Admixtures:

1. General: No admixtures may be used without specific approval of the ENGINEER.

2. Air Entraining Admixture: Conform to ASTM C260. Add air-entraining agent when required as indicated in ACI 301, paragraph 3.4.1 and as required herein.

3. Water Reducing Admixture: Conform to ASTM C494, Type A.

4. High Range Water Reducing Admixture (Super-plasticizer): Conform to ASTM C494, Type F or G.

5. Calcium Chloride: Calcium chloride or admixtures containing more than 0.50 percent chloride ions or thiocyanates are not permitted.

6. Certification: Written certification of compliance with the above requirements, including the chloride ion content of admixtures, will be required from the admixture manufacturer prior to mix design review by the ENGINEER.

G. Integral Fiber Reinforcement: 100% virgin polypropylene, collated, fibrillated fibers complying with ASTM C1116, Type III, tensile strength not less than 80 KSI, with graded lengths per manufacturer's recommendation, designed and manufactured for use in concrete by one of the following manufacturers:

1. Euclid Chemical CO., Cleveland, OH

2. Fibermesh Co., Chattanooga, TN

3. Forta Corp., Grove City, PA

H. Non-Slip Aggregate: Aluminum oxide abrasive grit, 100% passing #12 screen and 100% retained on #32 screen.

2.02 CURING MATERIALS AND CONCRETE TREATMENTS

A. Curing Materials:

1. Clear polyethylene sheeting complying with ASTM C171, minimum 4 mils (0.004") [1mm] thick. White opaque type may be used during hot weather concreting to reduce heat gain from solar radiation.

2. Coordinate selection of curing materials with finish flooring and flooring adhesive of setting material manufacturers to ensure compatibility.

B. White pigmented, liquid, membrane forming curing compound conforming to AASHTO M-148, Type II, Class B shall be used unless another method is approved by the ENGINEER in writing. If the use of a curing compound results in streaked or blotchy appearance, the method shall be stopped and a suitable means of curing used.

2.03 CONCRETE MIXES

A. Water/Cementitious Materials Ratio:

1. Water/cementitious materials ratio noted in concrete mix design categories is the ratio of weight of water to weight of cementitious materials (Portland cement plus fly ash or silica fume). The weight of water shall include all free water in the aggregates at the time of batching.

2. Unless otherwise indicated, all concrete shall have a maximum water/cemetitious materials ratio of 0.45 (minimum field compressive strength 4000 psi at 28 days).

B. Adjustment to Concrete Mixes:

1. Mix design adjustments may be requested by the CONTRACTOR when characteristics of materials, job conditions, weather, test results, or other circumstances warrant and as approved by the ENGINEER. No additional costs to OWNER will be allowed as a result of revised mix designs.

2. Laboratory test data for revised mix design and strength results must be submitted to and accepted by the ENGINEER before use in the Work.

3. The CONTRACTOR shall fully satisfy himself that his proposed mix designs possess the workability, finishability, and setting time characteristics necessary to achieve the specified concrete quality requirements in the hardened state.

C. Admixtures:

1. All concrete required to be air entrained shall contain an approved air entraining admixture.

2. All lightweight concrete and normal weight concrete to be placed by pumping, and with water/cementitious materials ratios below 0.49, shall contain a high range water-reducing admixture.

D. Polypropylene Fiber Reinforcement:

1. Where polypropylene fiber reinforcement is allowed for secondary reinforcing of slabs, paving, or flatwork, add fiber at the rate of 1.5 to 1.6 lbs/cy in accordance with manufacturer's instructions.

2. Fiber may be added at the batch plant or at the jobsite; if added at the jobsite, follow manufacturer's instructions for additional mixing time after addition of fibers.

3. If mix design specified the use of a superplasticizer, add and mix polypropylene fibers prior to adding the superplasticizer.

4. Use of polypropylene fiber reinforcement will indicate lower slump values if standard ASTM C143 slump test is used; ASTM C995 may be used to measure consistency and workability of fiber-reinforced concrete if approved by the ENGINEER and the testing and inspections agency.

2.04 SOURCE QUALITY CONTROL

A. Concrete Mix Designs: Prepare concrete mix designs in accordance with ACI 211.1 and 211.2.

B. Concrete Mixing: Proportion and mix concrete in accordance with ACI 304 and ASTM C94. Comply with referenced standards regarding number of drum revolutions, mixing speed, agitation speed, and addition of water after initial batching. Maximum time from initial batching to placement shall not exceed 1-1/2 hours.

C. Testing: Test fine and coarse aggregates for sieve analysis in accordance with ASTM C136.

PART 3 EXECUTION

3.01 EXAMINATION

A. Verification of Conditions: Examine areas and conditions under which the work of this

Section will be performed. Do not proceed with the work until unsatisfactory conditions have been corrected. Commencement of work implies acceptance of all areas and conditions.

3.02 PREPARATION

A. Preparation of Areas for Placement: Prior to mixing and placing concrete in a given section of the Work, the following items must be complete and all necessary inspections made for that section:

1. All formwork, bulkheads, construction joints, keyways, sleeves, inserts, plates, and other embedded work shall be in place and securely anchored.

2. Where indicated or required by finish flooring material warranties, under-slab vapor barrier is in place, with all seams and patches sealed with water-resistant tape.

3. All reinforcing, dowels, supports, and other accessories shall be complete and securely tied in place.

4. All materials and equipment for placing, finishing, curing, and protecting concrete, including temporary heaters, coverings, and enclosures shall be on hand at the jobsite.

5. Provide runways for wheeled equipment to protect formwork and reinforcing. Runways shall be in good condition, adequate to support all construction loads, and suitable and safe for the workmen.

6. Formwork and reinforcing shall be clean, with all debris, standing water, or other foreign matter removed.

B. Notification of Placement Schedule: Notify the ENGINEER and testing and inspections agency not less than 36 hours in advance of any scheduled placement of concrete in order to allow sufficient time for required inspections and correction of any defective formwork or reinforcing. Do not place concrete until all required inspections have been made.

C. Weather Conditions: Keep informed of forecast weather conditions for anticipated time of concrete placement. Do not schedule or commence concrete placement when inclement weather is likely to occur.

3.03 INSTALLATION

A. Concrete Placement:

1. Convey concrete from mixer as rapidly as possible.

2. Do not place concrete over standing water, saturated earth materials, or frozen surfaces.

3. Prevent separation of loss of ingredients.

4. Once placement has begun, continue without interruption until entire unit, section, or area between construction joints is complete. Assign a sufficient number of transit-mix trucks to ensure continuous placement without delays between trucks.

5. Provide all necessary hoppers, buckets, chutes, and other conveying equipment required for site and placement conditions. Use metal or metal lined chutes; provide baffles to prevent segregation where slope of chutes exceeds 1 vertical to 2.5 horizontal.

6. Do not allow concrete to free drop more than six feet except in drilled piers, Do not allow concrete to free drop more than 20 feet in drilled piers.

7. Place concrete in uniform horizontal layers not exceeding 12" in depth, maintaining generally level surface during placement, and compact each layer using mechanical vibrators, supplemented by hand-spading, rodding, or tapping forms as require to consolidate ingredients and to fill voids. Do not use vibrators to move concrete within forms.

8. Keep wet concrete off finished concrete and other materials or surfaces exposed to view. Provide protective coverings or wind screens as required by ambient conditions.

B. Consolidation:

1. Use equipment and procedures for consolidation of concrete in accordance with ACI 309.

2. Consolidation by mechanical vibration is required for all concrete except caissons or drilled piers below top 5 ft of shafts.

3. Provide internal vibrating equipment with a minimum frequency of 7000 rpm. Maintain one standby vibrator for each three vibrators used.

4. Begin vibration as soon as concrete is placed to accomplish continuous and thorough consolidation of concrete around reinforcing, inclusions, embedments, and against all form surfaces. Supplement mechanical vibration with hand-spading, rodding, or tapping of forms.

5. Insert and withdraw vibrators at uniform spacings to penetrate at least 6" into preceding layer such that each layer, including the top layer, is vibrated twice. Do not insert vibrators into lower layers of concrete which have begun to set.

6. Exercise caution to prevent over-vibration and resultant segregation of ingredients.

C. Grades and Levels: Provide means to continuously monitor finished grades and levels of horizontal concrete surfaces as concrete is being placed, in order to raise or adjust screeds or other depth-determining markers as required to accommodate deflection of formwork and maintain tolerances specified herein.

D. Tolerances:

1. Formed surfaces shall generally be within tolerances stated in ACI 301 unless otherwise noted herein. Tolerances shall be the maximum allowable dimensional departure from planes, points, or elevations established by the Drawings. Variations from plumb shall apply to the total building height. All work shall be constructed within the stated tolerances.

2. Slabs and Other Horizontal Surfaces:
   a. Finish elevation shall be within 1/4" of designated finish elevation at any point.
   b. Troweled surfaces and floated surfaces shall be level within 1/4" in any 10'-0", as determined by placement of a 10-ft straightedge placed anywhere on the surface in any direction.
   c. Screeded surfaces shall be level within 3/8" in any 10'-0", as determined by placement of a 10-ft straightedge placed anywhere on the surface in any direction.
   d. Sloped surfaces shall be constructed within same tolerances.

3. Columns, walls, and other vertical surfaces shall be plumb within 1/4" in any 10'-0" or 1" in the overall height of the building.

4. Sleeves and Blockouts in Walls:
   a. Size: +1/2", -0"
   b. Location:
      1) Minimum dimension 6" or larger: ±1/2"
      2) Minimum dimension less than 6": Locate in nearest open space between reinforcing, not less than 1" clear from nearest reinforcing bar.

5. Embedded Items: Unless noted otherwise on the Drawings, tolerances shall be as follows:
   a. Anchor Bolts:
      1) Adjacent anchor bolts in a group receiving a single fabricated setting piece: ±1/8".
      2) Location and alignment of anchor bolt groups from designated location and alignment: ±1/4".
   b. Embedded Plates and Weldments:
      1) Location: ±1" vertical, ±2" horizontal.
      2) Plumb and alignment: 1/4" in 12".

E. Pumping: Concrete may be placed by mechanical pumping with the prior approval of the ENGINEER. Provide special concrete design mix as required for placement by pumping. Pumping equipment, procedures, and mix design shall conform to ACI 304.2R.

F. Bonding New Concrete to Old:

1. Before concrete is deposited upon or against concrete that has taken its initial set or has hardened, thoroughly clean surface of existing concrete to remove all laitance and loose material.

2. Remove encrustations from formwork and reinforcing, and clean formwork with high-pressure water spray.

3. Moisten bonding surface of existing concrete just prior to depositing new concrete, and remove all standing water.

4. Chemical bonding agents are not permitted without the prior written approval of the ENGINEER.

G. Construction Joints:

1. Provide construction joints in monolithic concrete framing so that not more than 400 cy of concrete is placed in one day, and no side dimension of area placed exceeds 150 ft. Larger areas may be approved by the ENGINEER on an individual basis.

2. Locate construction joints so as not to impair the overall strength or integrity of the structure. Coordinate location of construction joints with the ENGINEER.

3. Make provisions for transfer of shear and other forces through construction joints. Provide horizontal keyways at center of slab or beam, and allow reinforcing to continue through the joint for splicing into reinforcement of next adjoining section. Provide additional reinforcing if directed by the ENGINEER.

4. Follow procedures for bonding new concrete to old, as specified herein, when placing concrete at previously constructed construction joints.

H. Isolation (Expansion) Joints: Unless otherwise indicated, form isolation joints with preformed compressible fillers as specified in Section 03 10 05. No part or portion of the filler material shall extend above the surface of the concrete.

I. Control Joints:

1. Install tooled or saw-cut control joints at locations as follows:
   a. Slabs on Grade: At structural grid lines and slab offsets, but not exceeding 15'-0" o.c. each way.
   b. Exterior Walks, Decks, Ramps: At spacing equal to width of walk or ramp, but not exceeding 6'-0" o.c. each way.
   c. Exterior Paving: At equal spacing not exceeding 15'-0" each way.
   d. Topping Slabs over Precast Concrete: Directly above each joint between precast members, plus 15'-0" o.c. in direction perpendicular to precast joints.

2. Control joints shall be not less than 1" deep, but shall not expose reinforcing.

3. Verify control joint locations with the ENGINEER prior to installing.

4. Install saw-cut control joints using "Soff-Cut" power saws as soon as possible after finishing and when the concrete will support the weight of the saw and the operator without disturbing the final finish.

3.04 FINISHING

A. Preparation for Finishing:

1. Strike slabs off true to required level at indicated elevations. Set edge forms and screeds accurately to elevations and contours of finished surfaces.

2. Construct slabs level to indicated elevations, except where drains occur or slopes are indicated. Maintain minimum slab thicknesses as indicated, increase thickness as required to compensate for form deflection, but not so much as to increase dead load of structure above acceptable tolerances defined in ACI 318.

3. Tamp surface of freshly placed concrete slabs with special tools to force aggregate away from surface, and screed with straight edge to produce a level, even, and dense surface, ready for floating or troweling.

4. On formed surfaces such as sides and soffits of beams and joists, faces of columns, and soffits of slabs, remove fins and rough edges, and patch or fill honeycombs or other minor voids with a paste of one part portland cement to two parts sand. Trowel to smooth finish, slush with adjacent surfaces. Rub with Carborundum bricks and water to remove all bulges or projections.

B. Finishes for Slabs and Non-Formed Surfaces:

1. Screeded Finish: Slabs scheduled or indicated to receive concrete toppings, or mortar setting beds shall receive no additional finish after screeding.

2. Float Finish: On slabs indicated or scheduled to receive deck coatings or elastomeric finishes, finish surface with wood or steel float to produce a true, even, and smooth finish, free of float marks or ridges. Remove excess water. Do not dust surface with dry cement. Just prior to final set, re-float surface to produce a finish similar in touch and appearance to fine sandpaper.

3. Steel-Troweled Finish: On slabs indicated or scheduled to receive resilient flooring, carpet, or similar finished flooring materials, or to be left unfinished, provide steel-troweled finish as follows:
   a. Finish surface with float as specified herein, then finish with steel trowel using circular motion, to produce a smooth, dense, non-dusting, and impervious surface, free from scoring marks of any nature.
   b. Finish exposed horizontal edges of slabs with an edging tool to provide a uniform, rounded edge with a radius not to exceed 1/4".
   c. Do not sprinkle surface with water or dust surface with dry cement during finishing.

4. Broom Finish:

a. Immediately after troweling, finish with hair broom to produce a uniform non-slip finish. Broom concrete perpendicular to long edge or to slope. Make striations straight, uniform, and even.
b. Use fine hair broom at interior ramps and slabs scheduled or indicated to receive thin-set tile finish.
c. Use medium or coarse hair broom at exterior stairs, walks, plaza, decks, balconies, roof walkways, pavements, ramps, and other surfaces requiring non-slip finish.

C. Finishes for Formed Surfaces: Interior and exterior surfaces which will be exposed to view in finished work, but which are not designated as architectural concrete shall be finished as follows:

1. After patching and filling of voids as specified herein, pre-dampen freshly hardened concrete, then apply slurry consisting of 1 part portland cement and 1-1/2 parts sand passing a #16 sieve over entire surface, using sponge rubber floats or clean burlap pads.

2. Remove excess slurry by scraping and rubbing with clean burlap, and allow finish to cure for not less than seven (7) days.

D. Aggregate Non-Slip Finish: At interior exposed concrete stairs and landings, apply non-slip aggregate to freshly troweled concrete and thoroughly work into surface. Distribute aggregate by hand broadcasting, providing uniform coverage at rate of 30 lbs per 100sf, or as directed by manufacturer.

E. Other Finishes: In addition to finishes scheduled above, all surfaces to receive waterproofing shall be free of voids, spalled areas, loose aggregate, and sharp protrusions. Remove all contaminants such as oil or grease, and clean surface of dust, dirt, and debris. Use patching materials and methods compatible with waterproofing and approved by the waterproofing manufacturer.

3.05 CURING AND PROTECTION

A. General: Protect all exposed surfaces of freshly placed concrete from premature drying, excessively hot or cold temperatures, washing rains, and mechanical injury.

B. Preservation of Moisture:

1. Exposed Surfaces: Apply curing compound that produces a moisture resistant membrane film on concrete surfaces to assure hydration of the cement immediately after finishing slabs or other non-formed concrete surfaces. As an alternative to a curing compound, sprinkle or spray horizontal surfaces with fine mist of clean water and immediately cover with polyethylene or other approved impervious sheeting to prevent evaporation. Leave coverings in place a minimum of seven (7) days.

2. Formed Surfaces: Where forms are exposed to sun or wind, minimize moisture loss by keeping forms continuously wet until they can be safely removed. If

forms are removed prior to seven days after placement of concrete, continue curing by application of liquid membrane-forming curing compound.

3. Cold Weather Curing: When the mean daily temperature is less than 40°F, provide temporary heaters and enclosures as required to maintain the temperature of the concrete between 50°F and 70°F for the required curing period.

4. Hot Weather Curing: When the mean daily temperature is more than 90°F, provide wind breaks, shading, or additional fog spraying or sprinkling as required to prevent flash setting, surface checking, or premature hardening of concrete.

C. Protection From Mechanical Injury: During the curing period, protect concrete from damaging mechanical injury, such as heavy shock, excessive vibrations, construction equipment, and materials, as well as from rain and running water. Do not load structures in a way to overstress the concrete until full design strength is attained.

3.06 CONCRETE TREATMENTS

A. General: Verify compatibility of concrete treatments with subsequent finish materials prior to application. Refer to QUALITY ASSURANCE in part 1 of this Section.

B. Concrete Sealer and Dustproofer:

1. At the option of the CONTRACTOR, sealer and dustproofer may be applied to fresh concrete, serving as a curing material, or to hardened concrete. If applied to hardened concrete, verify compatibility with curing materials and methods used.

2. Apply sealer and dustproofer in accordance with manufacturer's instructions. Comply with manufacturer's recommendations for coverage rates and number of coats.

3.07 MISCELLANEOUS CONCRETE REQUIREMENTS

A. General: Furnish and install all other concrete work indicated on the Drawings or specified in other Sections of the Technical Specifications as required to complete the Work, even though not specifically mentioned herein.

B. Anchors and Embeds: Install anchors and embeds furnished under other Sections, in accordance with approved shop drawings and/or setting instructions.

C. Equipment Bases (Housekeeping Pads):

1. Install concrete bases for all pumps, boilers, tanks, fans, transformers, floor mounted electrical equipment, etc., including anchor bolts and inserts, in accordance with setting instructions furnished by the installer of the equipment. Finish all bases in a workmanlike manner with a troweled finish.

2. The bases shall be located and sizes determined by the subcontractor furnishing the equipment, and approved by the ENGINEER.

3.08 REPAIR OF SURFACE DEFECTS

A. Formed Surfaces:

1. Allow ENGINEER to observe concrete surfaces immediately upon removal of forms.

2. Modify or replace concrete not conforming to required lines, details, and elevations.

3. Repair or replace concrete not properly placed resulting in excessive honeycombing and other defects. Patch, repair, or replace exposed architectural finished concrete as directed by the ENGINEER.

4. Patching of tie holes is not required.

5. Repair defects in structural concrete elements as follows:

a. Deep Defects Exposing Reinforcing: Chip to sound concrete and clean thoroughly to remove all loose concrete and dust. Apply thin coat of an approved epoxy adhesive. Form and pour, or dry pack with non-metallic, non-shrink grout, prior to development of tack-free condition of epoxy adhesive. Strip forms after grout has hardened and provide specified finish. Moist cure or apply specified clear curing and sealing compound immediately after finishing.

b. Defects Greater Than 1/2" Depth Not Exposing Reinforcing: Chip and clean to expose a fractured aggregate profile. Apply an approved epoxy adhesive. Dry pack using specified non-metallic, non-shrink grout prior to development of tack-free condition of epoxy adhesive. Provide specified finish and cure as specified above.

c. Defects Less Than 1/2" Depth and Tie Holes:

1) For concrete having a specified compressive strength of 5,000 psi or less: Chip and clean as specified above. Dry pack, finish, and cure as specified above.

2) For concrete having a specified compressive strength greater than 5,000 psi: Chip and clean as specified above. At CONTRACTOR's option, dampen surface and apply an approved epoxy mortar, followed by specified finish (no curing required); or apply thin coat of the specified bonding compound followed by dry pack, finish and cure as specified above.

d. Other equivalent repair procedures may be used subject to review and acceptance by the ENGINEER.

B. Unformed Surfaces:

1. Test unformed surfaces, such as monolithic slabs, for smoothness and verify surface place to tolerances specified for each surface and finish. Correct low and high areas as herein specified. Test unformed surfaces sloped to drain for trueness of slope, in addition to smoothness using a template having required slope.

2. Correct high areas in unformed surfaces by grinding or other methods acceptable to the ENGINEER after concrete has cured a minimum of 14 days.

3. Correct low areas in unformed surfaces during or immediately after completion of surface finishing operations by cutting out low areas and replacing with fresh concrete. Finish repaired areas to blend with adjacent concrete. Where acceptable to the ENGINEER, underlayment compound applied in accordance with the manufacturer's instructions, may be used.

4. Repair defective areas, except random cracks and single holes not exceeding 1" diameter, by cutting out and replacing with fresh concrete. Remove defective areas to sound concrete with clean, square cuts and expose reinforcing steel with at least 3/4" clearance all around. Dampen concrete surfaces in contact with patching concrete and apply bonding compound. Mix patching concrete of same materials to provide concrete of same type or class as original concrete. Place, compact, and finish to blend with adjacent finished concrete. Cure in same manner as adjacent concrete.

5. Repair isolated single holes not over 1" in diameter by dry-pack method. Cut out holes to sound concrete and clean of dust, dirt and loose particles. Dampen cleaned surfaces and apply an approved bonding compound. Mix dry pack, consisting of one part Portland cement to 2-1/2 parts fine aggregate passing an No. 16 mesh sieve, using only enough water as required for handling and placing. Place dry pack after bonding compound has dried. Compact dry pack mixture in place and finish to match adjacent concrete. Keep patched area continuously moist for not less than 72 hours.

6. Saw-cut all visible cracks in slabs-on-grade and parking decks to depth of at least 3/4" and fill with flexible epoxy structural sealant. Do not cut reinforcing.

3.09 FIELD QUALITY CONTROL

A. Evaluation and Acceptance Criteria: Evaluation and acceptance of work under this section will be in accordance with the provisions of ACI 301, Chapters 17 and 18.

B. Rejection of Concrete: Concrete which does not conform with the minimum requirements of the Contract Documents will be rejected.

C. Correction of Deficient Work: Assume responsibility for and bear all costs of correcting concrete work which does not conform to the specified requirements including, but not limited to, strength, durability, tolerances, and finishes. Correct deficient concrete by means acceptable to the ENGINEER.

D. Test and inspect concrete in accordance with building code requirements and Contract Documents. Required testing and inspections include, but are not limited to, the following:

1. Testing concrete for strength, slump, air content, temperature, and weight.

2. Making and testing concrete cylinders, including furnishing cylinder containers for specimens.

3. Inspection at the batching plant.

4. Transporting and storing of all specimens involved in testing and inspection. Test cylinders are to be transported to laboratory not later than twenty-four (24) hours after casting.

5. Inspection of mixing and placing of concrete at the site, including recording of amount and location of concrete placement, truck number and amount of water added to each load of concrete tested, time of transit, time mixed on job, time placement was completed, method of placing concrete, and any other pertinent information.

6. Submission of written reports of all tests and inspections in accordance with the submittal requirements of Part 1 of this Section.

E. Preparatory Inspections: Testing and inspection agency and ENGINEER must approve all formwork, reinforcing, embeds, and other conditions affecting concrete work prior to commencement of placing operations.

F. Batch Tickets: Retain copies of all transit-mix batch tickets until the completion of the Work. Submit batch tickets to the ENGINEER.

G. TESTING AND INSPECTION

1. Aggregates: Test fine and coarse aggregates at the source or ready-mix batch plant for sieve analysis in accordance with ASTM C136.

2. Field Tests: Test fresh concrete in accordance with the following test methods:

a. Sampling fresh concrete - ASTM C172

b. Making and curing test cylinders - ASTM C31

c. Compressive strength of test cylinders - ASTM C39
   1) Test one cylinders at 7 days for information.
   2) Test two cylinders at 28 days; acceptance criteria shall be based on the average strength of these two cylinders.
   3) Retain one cylinder for testing at 56 days if 28 days tests indicate lower than acceptable compressive strength.

d. Air content of air-entrained mixes - ASTM C173 or ASTM C231 (Do not use ASTM C231 method for lightweight aggregate concrete.

e. Slump - ASTM C143

3. Test Specimen Quantity:

a. Compressive Strength Tests:

1) Spread Footings: One (1) test per structure
2) Slabs-on-Grade and Pavements: One test for every 5000 sf.
3) Foundation and Retaining Walls: One test for every 250 lf of wall length for every two foot vertical rise in compacted fill depth. In no case shall less than eight tests be made
4) Structural Concrete: For each 100 cy or major fraction thereof of each type or category of general or structural concrete placed in any one day, gather one set of not less than four (4) test cylinders. Gather at least one set of four cylinders for each day's work if less than 50 cy of concrete is placed.
5) When more than 50 cy of concrete is placed in any one day, the interval between test samples shall be at least 30 cy in order to be representative of the entire day's operation.
6) Take samples at the point of discharge from the truck chute, except take samples of pumped concrete at the point of deposit in the field. Accurately mark all cylinders with date, reference to location in structure from which they were taken, and batch ticket number.

b. Slump Tests: Test one truckload of concrete delivered to the Site.

c. Air Content: Test one truckload of air-entrained concrete delivered to the Site.

d. ENGINEER or testing and inspection agency may require additional slump tests or cylinders should there be reasonable cause to suspect that concrete being placed does not comply with specified concrete quality.

4. Hardened Concrete Tests:

a. If the 28 day cylinder tests indicate compressive strengths below the acceptable limits for the specified design strength, the concrete represented by such tests shall be considered unacceptable and subject to removal pending testing of final cylinder at 56 days or additional testing of hardened concrete in place.

b. Consideration shall be given to the acceptance of such concrete if it can be demonstrated to the ENGINEER that the cylinder tests do not accurately represent the strength of the concrete in place, or that the structure is fully capable of carrying the loads for which it was designed.

c. Testing of hardened concrete may be accomplished by means of load testing in accordance with applicable building codes or by testing core samples in accordance with ASTM C42. Non-destructive tests such as ASTM C803 or C805 may be used to determine quality of concrete in question relative to concrete of known compressive strength, but not as an alternative to finite compressive strength tests.

d. All costs of additional testing of hardened concrete, plus the removal and replacement of non-complying concrete, if required, shall be the sole responsibility of the CONTRACTOR.

3.10 CLEAN-UP

A. Construction Cleaning: Remove all excess materials, concrete spills, formwork, shoring, and other debris from the jobsite. Do not use spilled concrete or debris from washout of trucks in filling or backfilling.

B. Truck Washout: Provide a designated area for washout and cleaning of concrete trucks following discharge of concrete in the Work. Do not allow washout on adjacent private or public property.

END OF SECTION 033005

SECTION 034005 - CIVIL PRECAST CONCRETE

PART 1 GENERAL

1.01 SCOPE

A. This section addresses precast concrete products.

1.02 QUALITY ASSURANCE

A. Standards (as applicable)

1. ASTM A185, Standard Specification for Steel Welded Wire Fabric, Plain, for Concrete Reinforcement, latest revision.
2. ASTM A615, Standard Specification for Deformed and Plain Billet-Steel Bars for Concrete Reinforcement, latest revision.
3. ASTM A996, Standard Specification for Rail-Steel and Axel-Steel Deformed Bars for Concrete Reinforcement, latest revision.
4. ASTM C33, Standard Specification for Concrete Aggregates, latest revision.
5. ASTM C94/C94M, Standard Specification for Ready-Mixed Concrete, latest revision.
6. ASTM C150, Standard Specification for Portland Cement, latest revision.
7. ASTM C206, Standard Specification for Finishing Hydrated Lime, latest revision.
8. ASTM C478, Standard Specification for Precast Reinforcement Concrete Manhole Sections, latest revision.
9. ASTM C494/C494M, Standard Specification for Chemical Admixtures for Concrete, latest revision.
10. ASTM C618, Standard Specification for Coal Fly Ash and Raw or Calcined Natural Pozzolan for Use as a Mineral Admixture in Concrete, latest revision.

1.03 SUBMITTALS

A. Shop Drawings: Submit detailed shop drawings for precast concrete structures.

B. Product Data: Submit Product Data from manufacturer's descriptive literature and specifications for all materials used in this Section.

C. Certifications: Certification that materials being provided meet the requirements of these specifications.

D. Project Record Documents: Submit upon completion of construction "as-built" documents in accordance with Section 017839, Project Record Documents.

1.04 DELIVERY, STORAGE, AND HANDLING

A. Deliver, store, and handle all materials to ensure installation in sound and undamaged condition.

B. Do not deliver precast sections to the job site until the concrete has attained at least 80% of specified design strength.

C. Precast concrete members shall be lifted and supported during manufacturing, stockpiling, transportation, and erection operations only at the lifting or supporting point, or both, as shown on shop drawings.

D. Transportation and on-site handling shall be performed with acceptable equipment and methods, as well as by qualified personnel.

E. Care shall be taken to avoid tensional stresses during transportation.

F. Place units so that identification markings are discernible.

G. Stack so that lifting devices are acceptable and undamaged.

PART 2 PRODUCTS

2.01 CONCRETE

A. All precast concrete requirements shall conform to ASTM C478.

B. All cement shall be Portland Cement.

1. Portland Cement shall conform to ASTM C150.

2. Portland Cement shall be Type I/II.

C. Air entraining admixtures shall conform to ASTM C260.

PART 3 EXECUTION

3.01 GENERAL

A. Examine each precast section upon arrival to the job site for cracks and other unsightly imperfections or structural defects.

1. Mark and remove all damaged sections from the job site.

3.02 INSTALLATION

A. Precast Sections

1. Set precast sections in accordance with the manufacturer's erection drawings.

END OF SECTION 034005

Appendix F
Asphalt Specifications

NOTE:

The specifications in this book are strictly to be used as an example as a work ng copy of specification that may be found on real life projects. Only portions of the specifications are shown in this book that the author felt were relevant to the subject matter discussed. The specifications included are not intended to be complete or a full representation of the project specifications.

REFERENCE:

Aims Community College
North Campus Refresh
Greeley, Colorado
Project Manual – Volume 2 or 2
July 29, 2016
Prepared by DLR Group

SECTION 321113 - STABILIZED SUBGRADE

PART 1 GENERAL

1.01 DESCRIPTION

A. This work consists of the contractor constructing one or more courses of a mixture of subgrade soil, approved stabilizing agent and water in substantial conformity with the design line, grades, thicknesses, and typical cross sections shown on the approved plans and the approved pavement thickness design.

B. The purpose of the work shall be to provide a structural section on which paving materials can be placed and to meet design specifications, while at the same time, protecting the underlying moisture conditioned subgrade soils.

1.02 SUBMITTALS

A. At least 15 days prior to commencing stabilization work, the CONTRACTOR shall furnish the following information to the OWNER'S REPRESENTATIVE:

1. The source and supplier of stabilizing agent and certifications, including purity of stabilizing agent, from the manufacturer's testing indicating that the stabilizing agent meets the appropriate requirements.

2. Description of the proposed construction equipment, construction methods, expected production rates and planned sequence of construction.

3. A mix design giving the percentage of stabilizing agent, source of the agent, properties and any special considerations.

4. Water Soluble Sulfate test results.

B. For each day's work, the CONTRACTOR shall furnish the following information to the OWNER'S REPRESENTATIVE by the following day:

1. Certified truck weight tickets of stabilizing agent, delivered or used at the site.

2. A summary of the amount of stabilizing agent used each day, areas stabilized and first mixed, areas second mixed and compacted, and areas with curing completed.

PART 2 PRODUCTS

2.01 STABILIZING AGENTS

A. The pre-approved stabilizing agents are listed in the table below. Various combinations of these materials may also be used, subject to a suitable mix design. Other agents may be used with prior written approval of the OWNER'S REPRESENTATIVE.

PRE-APPROVED STABILIZING AGENTS		
Agents	Must conform to requirements of	
Lime	ASTM	C 977, C 110
Fly Ash (C and F)	ASTM	C 618
Cement Kiln Dust	ASTM	D 5050
Portland Cement	ASTM	C 114

B. High-calcium quicklime shall conform to the requirements of ASTM C 977 and rate of slaking test shall produce a temperature rise of 20°C in 30 seconds and 35°C in 3 minutes per ASTM C110. Dolomitic quicklime, magnesia quicklime with magnesium oxide contents in excess of 4 percent of carbonated hydrated lime, shall not be used.

C. Fly ash may consist of Class C or Class F. Class F fly ash shall only be allowed in conjunction with lime or other stabilizing agents.

D. All stabilizing agents shall come from the same source as used in the design. If the source is changed, a new design must be submitted for the OWNER'S REPRESENTATIVE approval. Each lot of stabilizing agent furnished shall have the supplier's certificate of compliance.

2.02 WATER

A. Water used for mixing or curing shall be from a potable source. Non-potable water shall be tested in accordance with and meet the requirements of AASHTO T 26.

2.03 SUBGRADE

A. The subgrade to be stabilized shall be free of roots, sod, weeds, wood, construction debris, ice, snow, or other frozen materials, deleterious matter, and stones larger than 3 inches. Material in the stabilized zone shall have a soluble sulfate content of less than 0.5 percent. If the subgrade soils have a soluble sulfate content exceeding 0.5 percent, the mix design shall be addressed to the specific methodology used to prevent adverse sulfate reactions.

2.04 SEALANT

A. Emulsified asphalt is the only pre-approved sealant; no other sealant may be utilized without prior written approval of the OWNER'S REPRESENTATIVE. Emulsified asphalt used for a bituminous seal shall conform to requirements in the table below.

PRE-APPROVED AGENTS			
Sealant	Type	Requirements	
Emulsified Asphalt	SS-1	ASTM	D 977
Emulsified Asphalt	SS-1h	ASTM	D 977
Emulsified Asphalt	CSS-1	ASTM	D 2397

Emulsified Asphalt	CSS-1h	ASTM	D 2397

2.05 STABILIZED MIX DESIGN

A. Mix designs shall be performed under the supervision of and signed by a Professional ENGINEER registered in the State of Colorado. Laboratories shall be accredited by either AASHTO or A2LA.

B. Mix design shall comply with the following requirements in the table below:

STABILIZATION MIX DESIGN REQUIREMENTS

Stabilization Agent	Minimum pH Note 1 & 2	Maximum Plasticity Index (%) Note 3	Maximum Swell Potential (%) Note 4	Minimum Unconfined Compressive Strength (psi) Note 5 & 6
Lime	12.3	10	1.0	160
Fly Ash	N/A	10	1.0	160
Cement Kiln Dust	N/A	10	1.0	160
Portland Cement	N/A	10	1.0	160
Lime-Fly Ash	12.3	10	1.0	160

Note:

1. When lime is used, the pH should be no less than 12.3 as measured after completion of initial mixing with stabilizing agent and at ambient temperature.
2. Testing of pH is to be done in accordance with Eades-Grim pH test method (ASTM C 977 APPENDIX).
3. Plasticity Index is to be less than 10, per AASHTO T 89 and T 90 unless a greater value is acceptable by the OWNER'S REPRESENTATIVE.
4. Swell Potential to be less than 1.0 percent at 200 psf, ASTM D 4546.
5. Minimum of 160 psi ($M_R > 34{,}800$ psi, where $M_R = 10{,}000 + 124q_u$) in five (5) days of moist curing at 100°F (38°C). Testing is in accordance with ASTM D 1633 Method A for pozzolanic agents and ASTM D 5102 Procedure B for Hydrated Lime.
6. The design stabilizing agent percentage shall be increased by 0.5 percent to account for waste, inert materials, and construction variability.

PART 3 EXECUTION

3.01 EQUIPMENT

A. All equipment shall be subject to approval by the OWNER'S REPRESENTATIVE. All equipment and machinery shall be kept in good working order, free of leaks and properly muffled.

B. Dry Application Equipment

1. Equipment for spreading dry stabilizing agent shall be of an approved screw-type spreader box, mixer, or other semi enclosed equipment which is equipped with a metering device. Spreading of stabilizing agents by aggregate spreaders or motor-graders will not be allowed.

C. Slurry Application Equipment

1. A distributor or truck applicator shall be used and be capable of continuous agitation to keep the slurry mixture uniform. The applicator shall be capable of uniformly metering the stabilizing agent during application.

D. Mixing Equipment

1. Mixing equipment shall be of sufficient size to adequately mix the stabilizing agent into the soil and to pulverize the mixture according to the requirements of Section 3.02. The size of the mixer shall be adequate to mix and pulverize the mixture to a minimum depth of 12 inches in a single pass. Blades, discs, and similar equipment are not allowed without prior written approval of the OWNER'S REPRESENTATIVE.

E. Compaction Equipment

1. Compaction equipment shall be in good working order and of sufficient size and effective force to achieve the required compactive effort as prescribed in Section 3.02.

3.02 PROCESSING MATERIALS

A. It is the primary requirement of this specification to secure a completed subgrade structural section containing a uniform stabilized mixture. The mixture is to have a uniform density and moisture content, free from loose or segregated areas, well bound for its full depth, well cured, and with a surface suitable for placing subsequent courses.

B. It shall be the responsibility of the CONTRACTOR to regulate the sequence of his work, to use the proper amount of stabilizing agent, maintain the work, and rework the courses, as necessary, to meet the requirements.

3.03 APPLICATION

A. The subgrade shall not be treated when the ambient air temperature falls below freezing or the subgrade material is below 40°F, or when weather predictions suggest that subgrade material temperature may fall below 40°F within 24 hours, unless prior written approval of the OWNER'S REPRESENTATIVE has been issued. Prior to beginning any treatment, the subgrade shall also be constructed and finished to a smooth and uniform surface that is in conformity to the grade and typical section specified. Variation from the subgrade plan elevation specified shall not be more than ± 0.08 ft. The in-place density shall be at least 95% of maximum dry density as determined by ASTM D 698, Standard Proctor Density, and within 0 to 3% of optimum moisture content.

B. Stabilizing agent shall be applied at the minimum rate specified by the mix design for the depth of stabilized subgrade shown on the plans. The rate shall be determined from a design using the on-site soils and shall meet the requirements found in Item 5.5. Rate of application shall be verified using area/quantity calculations or testing of stabilized subgrade. Stabilizing agent shall be spread only on that area where the first mixing operations can be completed during the same working day. Lime slurry shall not be left exposed to the air for more than four hours without initial mixing.

C. The OWNER'S REPRESENTATIVE reserves the right to require variation of the rate of application of stabilizing agent from the mix design application rates during the progress of construction as necessary to maintain the desired characteristics of the stabilized subgrade.

D. Stabilizing agent shall be applied using the following methods:

1. Slurry Placement

a. The distribution of stabilizing agent shall be attained by successive applications over a measured section of subgrade until the proper amount of agent has been spread. The amount spread shall be the amount required for mixing to the specified depth, which will result in the percentage determined in the design. When quicklime is used in place of hydrated lime the amount of quicklime used will be determined by the certified lime purity for each load supplied as follows:

Quicklime delivered X % purity * 1.32 = A
Quicklime delivered *X % inert material = B
A + B – total hydrated lime available

2. Dry Placement

a. The amount of stabilizing agent spread shall be the amount required for mixing to the specified depth, which will result in the percentage specified by the design.

b. The stabilizing agent shall be distributed in such a manner that scattering by wind will be minimal. Agents shall not be applied when wind conditions, in the opinion of the OWNER'S REPRESENTATIVE, are detrimental to a proper application. The blended material shall be sprinkled or watered until moisture content is as specified in subgrade stabilization design.

c. After spreading of stabilizing agent and during mixing, water shall be added to hydrate the agent and to reduce dusting. The combination of stabilizing agent, soil and water shall be called the mixture.

E. High Sulfate Treatment

1. Where sulfates are over 0.5 percent and less than 1.0 percent the OWNER'S REPRESENTATIVE must address the method of treatment.

   a. *Note: When a double treatment of lime is required, the first 50 percent of the agent shall be placed, moisture treated and allowed to mellow or cure for up to three weeks, as determined by the OWNER'S REPRESENTATIVE. The last half of the lime shall then be applied.*

F. Mixing

1. No stabilization shall take place when precipitation may cause damage to the subgrade, as determined by the OWNER'S REPRESENTATIVE. Mixing shall be continuous. The full depth of the treated subgrade material shall be mixed with an approved mixing machine to the specified depth below the bottom of the pavement structure and/or curb. The mixing machine shall make a sufficient number of passes to adequately achieve 100 percent of the material passing the one-inch sieve and 60 percent passing the 1/4-inch sieve. Water shall be added to the subgrade during mixing to provide a moisture content of at least 3 percent above the optimum moisture of the mixture or as specified in subgrade stabilization design.

2. Mixing and remixing will be performed, as necessary, to assist the stabilizing agent-soil reaction and produce a homogenous mixture. Mixing and remixing shall continue until the combination of stabilizing agent and subgrade material is free of streaks or pockets of stabilizing agent.

G. Mellowing (Lime or Lime/Fly Ash Only)

1. The water content of the subgrade mixture shall be maintained at a moisture content above the optimum moisture content for a minimum of 2 days and until the PI meets the required subgrade stabilization design criteria. Remixing will be done as necessary to assist the reaction, as determined by the OWNER'S REPRESENTATIVE. Application of moisture shall be performed as necessary during the mellowing period; the material shall maintain a moisture content of at least 3 percent above the optimum moisture content. The stabilized material shall not be subject to traffic. If during the mellowing period the material is not in a semi-loose state, the chemical reaction process may slow down and, therefore, require additional time and/or mixing as determined by the OWNER'S REPRESENTATIVE.

F. Final Mixing (Lime or Lime/Fly Ash Only)

1. Final mixing of the treated subgrade shall not occur if the temperature of the soil to be stabilized is below 40°F. The treated subgrade shall be maintained at a temperature of 40°F or above until the treated material has been compacted.

2. The material shall be uniformly mixed by an approved method to meet the following requirements when tested dry by laboratory sieves:

Sieve Size	Minimum Percent Passing
1-inch sieve	100
No. 4 sieve	60

3.04 COMPACTION

A. Compaction of the mixture, for the full depth of the stabilized subgrade shown on the plans, shall begin as soon as practical after final mixing. Stabilized subgrade with cementitious stabilization agent shall be completed within 90 minutes of the time cementitious stabilization agent or water is applied. The field density of the compacted mixture shall be at least 95 percent of the maximum dry density of laboratory specimens prepared from samples taken from the treated subgrade material immediately prior to compacting. The specimens shall be compacted and tested in accordance with ASTM D 698 or ASTM D 558, as specified in subgrade stabilization design. The in-place field density shall be determined in accordance with ASTM D 1556, ASTM D 2167 or ASTM D 2922. The moisture content of the mixture shall be between 0 to 3 percent above the optimum moisture content. The optimum moisture content shall be determined in accordance with ASTM D 698 or ASTM D 558, as specified in subgrade stabilization design.

B. Initial compaction shall be done by means of a sheep foot or segmented wheel roller. Final compaction shall be by means of a smooth wheel or pneumatic tired roller. Areas inaccessible to a mechanical roller shall be compacted to the required density by other means suitable to the OWNER'S REPRESENTATIVE.

C. All irregularities, depressions, or weak spots which develop, as determined by the OWNER'S REPRESENTATIVE, shall be corrected immediately by scarifying the areas affected, adding or removing materials as required, and reshaping and recompacting by moisture conditioning and rolling. Adding additional stabilized material to an initial cured section, resulting in lamination and potential slip plane, is not allowed. The surface of the course shall be maintained in a smooth condition, free from undulations and ruts, until other work is placed thereon or the work is accepted. Should the material, due to any reason or cause, lose the required stability, density, and finish before the next course or pavement is placed, it shall be corrected and refinished at the sole expense of the CONTRACTOR, as directed by the OWNER'S REPRESENTATIVE.

3.05 FINISHING AND CURING

A. After the final layer of stabilized subgrade has been compacted, the shape of the surface shall be maintained by blading. The surface shall be smooth and conform to the required lines, sections, and grades, in accordance with the plans and thoroughly cured, or to within a minimum of 0.1 foot above the finished subgrade elevation to allow for trimming to final grade prior to placement of surface coarse. The completed section shall then be finished by rolling with suitable pneumatic tired equipment with sufficiently light effort to prevent hairline cracking.

B. Curing may be accomplished by periodic water application to maintain moisture content preventing sloughing or cracking in the surface of the stabilized subgrade to a depth no greater than 0.1 foot, or by the utilization of a bituminous seal. When bituminous seal

(Section 2.04) is utilized, the minimum application will be at the rate of 0.12 gallons per square yard, as directed and approved by the OWNER'S REPRESENTATIVE.

C. The completed section shall be cured for a minimum of 5 days before further courses are added or any traffic is permitted, unless otherwise permitted by the OWNER'S REPRESENTATIVE. The moisture cured duration may be reduced if a non-yielding surface is obtained to support construction traffic and either the next layer of stabilized soils are placed or the pavement layer is constructed, as approved by the OWNER'S REPRESENTATIVE.

D. If the surface of the finished layer is above the approved plan elevation tolerance specified in this section, the excess material shall be trimmed, removed, and disposed of. Any low areas will be replaced with the subsequent surface courses. No loose material shall be left in place after trimming. After trimming the stabilized subgrade surface shall be rolled again with a steel wheel or pneumatic tired roller to seal the surface.

3.06 TOLERANCES

A. Thickness

1. Stabilized zone thickness shall be verified by the use of phenolphthalein and shall be performed at intervals of approximately every 500-feet in each lane. When the measurement of the thickness is deficient by more than 1 inch from the plan thickness, two additional locations shall be measured randomly within the deficient area and used in determining the average thickness. When the average thickness is deficient by more than 1 inch, the entire area shall be reprocessed to meet the design parameters or the roadway design section must be re-evaluated (Section 3.07).

B. Grade

1. Prior to placement of surface course, any deviation in excess of 1/2 inch in 10 feet measured longitudinally shall be corrected. Variations in excess of this tolerance shall be corrected by the CONTRACTOR, at the CONTRACTOR's expense, in a manner satisfactory to the OWNER'S REPRESENTATIVE. Thickness requirements shall be met in areas corrected for grade.

C. Strength

1. The stabilized subgrade must develop a laboratory compressive strength of at least 160 psi at 5 days in accordance with Section 3.09, Schedule For Minimum Materials Sampling And Testing Table. Samples shall be molded from stabilized soil within 1.5 hours of final mixing with the material compacted per ASTM D 558 or ASTM D 698, as specified in subgrade stabilization design, at the field moisture content.

D. Stabilizing Agent Percentage

1. The percentage of stabilizing agent placed in the mixture shall be determined by the appropriate AASHTO chemical determination and/or by calculation of

stabilizing agent used for the volume of subgrade material stabilized. When the test result is deficient by more than 1 percent, the entire area shall be re-treated at the CONTRACTOR's expense.

3.07 CONFORMITY WITH PLANS AND SPECIFICATIONS

A. When thickness and/or strength criteria fail to meet design parameters, even after all possible attempts have been made to correct said deviations, remediation will be required as listed in table below. Evaluation of the roadway pavement section will be made and approved in writing by the OWNER'S REPRESENTATIVE. The pavement structural section shall be adjusted to compensate for any deficiency in the stabilized subgrade thickness and strength, at the CONTRACTOR's expense. Placement of subsequent surface coarse will not occur until the stabilized subgrade has been accepted in writing by the OWNER'S REPRESENTATIVE.

Deficiency	Remediation
< 25% of design thickness	Evaluate Roadway design section
> 25% of design thickness	remove and replace
< 25% of required strength	Evaluate Roadway design section
> 25% of required strength	remove and replace

3.08 MEASUREMENT

A. The area of stabilized subgrade shall be measured by the plan quantities completed, in place, and accepted. No separate measurement of depth or area, except as required for thickness testing shall be performed.

B. The quantity of stabilizing agent accepted and used shall be measured by the ton of fly ash, Portland cement, cement kiln dust, or hydrated lime used (or the calculated dry hydrated lime content of the lime slurry).

3.09 TESTING AND INSPECTION

A. Testing and inspection shall be performed in accordance with the following table:

SCHEDULE FOR MINIMUM MATERIALS SAMPLING AND TESTING

Test Type	Deficiency	Remediation
Sampling	AASHTO T 87	per 1,000 square yards
Sample Preparation	ASTM D 3551	
Maximum Dry Density and Optimum Moisture Content	ASTM D 698 (Lime) ASTM D558 (Cement)	As directed by the OWNER'S REPRESENTATIVE (minimum one per soil type)
In Place Soil Density	ASTM D 1556 ASTM D 2167 ASTM D 2922	One test for each 200 lane feet (not less than one test per day)
In Place Moisture Content	ASTM D 2216 ASTM D 3017	
pH	ASTM C 110 ASTM C 977 APPENDIX	One test per 1,000 square yards
Swell	ASTM D 4546 Method B	As directed by OWNER'S REPRESENTATIVE
Unconfined Compressive Strength (Lime)	ASTM D 5102 (Procedure B)	One set of four cylinders per 1,000 square yards. Sealed and moist cured at 100°F for 5 days
Compressive Strength Cementitious Agents	ASTM D 1633 (Method A)	One set of four cylinders per 1,000 square yards. Sealed and moist cured at 100°F for 5 days
Atterberg Limits	AASHTO T 89 & T 90	One test per 1,000 square yards
Thickness	As directed by OWNER'S REPRESENTATIVE	One test every 500 feet per lane
Percentage of Agent	AASHTO T 232 see section 5.9.4	One test per 1,000 square yards

3.10 PAYMENT

A. Stabilized subgrade found deficient in thickness shall be removed and replaced, or at the option of the OWNER'S REPRESENTATIVE. The pavement structural section shall be adjusted to compensate for any deficiency in the stabilized subgrade thickness and strength at the CONTRACTOR's expense as noted in Section 3.07.

END OF SECTION 321113

SECTION 321123 - AGGREGATE BASE COURSE

PART 1 GENERAL

1.01 SUMMARY

A. Section Includes:

1. Aggregate base course
2. Subgrade Compaction

1.02. RELATED SECTIONS:

A. Section 321113, Stabilized Subgrade.

B. Section 321216 - Hot Mix Asphalt Pavement (HMAP).

1.03 REFERENCES

A Standards (as applicable).

1. AASHTO T 27 - Sieve Analysis of Fine and Coarse Aggregates
2. AASHTO T 89 - Determining the Liquid Limit of Soils (ASTM D 4318)
3. AASHTO T 90 - Determining the Plastic Limit and Plasticity Index of Soils.
4. AASHTO T 96 - Los Angles Abrasion
5. AASHTO T 111 - Inorganic Matter or Ash in Bituminous Materials
6. AASHTO T 180 - Moisture-Density Relations of Soils Using a 4.54-kg (10-lb) Rammer and a 457-mm (18-in.) Drop
7. AASHTO T 190 - Resistance R-Value and Expansion Pressure of Compacted Soils, (ASTM D 2844)
8. AASHTO – T 191 - Density of Soil In-Place by the Sand-Cone Method
9. AASHTO T 238 - Density of Soil and Soil-Aggregate in Place by Nuclear Methods (Shallow Depth)
10. AASHTO T 239 - Moisture Content of Soil and Soil Aggregate in Place by Nuclear Methods (Shallow Depth
11. AASHTO T 294 - Resilient Modulus of Unbound Granular Base/Subbase

Materials and Subgrade Soils

12. ASTM D 1557, Standard Test Methods for Laboratory Compaction Characteristics of Soil Using Modified Effort (56,000 ft-lbf/ft^3(2,700 kN-m/m^3), latest revision.

13. ASTM D 2216 - Standard Test Method for Laboratory Determination of Water (Moisture) Content of Soil and Rock by Mass

1.04 SUBMITTALS

A. Aggregate base material annual report in accordance with Section 2.01, C.

B. Supplier Material Certification.

1.05 PROTECTION

A. Protect all subsurface sleeving and conduit.

PART 2 PRODUCTS

2.01 AGGREAGATE

A. Aggregate Fill Type: Provide aggregate base course in quantities necessary to conform with the drawings. Materials shall be sound, durable particle, free from clay, silt or organic materials. Material shall meet the following limits:

Sieve Size	%Passing by weight
3/4"	100
No.4	35-65
No.8	25-55
No. 200	3-12

B. The aggregate base material shall also meet the following requirements:

	Test Method	Requirement
R-Value	AASHTO T 190	Min. 78 at 300 psi
Plastic Index	AASHTO T 90	15 Max.
Liquid Limit	AASHTO T 89	30 Max.
Resilient Modulus, psi	AASHTO T 294	20,000 min.
Los Angles Abrasion %	AASHTO T 96	45 Max.

C. The aggregate base material shall be pre-qualified on an annual basis by an independent laboratory capable of performing the tests listed in this section. The annual report shall be prepared under the supervision of, stamped and signed by a Colorado Professional Engineer. Supplier must certify the material meets the requirements listed above.

D. Gradation and Atterberg Limits shall be performed on random samples taken from each lift in place at the project. The aggregate base course shall conform to the following range of tolerances from the approved gradation.

Properties	Tolerance
Passing No. 8 and larger sieves	±8%
Passing No. 30 sieve	±6%
Passing No. 200 sieve	±2%
Plastic Index	± 3

E. Should there be a change in source material or material placed at the project is not in conformance with the tolerances with 2.01, D. the ENGINEER shall suspend the use of such material until laboratory tests indicates the resilient modulus is acceptable. Material that exceeds the tolerances shall be removed and replaced by the CONTRACTOR at no expense to the OWNER.

2.02 SOIL STERILIZATION

A. The sterilization shall be with a pre-emergent herbicide agent, which shall be a soluble, dispensable or mixable in water and non-toxic to humans when applied per the manufacturer's recommendations.

PART 3 EXECUTION

3.01 EXAMINATION

A. Verify subgrade has been inspected, gradients and elevations are correct, and subgrade is dry.

3.02 PREPARATION

A. Establish a consistent subgrade depth in conformance with the dimensions shown on the drawings.

B. Compact subgrade to 95% Modified Proctor Density (ASTM D 1557) within a moisture content range of 2 percent below to 2 percent above optimum.

C. Determine moisture stability: Moisture stability is determined by R-value testing which shows a maximum 10 point difference in R-values between exudation pressures of 300 psi and 100 psi. Aggregate base course material shall be tested to determine compliance with these specifications prior to importation to the site.

D. Correct irregularities in substrate gradient and elevation by scarifying, reshaping, and re-compacting. This work will be paid for by the CONTRACTOR.

E. Do not place fill, grade or compact on soft, muddy, or frozen surfaces.

3.03 AGGREGATE PLACEMENT

A. Spread aggregate over prepared subgrade to the total compacted thickness shown on the drawings.

B. Place aggregate in maximum 6 inch lifts. Add small quantities of fine aggregate to coarse aggregate as appropriate to assist compaction. Water shall be applied during compaction in sufficient amount to assist in the compaction and prevent raveling.

C. Level and contour surfaces to elevations and gradients required to attain the specified finished surface elevations as shown on the drawings.

D. Use mechanical tamping equipment in areas inaccessible to compaction equipment.

3.04 COMPACTION

A. Compaction shall immediately follow aggregate placement. If compacted depth of the base exceeds 6 inches, the course shall be constructed in two or more layers of approximately thickness.

B. Each layer shall be compacted to a density of not less than 95 percent of maximum density in accordance with AASHTO T 180 (ASTM D 1557).

C. Moisture conditions shall be within 2 percent of optimum moisture content.

D. Testing shall be completed within 24 hours of the placement of the next paving course. CONTRACTOR will pay for retesting if next paving course is delayed beyond 24 hours.

3.05 PROOF ROLL

A. The CONTRACTOR shall proof roll the completed base course material to determine any soft, yielding or otherwise unacceptable areas. These areas shall be removed and replaced with no cost to the OWNER.

B. The proof roller equipment may be loaded trucks, loaders, rollers and shall be approved by the OWNER. Equipment weight shall not be less than 45 tons and capable of applying ground loads of not less that 18,000 pounds per axle unless otherwise approved. All tires shall be of equal size and diameter and shall have a tire pressure of at least 100 pounds per square inch. Tires shall be kept uniformly inflated within 5 pounds per square inch between tires. CONTRACTOR shall provide a means to verify tire pressure on the job at any time.

C. Rollers shall be operated between 2 and 6 miles per hour.

3.06 SOIL STERILIZATION

A. Soil sterilization shall be applied under all new asphalt paving unless waived by the ENGINEER on a case-by-case basis.

B. The agent shall be active for one year after application.

C. The applicator shall be certified by the U.S. Environmental Protection Agency and licensed in the State of Colorado as a "pesticide applicator" and shall be held responsible for any damage to plant growth outside of the roadway or to the pavement where such damage is attributable to carelessness or improper application of the agent.

D. The agent shall be applied to the aggregate base course no more than 3 days prior to paving. The agent shall not be used where it may contaminate water used for irrigation or drinking purposes.

3.07 TOLERANCES

A. Grade Tolerances

1. Maximum variation of ¼ inch in cross section
2. Maximum variation of ¼ inch measured with 16 foot straight edge measured longitudinally.

B. Density Tolerances

1. Density below specified minimum set in 3.04, B shall be corrected by the CONTRACTOR in a manner satisfactory to the ENGINEER at no cost to the OWNER.

C. Thickness Tolerances

1. Maximum variation of ½ inch in thickness. If the thickness is deficient or exceeds specified tolerance it shall be corrected by the CONTRACTOR in a manner satisfactory to the ENGINEER at no cost to the OWNER.

3.08 FIELD QUALITY CONTROL

A. Testing aggregate base course shall be provided in accordance with the following:

Test Type	Test Standard	Minimum Frequency of Tests
In Place Soil Density and Moisture Content	AASHTO T 191 ASTM D 2167 AASHTO T 239 AASHTO T 238 ASTM D 2216	One test for each 1,000 square yard (not less than one test per day). AASHTO T 191 and ASTM D 2216 shall be performed every 10th nuclear method density test
Atterberg Limits	AASHTO T89 & T90	One test per 2,000 tons
Moisture- Density Relationships	AASHTO T 180	One test per soil type
Gradation	AASHTO T27 and T111	One test per 2,000 tons
Thickness		One test per 5,000 square yards
Resilient Modulus	AASHTO T 294	Upon request by OWNER

END OF SECTION 321123

SECTION 32 12 16 - HOT MIX ASPHALT PAVEMENT (HMAP)

PART 1 GENERAL

1.01 SECTION INCLUDES

A. Plant Mixed Hot Mix Asphalt Pavement (HMAP)

B. Subgrade Compaction

1.02 RELATED SECTIONS

A. Section 321113, Stabilized Subgrade.

B. Section 321123 - Aggregate Base Course

1.03 SUBMITTALS

A. The CONTRACTOR shall submit all mixture designs, Certificates of Compliance and laboratory data to the ENGINEER for approval at least 7 calendar days before construction is to begin. The mix design (Proposed Design Job Mix) must be approved by the ENGINEER prior to the start of construction

B. Mixture designs shall be performed in a materials laboratory under the direct supervision of and shall be stamped and signed by a Professional ENGINEER licensed in Colorado and practicing in this field. In addition, the CONTRACTOR shall submit, as part of the mixture design, laboratory data documents to verify the following:

1. Source of Material
2. Gradation, specific gravity, source and description of individual aggregates and the final blend.
3. Aggregate physical properties.
4. Source and Grade of Performance Graded Binder.
5. Proposed Design Job Mix: aggregate and additive blending, final gradation shown on 0.45 power graph, optimum asphalt content.
6. Mixing and compaction temperature used.
7. Mixture properties determined at a minimum of four asphalt contents and interpolated at optimum and graphs showing mixture properties versus asphalt content.

C. The ENGINEER reserves the right to verify the CONTRACTOR mix design for each hot asphalt pavement grading utilizing materials actually produced and stockpiled. If mineral filler, RAP, and additive for the required laboratory tests, if required by the ENGINEER. The ENGINEER may request a Certificate of Compliance at any time on any material used.

1.04 QUALITY ASSURANCE

A. Pre-Paving Meeting: The ENGINEER may require a pre-paving meeting of all parties involved in supply, haul, lay down inspection, quality control and quality acceptance of HMAP. Areas of responsibility and contract names and numbers should be shared.

B. Change in Source or Grade

1. Should a change in the source of Lime occur, or more than one temperature grade change on either the high or low end of the Performance Graded Asphalt Binders - (Asphalt Cements) occur, a one point verification test (at optimum asphalt content) of the mix must be performed to verify that the applicable criteria shown on Minimum VMA% Table (Section 2.01, D.) or Superpave Mixture Properties (Section 2.06) is still met.

2. If testing shows non-compliance, a new Design Job Mix shall be established and approved by the ENGINEER before the new Performance Graded Asphalt Binders or Lime source if used. Any change in aggregate type or source will require a new mix design. The one point verification test may be performed on lab mixed samples or on plant mixed samples.

C. Mix Production Verification

1. Production verification shall occur prior to the start of the project. The production verification shall be performed by LABCAT Level C certified technicians to verify the volumetric properties of the mix. If the mix has been produced for another project within the last 90 days, data from that project can be submitted for this verification. Volumetric properties of the mix verification testing shall be with in the following tolerances compared to the Proposed Design Mix. The mix verification test reports shall be submitted to the OWNER prior to mix placement.

DESIGN MIX VERIFICATION TOLERANCES

Air Voids	+/- 1.2%
VMA	+/- 1.2%
Asphalt Cement Content	+/- 0.3%
Stability	Applicable minimum

2. The tolerances in this table for mix design verification only. See Section 3.08 for production tolerances.

1.05 PROTECTION

A. Protect all subsurface sleeving and conduit.

1.06 GEOTECHNICAL REPORT

A. The Contractor shall follow the specifications and recommendations of the geotechnical report prepared for this project titled Geotechnical Exploration Report Aims Community College, North Campus Refresh, Greeley, Colorado, EEC Project No. 1162022. The geotechnical report was prepared by Earth Engineering Consultants, 4396 Greenfield Drive, Windsor, CO 80550.

PART 2 PRODUCTS

2.01 AGGREGATE

A. Aggregate for HMAP shall be uniform quality, composed of clean, hard, durable particles of crushed stone, crushed gravel, or crushed slag. Excess of fine material shall be wasted before crushing. The material shall not contain clay balls, vegetable matter, or other deleterious substances and shall meet the following requirements:

AGGREGATE PROPERTIES

Aggregate Test Property	Coarse: Retained on #4	Fine: Passing the #4
Fine Aggregate Angularity, P-L 5113 Method A Traffic Level 1 & 2 Traffic Level 3 to 5	 - - - - - - - -	 40% Min. 45% Min.
Two Fractured Faces, CP-45 Top & Middle Lifts Bottom Lifts	 80% Min. 70% Min.	 - - - - - - - -
LA Abrasion, ASSHTO T 96	45% Max.	- - - -
Flat and Elongated (Ratio 5:1) %, AASHTO M 283	10% Max	- - - -
Adherent Coating (Dry Sieving) ASTM D 5711	0.5% Max.	- - - -
Sand Equivalent, (AASHTO - T 176	- - - -	45% Min.

B. Reclaimed Asphalt Pavement material (RAP) shall be used only where specifically allowed and shall be of uniform quality and gradation with a maximum size no greater than normal aggregate size of mix. Mixes shall not contain more than 20% RAP

C. The HMAP gradation for the Proposed Design Job Mix gradation shall be wholly within the control point gradation range set forth in the following table. The allowable job mix gradation for production shall be the design job mix gradation with the tolerances of Section 3.08, B. applied. The Proposed Job Mix and the final allowable job mix gradation for production shall report all sieve sizes listed in the following table.

GRADATION RANGE
(Percent by Weight Passing Square Mesh Sieve, AASHTO T 11 & T 27)

Mixture Grading Sieve Size	SX (1/2" nominal)		S (3/4" nominal)		SG (1" nominal)	
	Control Points	Restricted Zone*	Control Points	Restricted Zone	Control Points	Restricted Zone
1 ½"					100	
1"			100		90-100	
¾"	100		90-100		@	

½"	90-100		@		@	
3/8"	@		@		@	
#4	@		@		@	39.5
#8	28-58	39.1	23-49	34.6	19-45	26.8-30.8
#16	@	25.6-31.6	@	22.3-28.3	@	18.1-24.1
#30	@	19.1-23.1	@	16.7-02.7	@	13.6-17.6
#50	@	15.5	@	13.7	@	11.4
#200	2.0-8.0		2.0-7.0		1.0-7.0	

* The restricted zone is a guideline only. It is recommended that the mix design gradations go above the restricted zone boundaries, in the "fine" side.

** These limits shall include the weight at 1.0%.

@ These sieve sizes used to determine the final allowable Job Mix Formula (JMF) in accordance with Section 3.08.

D. VMA shall be based on tests of the Bulk Specific Gravity of the Compacted Mix (CP-L 5103) and Aggregate (T 84 & T 85), and the calculated according to CP-48. All mixes shall meet the minimum VMA specified in below

MINIMUM VMA, %

Nominal Maximum* Particle Size	Air Voids **		
	3.5%	4.0%	4.5%
1"	12.5	13.0	13.5
¾"	13.5	14.0	14.5
½"	14.5	15.0	15.5

* Nominal Maximum Particle Size is defined as one sieve size larger than the first sieve to retain more than 10%, but shall not exceed the 100% passing size. The Nominal Maximum Particle Size can vary during mix production even when the 100% passing size is constant.

** Minimum VMA criteria apply to the mix design only. The minimum VMA criteria shall be linearly interpolated based on actual air voids. See Section 3.08 for tolerances.

2.02 PERFORMANCE GRADED ASPHALT BINDERS - (ASPHALT CEMENTS)

A. The CONTRACTOR shall provide to the ENGINEER acceptable "Certifications of Compliance" of each applicable asphalt cement grade from the supplier. Upon non-conformance with the specifications, the asphalt cement may be rejected as directed by the ENGINEER. When production begins, the CONTRACTOR shall, upon request, provide to the ENGINEER a one quart can of each specified asphalt cement. Additionally, when requested, the CONTRACTOR shall provide the refinery test results that pertain to the asphalt binders used during production.

B. The Asphalt Binder shall meet the requirements of the Superpave Performance Graded Binders (PG) as presented in the table below:

PROPOERTY OF PERFORMANCE GRADED BINDERS

Property	PG 58-28	PG 64-22	PG 76-28
Traffic Level: Recommended Usage	1 and 2	2 through 5	3 through 5

Flash Point Temperature, °C, ASSHTO T 48	230 Min.	230 Min.	230 Min.
Visocity at 135°C, Pas, ASTM D 4402	3 Max.	3 Max.	3 Max.
Dynamic Shear, Temperature °C where C* /Sin δ @10 rad/sec. ≥ 1.00 Kpa, ASSTH TP 5	58°C	64°C	76°C
Rolling Thin Film Oven Residue Properties, AASHTO T 240			
Mass Loss, %, AASHTO T 240	1.00 Max.	1.00 Max.	1.00 Max.
Dynamic Shear, Temperature °C where G* /Sin δ @10 rad/sec. ≥ 2.20 Kpa, ASSTH TP 5	58°C	64°C	76°C
Elastic Recovery1, 25°C, % Min.*	N/A	N/A	50 Min.
Pressure Aging Vessel Residue Properties, Aging Temperature 100°C AASHTO PP1			
Dynamic Shear, Temperature °C where G* /Sin δ @10 rad/sec. ≥ 2.20 Kpa, ASSTH TP 5	19°C	25°C	28°C
Creep Stiffness, @ 60 sec. Test Temperature in °C, AASHTO TP 1	-18°C	-12°C	-18°C
S, Mpa, AASHTO TP 1	300 Max.	300 Max.	300 Max.
m-value, ASSHTO TP 1	0.300 Min.	0.300 Min.	0.300 Min.
**Direct Tension Temperature in °C, @ 1.0 mm/min., where Failure Strain >1.0%, ASSHTO TP 3	-18°C	-12°C	-18°C

* Elastic Recovery by Task Force 31, Appendix B Method

** Direct tension measurements are required when needed to show conformance to AASHTO MP 1

2.03 ADDITIVES - HYDRATED LIME

A. Lime shall be added at the rate of 1% by dry weight of the aggregate and shall be included in the amount of material passing the No. 200 sieve. Hydrated lime for aggregate pretreatment shall conform to the requirements of ASTM C 207, Type N. In addition, the residue retained on a 200-mesh sieve shall not exceed 10% when determined in accordance with ASTM C 110. Drying of the test residue in an atmosphere free from carbon dioxide will not be required.

B. Hydrated may be substituted with an approved anti-stripping additive.

2.04 TACK COAT

A. The emulsified asphalt, for Tack Coat shall be CSS-1h or SS-1h and conform to ASSHTO M208 or M140, respectively.

2.05 MIX DESIGN AND PLANT PRODUCED MIXTURE REQUIREMENTS

A. The mix design materials shall be those listed in this specification and used for the project. No substitutions are allowed during production.

B. Grading SG (1-inch nominal aggregate) shall only be designed using the 150mm Superpave molds. Hveem Stability and Lottmans are not required for Grading SG mixtures. Grading S and SX shall be designed using 100mm Superpave molds when using the Superpave design method.

2.06 SUPERPAVE MIXTURE DESIGN METHOD

A. The CONTRACTOR shall submit a Proposed Design Mix for each mixture required by the Contract. The design shall be determined using the Colorado Procedure CP-L 5115 for the Superpave Method of Mixture Design. Guidance is provided in "Superpave Level 1 Mix Design" SP-2 published by the Asphalt Institute. Mix designs shall meet the following requirements.

SUPERPAVE MIXTURE PROPERTIES

Traffic Level	0	1 & 2 & 3	4 & 5
Traffic Loading, Total 18 kip ESALs Over Design Life (Usually 20 Years)	Non Vehicular or Paths	1) <300,000 2) 300,000 to <1.0 Million 3) 1.0 to <3.0 Million	4) 3.0 to 10.0 Million 5) > 10.0 Million
Initial gyrations, $N_{initial}$ (For Information Only)	6	7	8
$N_{initial}$ Air Voids	>8.5	>9.5	>11.0
Design gyrations, N_{design} (Air Voids: 3.5% to 4.5%) (See Note 1, 2)	50	75	100
Hveem Stability CP-L 5106 (Grading S & SX only) (See Note 1,2)	N/A	28 Min.	30 Min.
Voids filled with Asphalt, MS-2	70-80	65-78	65-75
Lottman, Tensile Strength Ratio, % Retained, CP-L 5109, Method B	80 Min.	80 Min.	80 Min.
Lottman, CP-L 5109 Dry Tensile Strength, psi	30 Min.	30 Min.	30 Min.
VMA %, CP-48 (See Note 1,2)	Minimum VMA criteria applies to the mix design only (2.01, D.). The minimum VMA criteria shall be linearly interpolated based on actual air voids.		

Note 1: Maximum Theoretical Specific Gravity of mix by cp-51.
Note 2: Refer to Section 3.09 for production tolerances.

PART 3 EXECUTION

3.01 EQUIPMENT

A. MIXING PLANT

1. The mixing plant shall be capable of producing a uniform material, have adequate capacity, and be maintained in good mechanical condition. Defective parts shall be replaced or repaired immediately if they adversely affect the proper functioning of the plant or plant units, or adversely affect the quality the HMAP.

2. Dust, smoke, or other contaminants shall be controlled at the plant site to meet all air quality requirements in the "Colorado Air Quality Control Act," Title 25, Article 7, CRS and regulations promulgated thereunder.

3. Acceptable safety equipment, approved by the ENGINEER, shall be provided by

the CONTRACTOR to accommodate sampling and testing.

B. HAULING EQUIPMENT

1. Trucks used for hauling HMAP material shall have tight, clean, smooth beds thinly coated with a minimum amount of paraffin oil, lime solution, or other approved release agent. Petroleum distillates such as kerosene or fuel oil will not be permitted. Each truck shall have a cover of canvas or other suitable material to protect the mixture from the weather and excessive temperature loss or cooled layers of mix in truck.

C. BITUMINOUS PAVERS

1. Self-propelled pavers shall be provided for full lane width paving, and shall be equipped with a screed assembly, heated if necessary, capable of spreading and finishing the HMAP material in full lane widths applicable to the typical section and thicknesses shown in the Contract. Pavers used for shoulders, patching and similar construction, not requiring fine grade control, shall be capable of spreading and finishing courses of HMAP material in widths shown in the Contract.

2. The paver's receiving hopper shall have sufficient capacity for a uniform spreading operation and shall have an automatic distribution system that will place and spread the mixture uniformly in front of the screed.

3. The paver shall be capable of operating at forward speeds consistent with uniform and continuous laying of the mixture. Stop and go operations of the paver shall be avoided. The screed or strike-off assembly shall produce the specified finished surface without tearing, shoving, or gouging the mixture. Self-propelled pavers shall be equipped with automatic screed controls with sensors capable of sensing grade from an outside reference line, and maintaining the screed at all the specified longitudinal grade and transverse slope. The sensor shall be constructed to operate from either or both sides of the paver and shall be capable of working with the following devices when they are required for the situation:

(1) Ski-type device at least 30 feet in length;
(2) Short ski or short shoe;
(3) Adequate length of control line and stakes, if no other type of geometric control is present.

4. The controls shall be capable of maintaining the screed at the specified transverse slope within plus or minus 0.1 percent. Automatic mode should be used where possible. If the automatic controls fail or malfunction, the equipment may be operated manually for the remainder of the normal working day, provided specified results are obtained.

5. If the CONTRACTOR fails to obtain and maintain the specified surface tolerances, the paving operations shall be suspended until satisfactory

corrections, repairs, or equipment replacements are made.

6. Placement of HMAP on a waterproofed bridge deck shall be accomplished with equipment that will not damage the membrane or protective covering.

3.02 MANUFACTURE

A. Preparation of Aggregates

1. Heating and drying of the aggregates shall be accomplished without damaging the aggregate. Hydrated lime shall be added to achieve complete and uniform coating of the aggregate, in accordance with one of the following methods:

(a) Lime Slurry Added to Aggregate: The hydrated lime shall be added to the aggregate in the form of a slurry and then thoroughly mixed in an approved pug mill. The slurry shall contain a minimum of 70 percent water by weight.

(b) Dry Lime Added to Wet Aggregate: The dry hydrated lime shall be added to wet aggregate (a minimum of three percent above saturated surface dry) and then thoroughly mixed in an approved pug mill.

2. The lime-aggregate mixture may be fed directly into the hot plant after mixing or it may be stockpiled for not more than 90 days before introduction into the plant for mixing with the asphalt cement. The hydrated lime may be added to different sized aggregates and stockpiled by adding 75 percent of the lime to the aggregate passing the No. 4 sieve and 25 percent to the aggregate retained on the No. 4 sieve.

B. Mixing

1. The dried aggregates and asphalt cement shall be combined in the mixer in the quantities required to meet the design job mix. The materials shall be mixed until the aggregate is completely and uniformly coated, and the asphalt cement is uniformly distributed throughout the aggregate Baghouse fines shall be fed back to the mixing plant in a uniform and continuous manner to maintain uniformity in the mixture. The Baghouse, fines feeder, auger, and related equipment, shall be in good working condition and operated in accordance with manufacturer's recommendation. If the ENGINEER determines that non-uniform operation of the equipment is detrimental to the mixture, it may suspend all paving operations until the CONTRACTOR takes appropriate action.

2. The minimum temperature of the mixture when discharged from the mixer shall be as shown in the following table:

TABLE 9.6.2.1 – MIXTURE DISCHARGE TEMPERATURES

Asphalt Grade	Minimum Discharge Temperature	Maximum Discharge Temperature

PG 58-28	275° F	305° F
PG 64-22	290° F	320° F
PG 76-28	320° F	350° F

3. HMAP mix shall be produced at the lowest temperature within the specified temperature range that produces a workable mix and provides for uniform coating of aggregates (95 percent minimum in accordance with AASHTO T 195), and that allows the required compaction to be achieved.

4. HMAP mix may be stored provided that any and all characteristics of the mixture are not altered by such storage. If storing or holding of the mixture causes segregation, excessive heat loss, or adversely affects the quality of the finished product, corrective action shall be taken. Unsuitable mixture shall be disposed of at the CONTRACTOR's expense.

5. When placing hot bituminous mixture over bridge decks covered by waterproofing membrane, the minimum temperature of the mixture, when rolling operations begin, shall be 250° F. The job mix temperature may be increased up to 30° F to obtain this temperature.

C. Hauling

1. Each truck shall use covers to protect the mix during transport in cold weather, dust storms or if precipitation is eminent. The ENGINEER can reject any mix which shows an excess or deficiency of asphalt cement, damage due to burning or overheating, an improper gradation, or thermal segregation with cold areas 30° F below the minimum discharge temperature.

3.03 TACK COAT

A. Prior to placement of HMAP, a tack coat shall be applied to all existing concrete and asphalt surfaces. The material shall be in accordance with Section 2.04 Tack Coat. The emulsified asphalt shall be diluted 1:1 with water and applied at 0.10 ± 0.01 gallons per square yard of diluted material. The ENGINEER may direct other application rates to match the age of condition of the surface.

B. The surface to receive the tack coat shall be dry and cleaned by sweeping, or an other approved method, until dust, debris, and foreign matter are removed. The tack coat shall then be applied uniformly by squeegee, brooms, or distributor. Prior to paving, all water must have evaporated from the tack coat. Contaminated areas shall be cleaned and tack coat shall be reapplied.

3.04 PLACEMENT

A. Hot mix asphalt shall be placed only on approved, properly constructed surfaces that are free from loose material, water, frost, snow or ice. The hot mix asphalt and tack coat shall be placed in accordance with the temperature limitations of Minimum Air And Surface Temperatures Limitations For Mix Placement Table below and only when weather conditions permit the pavement to be properly placed and finished as determined by the ENGINEER. Placement temperature as stated shall be increased by 5° F for each 10 miles per hour wind velocity to a maximum increased placement temperature of 70° F.

MINIMUM AIR and SURFACE TEMPERATURES
LIMITATIONS for MIX PLACEMENT TABLE

Compaction Layer Thickness	Top Layer of Pavement*		Lower Layers*	
	PG 58-28 PG 64-22	PG 76-28	PG 58-28 PG 64-22	PG 76-28
<2 inches (not recommended)	60° F	75° F	N/A	N/A
2 inches to <3 inches	50° F	65° F	40° F	50° F
3 inches or more SG mix only	50° F	50° F	40° F	40° F

*Air temperature is taken in the shade. Surface temperature is taken on the subgrade or base. The ENGINEER may not waive the above temperature limitations for PG 76-28.

B. The mixture shall not be placed at a temperature lower than 245° F for mixes containing PG 58-28 or PG 64-22 asphalt, and 290° F for mixes containing polymer modified asphalt. Mix which is too cold or damaged by weather, will be rejected.

C. The mixture shall be laid upon an approved surface, spread and struck off to obtain the required grade and elevation after compaction. The minimum lift thickness shall be at least three times (preferably four times) the nominal particle size. The mixture should be placed approximately 25 percent thicker than the existing surrounding mat thickness to account for compaction. Raking is discouraged and will not be allowed except to correct major problems of grade and elevation. Casting or raking that causes any segregation will not be permitted.

D. On areas where the use of mechanical spreading and finishing equipment is impracticable, the mixture shall be carefully dumped, spread, raked, screeded, and luted by hand tools to the required compacted thickness plus 25 percent. Carefully move or minimally work the HMAP mix with the use of rakes, lutes, or shovels to avoid segregation. Mixtures made with modified asphalt cement require more rapid completion of handwork areas than for normal mixtures. Hauling and placement sequences shall be coordinated so that the paver is in constant motion. Excessive starting and stopping shall not be allowed. A construction joint shall be placed any time the paver stops, and the screed drops enough to cause a surface dip in violation of Section 3.08, A., Surface Tolerances; or the mat temperature falls below that allowed in Section 3.07, A. Compaction. Bituminous pavers shall be used to distribute the mixture either over the entire width or over such partial width as may be practicable. Echelon paving will be permitted.

3.05 LONGITUDINAL JOINTS

A. The longitudinal joints in both a new pavement and an overlay pavement layer shall offset the joint in the layer immediately below by a minimum of 6 inches. The joints in any pavement layer shall not fall in a wheel track or path. The joints in the top layer of new pavement, not built on top of an existing pavement, shall be located on lane lines, or as shown on the plans. Longitudinal joints shall be minimized, where feasible, with wide paving pulls or echelon paving. Joints shall not cross any centerline, lane line, or edge unless approved by the ENGINEER. The CONTRACTOR shall submit, prior to paving, a joint plan and pavement marking plan showing locations and the methods to establish a field control line. The ENGINEER must approve such plans prior to paving. The CONTRACTOR shall use a continuous string line to delineate longitudinal joints during paving as shown on the joint plan. All string lines shall be removed at the end of each day's paving.

B. The free edge of the paved pass shall be laid as straight as possible, to the satisfaction of the ENGINEER. This joint, if cold, shall be tack coated prior to placement of adjacent paving.

C. The new compacted mat shall overlap the previously placed mat no more than 1.5 inches. Excess overlap or thickness shall not be raked or cast onto the new mat, but shall be wasted by pulling back and removing. The hot edge shall be blocked or bumped in a smooth line consistent with the previous longitudinal edge. Minor raking will only be allowed to correct major grade problems or provide mix around manholes and meter covers. The longitudinal joint shall be rolled from the hot side and overlap the joint by approximately 6 inches on the cold side.

3.06 TRANSVERSE JOINTS

A. The CONTRACTOR shall submit, prior to paving, a joint plan showing locations and the methods to be used to construct transverse joints. The ENGINEER must approve such plans prior to paving. Placing of the HMAP shall be continuous with a minimum of transverse joints, and rollers shall not pass over the unprotected end of a freshly laid mixture. Transverse joints shall be formed by cutting back on the previous run to expose the full depth of the course. Tack coat material shall be applied to contact surfaces of all joints just before additional mixture is placed against the previously compacted material.

B. The end of transverse joints shall be located so they will be constructed with a full head of mix in front of the screed. When butt joints are constructed, runoff boards shall be used to support the roller on the downstream side of the joint. All tapered sections, rounded edges and segregated areas shall be removed to achieve a vertical face at the butt joint before paving is restarted.

C. When a tapered joint is required for traffic access, the ramp shall be removed back to a full depth from the segregated section before paving is restarted.

D. When restarting paving operations, the paver screed shall be placed on the starter block on the completed side of the transverse joint. The starter block should be approximately 25% of the thickness of the existing completed mat, so that adequate grade and compaction can be achieved on starting the paving operation. The screed should be nulled (angle removed) when on starting blocks and an up angle of attack set. Proper head of mix should be introduced into the paver prior to starting. The new compacted (downstream) side of the joint shall not be allowed except to correct major grade problems. The surface tolerance at the transverse joint must be verified with a 10-foot straight edge before the paver is more than 100 feet from the joint.

3.07 SEGREGATION

A. The asphalt mixture shall be transported and placed on the roadway without segregation. All segregated areas shall be removed immediately and replaced with specification material before the initial rolling. If more than 50 square feet of segregated pavement is removed and replaced in any continuous 500 linear feet of paver width laydown, operations shall be discontinued until the source of the segregation has been determined and corrected.

B. The ENGINEER will visually determine areas that are segregated, and may also use density and gradation measures to help in this determination. The ENGINEER will visually determine the extent of the segregation. The CONTRACTOR will not be allowed additional compensation for correction of segregated areas.

3.08 COMPACTION

A. The temperature of the mixture immediately behind the screed shall be at least 245° F for PG 58-28 or PG 64-22 binder and 290° F for PG 76-28 binder. The breakdown compaction shall be completed before the mixture temperature drops 20° F.

B. The HMAP shall be compacted by rolling. The number, weight, and type of rollers furnished shall be sufficient to obtain the required density and surface texture while the mixture is in a workable condition. Compaction shall begin immediately after the mixture is placed and be continued until the required density is obtained. Final compaction shall be obtained using steel wheel rollers.

C. If the required density is not achieved and the surface temperature falls below 185° F, or there is obvious surface distress or breakage, no further compaction effort will be permitted unless approved by the ENGINEER. Price Reduction criteria in Section 9.14 shall still apply in such cases. The criteria for mixtures containing PG 76-28 asphalt cements shall be 235° F. These minimum compaction temperatures may be adjusted according to the asphalt cement supplier recommendations. Adjusted minimum compaction temperatures must be shown on the approved mix design or on other asphalt cement supplier documents, to be available on the job site. Pavement operations shall be suspended when density requirements are not met, and the problem shall be resolved prior to continuing paving operations.

D. All roller marks shall be removed with the finish rolling. Use of vibratory rollers with the vibrator on will not be permitted on bridge decks.

E. The CONTRACTOR shall establish a rolling pattern or procedure during the beginning of paving operations, which will achieve the required compaction and surface tolerances. This procedure may be re-evaluated by the CONTRACTOR and ENGINEER throughout the paving operations.

F. All HMAP paving shall be compacted to between 92.0 and 96.0 percent of Maximum Theoretical (Rice) Density, (CP-51: Maximum Specific Gravity of Bituminous Paving Mixtures). The average (mean) of the 3 most recent production CP-51 Rice values shall be used in calculating Relative Compaction according to CP-44.

G. The CONTRACTOR shall core the pavement, as required by the ENGINEER, for field density tests in accordance with Colorado Procedure 44, Method B (AASHTO T 230), or for field calibration of nuclear density equipment in accordance with the Appendix of Colorado Procedure 81 (ASTM D 2950). At a minimum, cores for nuclear density equipment calibration shall be taken at the beginning of placement of each pavement layer or change of mixture materials or gradation. Untested areas during placement will also require cores to be taken to verify compaction.

H. Along forms, curbs, headers, walls, and all other places not accessible to the rollers, the mixture shall be thoroughly compacted with mechanical tampers.

I. Any mixture that becomes loose and broken, mixed with dirt, or is in any way defective, shall be immediately removed and replaced with fresh not mixture and compacted to conform with the surrounding area.

3.09 PRODUCTION TOLERANCES

A. Surface Tolerances

1. The variation between any tow contacts with the surface shall not exceed 3/16 inch in 10 feet for full lane width paving. For patching surface tolerances the variation shall not exceed ½" in 10 feet. Irregularities exceeding the specified tolerance shall be corrected at the CONTRACTOR's expense. Transverse measurements for variations shall exclude breaks in the crown sections.

2. There shall be no allowance for standing water on the final pavement surface. After compaction, the asphalt pavement adjoining the concrete curb and gutter pan shall be between ¼" and ½" higher than the lip of the concrete pan.

B. Job Mix Formula Tolerances

1. Production test results that deviate from the design job mix by more than shown in the following table are subject to Section 3.09:

JOB MIX FORMULA TOLERANCES

Item	
Passing No. 3/8" and Larger [1]	±6%
Passing No. 4 and No. 8	±5%
Passing No. 30 to No. 50	±4%
Passing No. 200 [2]	±2%

Air Voids	±1.2%
VMA	±1.2%
Hveem Stability	See footnote #3
Asphalt Content	±0.3%
Asphalt Content, Mixes with >10% RAP	±0.4%

1. There is 1.0 percent tolerance for the maximum sieve size.
2. Mixes with passing No. 200 sieve material produced over 7.0 percent are allowed only when the above Air Voids and VMA tolerances are still met.
3. Hveem Stability must meet the minimum value specified in table 9.3.2.

2. When disagreements concerning determination of specification compliance occur, only valid tests from both the ENGINEER and CONTRACTOR will be considered. The ENGINEER shall determine validity. Generally, valid tests are those in which sampling and testing have been performed according to referenced procedures and the results are within stated precision statements. When disagreements occur with asphalt content and gradation tests results, solvent extracted aggregate testing shall take precedence over burnoff oven extracted aggregate, which shall take precedence over cold feed belt testing.

3.10 CONFORMITY WITH PLANS AND SPECIFICATION

A. General

1. All work performed and all materials furnished shall conform to the lines, grades, cross sections, dimensions, and material requirements, including tolerances, shown in the contract.

2. For those items of work where working tolerances are not specified, the CONTRACTOR shall perform the work in a manner consistent with reasonable and customary manufacturing and construction practices.

3. When the ENGINEER finds that the materials furnished, the work performed, or the finished product does not conform with the contract, but that reasonably acceptable work has been produced, the ENGINERR shall determine the extent of the work to be accepted and remain in place. Cost reduction, when allowed, shall be accomplished by adjusting pay quantities as indicated herein and applying contract unit prices to the reduced quantities. If the work is to be accepted, the ENGINEER shall:

   a. Document the basis for acceptance by "Cure Notice" which may provide for an appropriate adjustment in the payment quantity for such work or materials not otherwise provided for in this section.

   b. Notify the CONTRACTOR in writing that the payment may be adjusted in accordance with this section when "P" is 25 or less, or require appropriate remediation being performed.

c. In lieu of cost (quantity) adjustment, permit correction or replacement of the finished product provided the correction or replacement does not adversely affect the work or the ENGINEER.

4. When the ENGINEER determines that the material furnished, work performed, or the finished product is not in conformity with the contract and has resulted in inferior or unsatisfactory product, the finished product or materials shall be removed and replaced or otherwise corrected by, and at the expense of, the CONTRACTOR.

5. Materials shall be sampled and tested by a qualified testing laboratory in accordance with the sampling, testing schedules, and procedures contained in the Section 3.11 Testing and Inspection. The approximate maximum quantity represented by each sample shall be as set forth in the testing schedule. An additional number of samples, in relation to the quantity of materials represented, may be selected and tested at the ENGINEER's discretion. The quantity represented by five consecutive random samples shall constitute a lot, whenever production schedules and material continuity permits. When it is necessary to represent short production runs, significant material changes, or other unusual characteristics of the work, the ENGINEER may establish a lot consisting of the quantity represented by any number of consecutive random samples from one to seven inclusive. Testing results that are determined to have sampling or testing errors, as determined by the ENGINEER, shall not be used.

3.11 COST REDUCTION FORMULA

A. Materials or work shall only be evaluated for price adjustment when deviations from specifications occur on any of the individual tests for the lot. The several individual test values shall be averaged and the percentage of cost (quantity) reduction for the lot shall be determined by applicable formula. This shall apply only when a cost reduction factor for the element is listed in Table 3.10,C., 1., Table of Price Reduction Factors

B. When the Lot is represented by Three through Seven Tests

1. The formula in (a) and (b) below shall be used.

a) $P = (X_n + \bar{a}R - T_u) * F$ Shall be used if a maximum limit only is specified or; when the average of the several test values is above the mid point of the specification band or above the job-mix formula value.

b) $P = (T_L + aR - X_n) * F$ Shall be used if the minimum limit only is specified or; when the average of the several test values is below the mid point of the specification band or below the job-mix formula value.

C. When the lot is represented by fewer than three tests.

1. The materials shall be evaluated for cost (quantity) reduction by the following procedure: Lots represented by two tests shall be divided into two separable lots

represented by one test each, as determined by the ENGINEER. Each lot that deviates from the specifications shall be cost adjusted by one of the following formulae.

c) $P = 0.76 * (T_o - T_u) * F$ When a maximum limit only is specified or the test value is above the maximum specified limit.

d) $P = 0.76 * (T_L - T_o) * F$ When a minimum limit only is specified or the test value is below the minimum specified limit.

Where:

P is the percentage of reduction in payment quantity.

X_n is the average of the several test values from samples taken from the lot, with "n" indicating the number of values

a is a variable factor to be used if "n" changes according to the following:

when "n" is	"a" equals
3	0.45
4	0.38
5	0.33
6	0.30
7 or greater	0.28

R is the difference between the highest and lowest values in the group of several test results from the lot.

T_u is the upper or maximum tolerance permitted by the specifications.

T_L is the lower or minimum tolerance limit permitted by the specifications.

T_o is the test value of the test that deviates from the specifications.

F is the cost reduction factor to be applied for each element as shown in the following table:

TABLE OF PRICE REDUCTION FACTORS

ELEMENT	FACTOR "F"
100 percent size sieve	1

1/2 inch sieve and larger	1
3/8 inch sieve, #4, #8, #30 and #50 sieves	3
No. 200 sieve	6
Density of Hot Mix Asphalt	8
Asphaltic Cement Binder content (all asphalt-aggregate mixtures)	20
Total air voids	30
Voids in mineral aggregate	20
Stability	5

2. If "P" is less than ten or a negative quantity, the material shall be accepted as being in conformity. In cases where one or more elements show a positive "P" value, such positive values shall be added and the resulting sum shall be used to determine whether the material is in conformity. If the total "P" value is between 10 and 25, the ENGINEER may require correction or may accept the material at a reduced cost. If "P" is greater than 25, the ENGINEER may:

    1) require complete removal and replacement with specification material at no additional cost to the OWNER;

    2) require corrective action to bring the material into conformity at no additional cost to the OWNER;

    3) where finished product is found to be capable of performing the intended purpose and the value of the finished product is not affected, permit the CONTRACTOR to leave the material in place with an appropriate cost adjustment to be based on the ENGINEER's evaluation but not less than that which would have occurred had an adjustment been made where "P" = 25.

3. If asphaltic cement content, aggregate sieve analysis, or compaction deviates from the specification requirements and the total "P" is three or greater, the reduction shall apply to the contract cost (quantity) multiplied by 0.60 for aggregate base course and Hot Bituminous Pavement mixtures.

4. The CONTRACTOR shall not have the option of accepting a cost reduction in lieu of intentionally producing material not meeting specification. Continued production of non-specification material shall not be permitted. Material that is defective as identified by visual inspection shall be isolated and rejected without regard to sampling sequence or location within a lot.

D. Cost Reduction for Thickness Deficiencies:

1. If the cores indicate a thickness deficiency, additional cores will be taken by the CONTRACTOR to be given to the OWNER so that price reductions can be determined per Lot. A Lot encompasses 250 lineal lane feet or the quantity between tests, and a price reduction shall be determined at the unit cost of the HMAP. The following chart is for new construction only. It is not to be used for overlays.

PRICE REDUCTION – THICKNESS

% Price Reduction	0%	10%	20%	30%	40%	Remove & Replace
Thickness Deficiency	0 - 0.50"	0.51" - 0.75"	0.76" - 1.00"	1.01" - 1.25"	1.26" - 1.50"	1.51" or greater

2. Any deficiency in the total thickness of the pavement shall not exceed ten (10%) percent for any one sample with the average deficiency for all samples not to exceed seven and one-half (7.5%) percent. Final decision for the correction of the deficiencies shall not be made until a pavement evaluation has been made by an independent testing laboratory at no cost to the OWNER.

3.12 TESTING AND INSPECTION

A. If any materials furnished or work performed by the CONTRACTOR fails to fulfill the specification requirements, such deficiencies shall be reported to the ENGINEER and the CONTRACTOR immediately. Preliminary written field reports of all tests taken and observation results shall be given to the CONTRACTOR immediately after they are performed. Field reports shall be forwarded to the ENGINEER no later than 1 week following the testing.

B. Reports of all tests taken, including failing tests, shall be reported to the ENGINEER, to the OWNER and to the CONTRACTOR no later than 1 week following the sampling. Density test results will be given in writing at the time the testing occurs.

C. Testing of Hot Mix Asphalt Pavement shall be performed in accordance with Schedule For Minimum Materials Sampling And Testing below. The tests shall be performed under the general supervision of and signed by a Professional ENGINEER registered in the State of Colorado. Laboratories shall be inspected by either AASHTO or accredited A2LA or equivalent in the elements listed below. Technicians taking samples and conducting compaction tests must have a LABCAT Level A certification or equivalent. Technicians conducting tests of asphalt content and gradation must have a LABCAT Level B certification or equivalent. Technicians performing voids testing must have a LABCAT Level C certification or equivalent.

SCHEDULE FOR MINIMUM MATERIALS SAMPLING AND TESTING
HOT MIX ASPHALT PAVEMENT

Test	Standard*	Minimum Frequency
Sampling	AASHTO T168, ASTM D 979 and ASTM D3665	One test for each day
Density	AASHTO T 166, T 238, T 230 or CP-44, CP-81, CP-82	One test for each 250 lineal feet per Lane
Thickness (Core)	ASTM D3549	One test for each 1000 lineal feet per Lane, 2
Air Voids & VMA	AASHTO T 166 or CP-44 & CP-48	One test for each day
Gradation	AASHTO T 27, T 11 or CP-31A, CP-31B	One test for each day
Hveem/Marshall Stability As Applicable	AASHTO T 245 or CP-L 5106	One test for each day
Asphalt (AC) Content	AASHTO T 164 or CP-L 5120 or other methods agreed upon between ENGINEER and CONTRACTOR	One test for each day
Maximum Theoretical Specific Gravity (Rice)	AASHTO T 209 or CP-51	One test for each day
Lottman Stripping, TSR & Dry Density	AASHTO T 283 or CP-L 5109, Method B	As requested by the ENGINEER

* Agency may determine the method used (CP vs. AASHTO).

D. Inspectors shall be responsible for checking temperatures of mix in the truck and on pavement, segregation, rolling patterns and other construction means and methods that affect the performance of the pavement system. The CONTRACTOR shall provide assistance in sampling and testing at all facilities and at the job site.

E. The Schedule For Minimum Materials Sampling And Testing is to be considered quality control testing, and is to be performed by the CONTRACTOR at no cost to the OWNER.

END OF SECTION 321216

SECTION 321313 - PORTLAND CEMENT CONCRETE PAVEMENT (PCCP)

PART 1 GENERAL

1.01 DESCRIPTION

A. This work shall consist of a pavement and ancillary structures composed of Portland Cement concrete, with or without reinforcement as specified, constructed on a prepared subgrade in accordance with these specifications and in conformity with the lines, grades, thickness and typical cross section shown on the plans.

B. At the option of the CONTRACTOR, and with the OWNER's concurrence, Portland Cement concrete pavement (PCCP) may be constructed with equipment utilizing stationary side forms or by the use of slip-form paving equipment.

1.02 SUBMITTALS

A. TRIAL MIX DESIGN, PERFORMANCE HISTORY, AND AGGREGATES

1. The CONTRACTOR shall submit a laboratory Trial Mix Design (proportions, sources and strength data), aggregate data, and certifications for cementious materials and admixtures for each class of PCC being placed on the project. The trial mix design test data shall show the mix design material sources and proportions, actual tested slump, air content, unit weight, water/cement ratio, and laboratory 7 and 28-day compressive strength results. The laboratory Trial Mix Design mix proportions must produce at least 5,000 PSI in the 28-day laboratory compressive strength test (minimum 1,000 PSI over-design from the required 4,000 PSI field strength). The laboratory Trial Mix Design shall also achieve a flexural strength of 700 PSI at 28 days.

2. In lieu of laboratory Trial Mix Design data, a Field Performance History shall be accepted as proof of potential for performance provided that it meets the strength criteria set forth in ACI 301 for acceptance of concrete mixes. Each Field Performance History shall establish the mix proportions and sources of all ingredients, and data on actual field strengths, slump, air content, unit weight.

3. Aggregate test data shall include gradations, minus No. 200, sand equivalent, fineness modules, specific gravity, absorption, soundness, LA abrasion per the requirements of 2.04 and 2.05. Should the ASTM C1260 (Potential Alkali Reactivity of Aggregates) be greater than 0.10%, the test shall be repeated using the revised job mix materials to establish that the Potential Alkali Reactivity is less than or equal to 0.10% for the job mix. The CONTRACTOR shall be responsible for the design mix proportions and subsequent adjustments necessary to produce the specified concrete.

4. The CONTRACTOR shall submit a new trial mix design when a change occurs in the mix proportions, source or type of cement, fly ash or aggregate, or failure of field tests to meet the specifications. All Trail Mix Designs or Performance

Histories shall be stamped and dated by a Professional ENGINEER licensed in Colorado. All such Designs or Histories shall be updated or re-conformed each calendar year.

B. CURING COMPOUND AND SEALER

1. The CONTRACTOR shall submit manufacturer's literature that shall include surface preparation, application instructions, recommendations and storage and handling requirements.

2. The CONTRACTOR shall deliver the approved product in factory sealed and numbered container. Store drums in a cool dry area and protected from freezing.

C. APPLICATOR QUALIFICATIONS

1. The applicator must provide prior experience applying specified product or similar products, or have manufacturer's representative on site ensuring that preparation and application are performed correctly.

PART 2 PRODUCTS

2.01 MATERIALS

A. CDOT Class "P" Concrete as per Colorado Department of Transportation Standard Specifications for Road and Bridge Construction, subsection 601.02 and 601.03. The concrete shall be composed of Portland Cement (with or without) fly ash, fine and course aggregates, admixtures, and water. The mix shall be generically known as "MGPEC PCCP Mix".

2.02 PORTLAND CEMENT

A. Portland Cement shall be Type I-II or Type II "medium alkali" (with a maximum alkali content of 0.8%) and shall conform to ASTM C 150.

2.03 FLY ASH

A. Fly ash for concrete shall conform to the requirements of ASTM C 618, Class C or Class F. All chemical requirements of ASTM C 618 Table 1-A shall apply, with the exception of footnote A.

B. The fly ash shall be limited to a maximum of 25 percent by weight of total cementious material, when used with Portland Cement.

2.04 FINE AGGREGATE

A. Fine aggregate shall be hard, durable and uncoated particles of natural or manufactured sand or a combination thereof. It shall be essentially free from frozen material, salt, alkali, vegetable matter or other objectionable material and conform to the requirements of AASHTO M6, including Class A criteria for deleterious substances; or conform to the requirements of ASTM C-33 using Class 4S criteria. The minimum sand equivalent, as tested in accordance with Class 4S criteria. The minimum sand equivalent, as tested in accordance with AASHTO T 176 shall be 80, unless otherwise specified. The fineness modulus shall be between 2.50 and 3.50, unless otherwise approved. ASTM C1260 (Potential Alkali Reactivity of Aggregates) testing is also required.

2.05 COURSE AGGREGATE

A. Course aggregate shall consist of durable particles of gravel, crushed stone or combinations thereof and shall be free from frozen material, salt, alkali, vegetable matter or other objectionable material either free or as an adherent coating, and conform to the requirements of AASHTO M 80 using Class A criteria; or ASTM C-33 using Class 4S criteria. ASTM C1260 (Potential Alkali reactivity of Aggregates) testing is also required.

B. Grading No. 57 or 467 may be used where the pavement thickness exceeds six (6) inches and machine pavers are used. Grading No 67 shall be used when the pavement thickness is six (6) inches or less.

C. Course aggregate shall conform to the grading shown in the following Table

TABLE 2.05
PCCP AGGREGATE GRADATIONS

SIEVE SIZE	No 67 (3/4" Nominal) % Passing	No 57 (1" Nominal) % Passing	No 467 Combined Grading % Passing
2 inch	-	-	100
11/2 inch	-	·100	95 to 100
1 inch	100	95 to 100	-
¾ inch	90 to 100	-	35 to 70
½ inch	-	25 to 60	-
3/8 inch	20 to 55	-	10 to 30
No 4	0 to 10	0 to 10	0 to 5
No 8	0 to 5	0 to 5	-
No 200	-	-	-

2.06 WATER

A. Water for use in PCCP and for curing shall be free from oils, acids, organic matter or other deleterious substances and shall not contain more than 1,000 parts per million of chlorides as CL nor more than 1,000 parts per million of sulfates as SO_4. Potable water shall not require testing, but water from other sources shall be sampled and tested before use in PCPP. Tests shall be made in accordance with AASHTO T 26.

2.07 ADMIXTURES

A. Admixtures for air entertainment, water reducers, or other uses shall be in a liquid state and conform to the requirements of ASTM C 260 or C494.

B. Calcium chloride is not allowed for normal use. It may be allowed (up to 2%) only for a fast track mix (open to traffic less than 24 hours), and only when required rebar or dowels is of non-corrosive materials, upon approval by the OWNER. The OWNER shall consider the possible detrimental effects of the additional calcium to any on-site soil containing soluble sulfates. This may result in the lessening of future alkali reactivity protection.

2.08 PROPORTIONING OF PORTLAND CONCRETE (PCC)

A. The proportions of materials to be used for mix design and field placement shall produce a workable PCC having:

1. A slump of:

- Of 2" to 4" (3"+ 1') for hand pours
- Or 1" to 2" (1.5 + ½") for machine placement

2. With entrained air content of:

- 5.0% to 8.0% (6.5% +/-1.5%) When using size #67 maximum coarse aggregate (3/4" nominal maximum size) in mix, or
- 4.5% to 7.5% (6.0% +/-1.5) when using #57 or #467 maximum coarse aggregate in mix (1.0" or 1.5" nominal maximum size).

B. Minimum total cementious content shall be at least 610 pounds per cubic yard with a maximum of 25 percent ash substitution based on weight and a maximum water to cement ration of 0.45.

C. The Portland Cement concrete mixture shall conform to ACI 211 and 301 criteria. A minimum of 55% course aggregate to total aggregate shall be used in the mix. If the mix contains a minimum or 55% of size #467 course aggregate, then load transfer dowels in transverse joints can be eliminated.

D. The field mix shall achieve a minimum 28-day compressive strength of 4,500 PSI. The field mix shall achieve a flexural strength of 600 PSI at 28 days. Project produced field mixes shall not normally have flexural strength as a spec, but may occasionally be tested to flexural strength. Some projects, by special provision by the OWNER, may require the 600 PSI flexural strength criteria be met on project produced mixes.

2.09 JOINT SEALANT

A. The joint sealant for all sawed longitudinal and transverse joints shall be a Dow Corning 888 or 890SL gray silicone joint sealant or an approved equal. Blocking medium shall be used below proposed joint sealant to keep the final sealant dimensions per the manufactures recommendation, by filling the lower part of the sawn joint reservoir to ensure a nominal sealant depth. The blocking medium shall be a commercial grade non-absorbing bond breaking cord (backer rod) compressible material in the form of an expanded closed-cell polyethylene foam backer rod or non-plastic rope that is compatible with the joint sealant material.

2.10 EXPANSION JOINT

A. The polyethylene expansion joint shall be black, flexible, low density, expanded extruded polyethylene plank formed by the expansion of polyethylene base resin, extruded as a multicellular, closed cell, homogeneous foamed polyethylene. Laminations shall not be permitted.

B. The joint material shall conform to the following physical property requirements:

1. Compression, psi; when tested in accordance with ASTM Designation D 1056 except that compressive strength shall be determined at 10 percent and 80 percent deflection:

Less than 10 psi	at 10 percent deflection
Less than 125 psi	at 80 percent deflection

2. Water absorption when tested in accordance with ASTM Designation C 272 using conditioning procedure 4.1.1 at a temperature of 50° + 3 degrees Centigrade:

Less than 0.5% by volume

3. Density when tested in accordance with ASTM Designation D 1564:

2.6+0.2 pounds per cubic foot

4. Size: Thickness shall be 1 inch, + ½ inch, -0 inch and full depth of the slab. Width shall be sufficient to fill in the joint without laminating to within ½ inch below the finished surface of the PCCP.

a. Regardless of the type of joint filler used, the joint shall be clean and free from all loose material, dirt, dust, grease, oil, or other foreign matter and shall be smooth and surface dry prior to installation of the filler.

2.11 CURING COMPOUND OR COMBINATION CURE/SEALER

A. Allowable curing compound types and specification shall vary depending upon when an expected snow or freeze condition may occur, or when de-icing materials will be soon used.

B. For PCCP placed between April 1 through September 14:

1. For normal PCCP, related vertical and flatwork, white pigmented curing compound conforming to ASTM C-309 Type 2 (white pigmented dye) shall be used unless another method conforming to ACI 308, Chapter 2, is approved by the OWNER. For colored concrete, products must meet ASTM C-309 Type 1 or 1-D.

C. For PCCP placed between September 15 and March 31:

2. A combination cure-sealer shall be used for PCCP and other vertical or flatwork placed during these dates, or when the OWNER predicts an event where they expect to receive snow, freezing conditions and/or the need for use of de-icing materials within 28 days after concrete is placed. Provide adequate texture to surfaces prior to applying the cure-seal, as the solvent based product has a high gloss finish and can pose visual distractions to drivers at night time if applied to smooth concrete surfaces.

D. The combination cure-seal for PCCP, related vertical and flatwork products must meet ASTM C-1315 Type II, Class A (pigmented, non-yellowing). For colored concrete, products must meet ASTM C-1315 Type I, Class A (clear non-yellowing). The compound must be an acrylic copolymer type, non-freezing solvent based, with a minimum of 25% solids content. Compound must be VOC compliant in accordance with EPA 40 CFR Part 59. The final gloss appearance will serve as proof of application.

E. The CONTRACTOR shall use a sealer that when applied according to manufactures recommendations will not adversely affect the skid resistance of the pavement.

PART 3 EXECUTION

3.01 CURING COMPOUND AND CONCRETE SEALER APPLICATION

A. The CONTRACTOR shall be responsible to protect the concrete being cured for the elements, traffic, and vandalism. Those surfaces covered by forms shall be similarly treated after the forms are removed. Inadequate protection by the CONTRATOR shall be cause for suspension of concreting operations and replacement of the affected concrete at no expense to the OWNER.

B. The CONTRACTOR shall examine concrete surfaces that are to receive curing and sealing compound. The CONTRACTOR shall notify OWNER if surfaces are not acceptable. Unacceptable conditions include, but are not limited to, cracked, curled, or spalled concrete surfaces. The CONTRACTOR shall not begin surface preparation or application until unacceptable conditions are corrected.

C. Clean surface of substances that might interfere with penetration or performance of concrete curing and sealing compound. Remove oil, existing curing compounds, laitance, and other substances that could prevent adhesion or penetration of concrete sealers. If necessary, the CONTRACTOR shall clean surface with a concrete cleaner. Protect adjoining work, including sealant bond surfaces, from spillage or blow-over of concrete sealer. Cover adjoining and nearby surfaces comprised of aluminum and glass if there is the possibility of concrete sealer being deposited on surfaces. Cover live plants and grass.

D. Curing and Cure-sealing compound should be applied to freshly placed concrete as soon as the surface water has dissipated and/or immediately upon removal of formwork. Apply compound at uniform coverage rate in accordance with manufacturer's instructions with a low-pressure sprayer. If using perform pavement markings, apply pavement markings first, cover pavement marking, and then apply sealer. Do not dilute curing and sealing compound. Protect horizontal surfaces from traffic until curing and sealing compound has cured.

3.02 EQUIPMENT

A. GENERAL

1. Equipment and tools necessary for handling materials and performing all parts of the work must have adequate capacity and be in good mechanical condition. This equipment shall be on the site, available for inspection and testing, before paving operations are started. All equipment, tools, and machinery shall be maintained in a satisfactory working condition.

2. The CONTRACTOR shall provide equipment of such capacity that the paver shall operate continuously or at a constant rate of production insofar as feasible. In the event that any piece or equipment does not have sufficient capacity to keep pace with the other operations, the OWNER may limit the size of the batch or otherwise limit the rate of production to prevent poor workmanship, overloading of equipment, or frequent delays.

3. Any equipment operating entirely or partially on the pavement, regardless of the age of the pavement, shall be equipped so that only rubber-tired wheels shall come in contact with the pavement.

4. Equipment shall be approved by the OWNER. All equipment and machinery shall be kept in good working order, free of leaks and properly muffled. All taxes, licenses and fees shall have peen paid and proper licenses and permits shall be posted as required by law.

B. VIBRATORS

1. The rate of vibration shall be not les than 3,500 vibrations per minute for surface vibrators and 5,000 vibrations per minute for internal vibrators. For hand vibration, the amplitude of vibration shall be sufficient to be perceptible on the surface of the concrete more that one (1) foot from the vibrating element. The CONTRACTOR shall furnish a tachometer or other suitable device for measuring and indicating the actual frequency of vibrations.

2. Vibrators shall not rest on new pavements or side forms. Power to vibrators shall be so connected that vibration shall cease when the forward or backward motion of the machine is stopped.

C. CONCRETE SAW

1. The CONTRACTOR shall provide sawing equipment adequate in number of units and power to complete the sawing with a diamond edge saw blade or an abrasive wheel to the required dimensions. The CONTRACTOR shall provide at least one standby saw that is also in good working order. An ample supply of saw blades shall be maintained at the site of the work at all times during sawing operation.

D. FORMS

1. Forms shall be of such section and sufficient rigidity, both in the form and in the interlocking connection with the adjoining forms, that springing shall not occur under the weight of the subgrading and paving equipment or from the pressure of the pavement when placed. The CONTRACTOR shall provide sufficient forms so that there shall be no delay in placing the pavement due to lack of forms. Form sections shall be straight, free from warps, bends, indentations, or other defects. Defective forms shall be removed from the work.

E. SLIP-FORMS

1. Slip-form paving equipment shall be equipped with traveling side forms of sufficient dimensions, shape, and strength to support the PCCP laterally for a sufficient length of time during placement to produce pavement of the required cross section.

2. No abrupt changes in longitudinal alignment of the pavement shall be permitted. The horizontal deviation shall not exceed 0.1 foot from the alignment established by the OWNER.

3.03 CONSTRUCTION

A. PREPARATION OF GRADE

1. The subgrade shall be moist at the time of placing PCC. Sprinkling shall be such that mud and pools of water shall not be formed At the time of placing the PCC, the grade shall not be muddy, soft, yielding, or frozen. Subgrades shall be prepared, proof rolled and accepted in accordance with Item 3 Embankments, MGPEC Pavement Design Standards and Construction Specifications.

B. SETTING FORMS

1. Before placing side forms, the underlying material shall be at the proper grade. Side forms shall have full bearing upon the foundation throughout their length and width of the form base and shall be placed to the required grade and alignment of the edge of the finished pavement. They shall be so supported that they shall not deviate vertically at any time more than 1/8 inch from the grade established by the OWNER.

2. The maximum vertical deviation of the top of any individual side form, including joints, shall not exceed 1/8 inch from a 12-foot straightedge, nor shall the inside face vary more that ¼ inch from a 12-foot straightedge. Stake pockets and interlocking devices shall be in such condition that they shall prevent movement of the form.

3. Side forms shall be staked firmly by means of steel stakes at each end of the section and at intermediate points not more that 5 feet apart and shall be so designed that stakes may be driven through the base of the form. Forms shall be

provided with means for locking stakes in position. Side form sections shall be laid with an expansion gap of approximately 1/8 inch. The stakes used in staking side forms shall be of sufficient length so that the side forms shall be held firmly in place. Any lateral movement of forms greater than ¼ inch while supporting moving equipment shall be considered as evidence that the steel stakes do not hold the side forms firmly in place and the CONTRACTOR at his expense shall provide longer stakes. Immediately in advance of placing pavement and after all subgrade operations are completed, side forms shall be trued and maintained to the required line and grade for a distance sufficient to prevent delay in placing the pavement. Side forms shall remain in place until the day after placing the pavement, and in all cases until the edge of the pavement no longer requires the protection of the forms. When allowed, if the CONTRACTOR removes forms prior to seven days, he shall apply curing compound to the exposed surface.

4. Side forms shall be thoroughly cleaned and oiled each time they are used and before pavement is placed against them.

3.04 REINFORCEMENT

A. PCCP shall be reinforced at structure approaches and other locations as shown on the plans or directed by the OWNER, and as specified in this section. Any PCCP that carries heavy traffic loads, 1,000,000 ESALS or greater, shall have smooth dowels inserted in the transverse joints. Also, use if the PCC mix has less that 55% course aggregate to total aggregate.

Pavement Thickness (T)	Tie Bar Size	Dowel Bar Diameter
$T \leq 8$ in.	No. 4	1 in.
8 in. $\leq T \leq$10in.	No. 5	1.25 in.
10 in.$\leq T \leq$14 in	No. 6	1.5 in.

B. Bar reinforcement shall be held accurately and firmly in position during the placing and compacting of the PCCP without sagging by means of supporting devices which shall be left in place. The supports shall be specifically manufactured for the purpose and each support shall be capable of supporting a vertical load of 200 pound. Spacing of tie bars shall typically be 30 inches on center and dowel bars at 24 inches on center, unless otherwise shown on the plans. Vertical placement shall be at mid thickness.

3.03 MIXING

A. All PCCP shall be homogeneous and thoroughly mixed in appearance, and there shall be no lumps or other evidence of undispersed cement. Concrete shall be plant mixed concrete, which meets the requirements of AASHTO M 157.

3.04 PLACING PCCP

A. PCCP shall not be placed on frozen ground, nor shall it be mixed or placed when air temperatures are expected to fall below 35°F unless adequate means are taken to insure the surface of the concrete does not fall below 40°F until the concrete meets a minimum strength of 2,500 psi.

B. The CONTRACTOR shall make adequate advance arrangements for preventing delay in delivery and placing of the PCCP. In interval of more than 45 minutes between placing of any two consecutive batches or loads shall constitute cause for stopping the paving operation. The CONTRACTOR shall make a construction joint at his expense and as directed by the OWNER, in the concrete already placed. Resumption of paving operations must be approved by the OWNER, after adjustments to the operations by the CONTRACTOR.

C. Unless otherwise specified, PCCP shall be placed in 12-foot traffic lane widths separated by construction joints as shown on the plans, or, at the option of the CONTRACTOR and with the OWNER concurrence, the PCCP may be placed monolithically two or more lanes wide without a construction joint, but with a longitudinal weakened plane joint at each traffic lane line. Any PCC showing improper proportions of materials, including water, shall not be used in the pavement and any such unsatisfactory PCCP shall be removed and disposed of by the CONTRACTOR at his expense.

D. The CONTRACTOR shall protect freshly placed PCCP from damage by any cause; see Section 3.09 Curing for additional information. Any damage that had occurred shall be repaired at the CONTRACTOR's expense. Concrete work shall be adequately barricaded in all directions to protect the work. Equipment that damages the subgrade or base shall not be allowed.

E. Expansion joint material shall be protected while depositing fresh PCC.

3.05 SPREADING, COMPACTING, AND SHAPING

A. FORM CONSTRUCTION

1. The PCC shall be distributed uniformly with a mechanical spreader. The placed PCC shall be vibrated, screeded, and tamped by a machine or machines. Any delay in excess of 15 minutes in vibrating, screeding, and tamping shall constitute cause for stopping the placement of PCC until the machines performing such work are again in proper position in the paving train.

2. The screeds shall be adjusted to an elevation slightly above grade so that when properly consolidated and finished, the completed surface of the pavement shall be at the established grade, true to the cross section shown on the plans, and free from porous areas. The tops of the forms or the adjacent pavement and the contact surface of the crawler tracks or wheels shall be kept clean by effective devices attached to the machine. The travel of the machine shall be maintained true without lift, wobble, or other variation tending to affect precision screeding. During each pass of the machine, a roll of PCC shall be maintained ahead of the front screed for the entire width of pavement being placed and except when making an expansion joint, the machine shall not be operated beyond that point where the roll of PCC can be maintained.

3. PCC required to be placed in widths less than a traffic lane may be compacted and shaped by a powered mechanical compacting and shaping machine, supplemented by hand methods as necessary. Where hand compaction is

performed, the tamper shall be constructed of heavy plank which length exceeds the width of pavement by a minimum of 1 foot; shall be covered with a heavy strip of metal for a tamping surface; and shall be stiffened adequately to maintain the required shape during use. For PCC production in excess of 40 cubic yards per hour, and where all compaction is performed by hand methods, not less that two tampers shall be used.

4. The hand tamper shall be used with a combined tamping and longitudinal motionraising it from the side form and dropping it to consolidate the PCC.

   a. A surplus of PCC shall be kept in front of the hand tamper and tamping shall continue until the required cross section is obtained and the mortar flushes slightly to the surface.

   b. Where hand compaction is performed on grades in excess of 5 percent, a light board constructed similar to the heavy tamper shall be used following the heavy tamper or tampers to correct any displacement caused by gravity flow of the PCC.

B. SLIP-FORM CONSTRUCTION

1. The CONTRACTOR shall establish references at reasonable intervals or both sides of the roadways, for the line and grade control of the placing operations. The CONTRACTOR shall furnish, place, maintain, remove and dispose of such supports, wire devices, and materials as may be required to provide continuous line and grade reference controls to the placing machine or paver. The slip-form paver shall be equipped with a control system, which shall automatically sense and simultaneously control the laying or trimming of the materials to the specified longitudinal and lateral grade from both sides of the roadway. The control systems shall be automatically actuated from an independent line, and grade control reference through a system of mechanical sensors or sensor directed devices, which shall maintain the equipment at the proper transverse slope and at the proper elevation to obtain the required thickness and surface. The material placed shall be subject to the smoothness and thickness tolerances.

2. Slip-form paving equipment shall spread, consolidate, screed and float-finish the freshly placed PCC in such a manner that a minimum of finishing with a hand float, as specified herein, shall be required to provide a dense and homogeneous pavement. The PCC shall be distributed uniformly into final position by the slip-form paver without delay.

3. The PCCP, for the full paving width, shall be effectively consolidated by internal vibration with transverse vibrating units or a series or equally spaced longitudinal vibrating units. If a series of longitudinal vibrating units are used, they shall be equally spaced at intervals not to exceed 30 inches, measured center to center.

4. When PCC is being placed adjacent to an existing pavement, that part of the equipment which is supported on the existing pavement shall be equipped with protective pads on crawler tracks or rubber-tired wheels on which the bearing

surface shall be offset to run a sufficient distance from the edge of the pavement to avoid breaking or cracking the pavement edge.

5. At locations inaccessible to slip-form paving equipment, PCCP shall be placed by methods and equipment conforming to the requirements for placing PCC in widths less than a traffic lane as specified in 3.05, A. Locations inaccessible to the slip-form paving equipment shall be finished by the hand float method.

3.06 FINISHING

A. GENERAL

1. Placing of PCC shall cease at such time that finishing operations can not be completed during daylight hours or when sufficient illumination that duplicates daylight is available. Necessary workmen shall remain at work long enough to complete the finishing and curing of the pavement.

B. HAND FLOAT METHOD

1. The surface of the PCCP shall be finished smooth and true to grade with two wooden or magnesium floats 8 feet long, 1 inch thick and 4 inches wide, rigidly ribbed and with adjusting screws between the rib and float bars at not more that 2 foot center, to ensure a true and flat surface on the under side at all times. Each float shall be operated from the side of the pavement and the float shall be parallel with the centerline of the pavement. The edge of the float shall be used to cut down all high areas, and the material so removed shall be floated into the depressions until a true surface is obtained. Each successive passage of the float shall just lap the previous path. Upon completion of the passage, the float shall be brought back and the overlap between the two passages smoothed.

2. The floats shall be operated as far back of the tamping machine as the PCC remains workable and the number of passes shall be sufficient to remove all perceptible inequities.

C. EDGING

1. After the preliminary finishing has been completed, the edges of an initial pavement lane shall be rounded to a ½ inch radius. Transverse construction joints, expansion joints, and joints adjacent to an existing pavement shall be rounded to a ¼ inch radius.

D. TEXTURE

1. In advance of the curing operations, the pavement shall be textured with a drag strip of burlap or other device, which shall produce scoring parallel to the centerline. The burlap drag shall consist of one or more pieces of burlap fastened to a cross member riding on the subgrade or side forms by means of wheels or skids to form a continuous strip of burlap the full width of the pavement. Drags shall be maintained clean and free from encrusted mortar. Drags that cannot be cleaned shall be discarded and new drags substituted. Complete pavement that,

in the opinion of the OWNER, is found to have a surface texture that does not provide satisfactory skid resistance shall be ground or scored by abrasive means to provide a surface texture satisfactory to the OWNER, at the CONTRACTOR's expense.

E. TINING

1. The OWNER may require tining to form groves on The PCCP surface, as shown on design plans. In advance of curing operations, the PCCP shall be first textured with a drag strip of burlap or similar device. If the OWNER requires, tining with a mechanical spring steel tine device, will be required and which shall form grooves as shown on the plans. In lieu of Plan details, the tines shall be rectangular in cross section, 3/32 to 1/8 inch wide, and 4 to 5 inches long. Tines shall be spaced 3/4 inches center to center and be of sufficient thickness and resilience to result in grooves 1/8 to 1/4 inch deep in the finished PCCP. The speed of the tine machine shall be slow enough so that the tines shall penetrate the surface to the desired depth, yet fast enough so, the machine can keep up with paving operations. A 1-inch gap shall be left between each tine strip to prevent overlapping the tined surface and producing a weak surface area.

3.07 JOINTS

A. GENERAL

1. Joints in pavement shall be designated as longitudinal and transverse construction joints, transverse expansion joints, and longitudinal and traverse contraction joints, and shall be constructed as shown on the plans and in accordance with the following provisions.

2. Construction joints are those made by placing fresh PCC against hardened PCCP at planned locations. PCCP on both sides of longitudinal construction joints shall be connected with tie bars as shown on the plans.

3. All transverse joints shall be constructed at the angle to the centerline of the pavement shown on the plans and the faces of all joints both transverse and longitudinal shall be perpendicular to the surface of the pavement.

4. All sawed joints shall be clean, free of all foreign material, and sealed after completion of shoulder work and prior to acceptance of the project or portion of the work.

5. Tie bars shall be required for all longitudinal joints and shall be placed as specified on the plans.

B. JOINT LAYOUT

1. The CONTRACTOR shall propose a joint layout for approval by the OWNER. Panel width should match lane width, with joints on lane lines whenever possible. Recommended panel width is 12 feet, and maximum panel width should be 15 feet.

C. WEAKENED PLANE JOINTS

1. Sawed TRANSVERSE weakened plane joints shall be formed by cutting the pavement with a power driven saw at the locations shown on the plans. The grooves for transverse weakened plane joints shall be saw cut to the dimensions shown on the plans for "transverse weakened plane joint." (Recommended saw depth of 1/3 the pavement thickness for joints with steel reinforcement, and 1/4 the pavement thickness for joints without steel reinforcement.)

2. Sawed LONGITUDINAL weakened plane joints shall be formed by cutting the pavement with a power driven saw at all lane lines. The grooves for longitudinal weakened plane joints shall be saw cut to the dimensions shown on the plans for "longitudinal weakened plane joint".

3. The sawed joint shall go through the pavement edge at full depth of cut. The time of sawing shall be determined by the CONTRACTOR to prevent uncontrolled cracking and raveling from the sawing. Every other planned transverse weakened plane joint in the initial lane of PCCP and the first joint immediately after the transverse joint shall be sawed within 24 hours after the PCCP has been placed, unless otherwise authorized by the OWNER. The remaining longitudinal and transverse weakened plane joints shall be saw cut before uncontrolled cracking occurs and within 48 hours and before permitting the CONTRACTOR's traffic or public traffic to use the pavement.

4. In succeeding lanes of the PCCP, transverse joints opposite those, which have opened in the initial lane, shall be sawed within 24 hours after the PC has been placed.

5. No transverse sawing shall be done within 5 feet of transverse uncontrolled cracks. If a transverse uncontrolled crack falls within 5 feet of the location of a proposed sawed joint, the sawed joint shall be omitted. Non-conforming sawed joints shall not be paid for. Cracks shall be routed and sealed at the CONTRACTOR's expense.

6. When the pavement is cured by means of a curing seal, all portions of the seal which have been disturbed by sawing operations shall be restored by spraying the areas with additional approved curing seal.

D. JOINT SEALING

1. Sawed longitudinal, transverse weakened plane joints and transverse cracks shall be sealed with joint sealant material meeting the requirements of Section 2.09, Joint Sealant. The joints shall conform to the details shown on the plans. After saw cutting, the joint shall be cleaned by low-pressure water flushing to remove the slurry. Prior to sealing, the joint shall be cleaned per joint sealant manufacturer's recommendations. Immediately prior to sealing, joints shall be blown using a minimum air pressure of 100 psi. All joints shall be clean and dry prior to sealing. Prior to placing the joint sealant, a commercial quality non-absorptive blocking medium or bond breaker cord (backer rod) shall be placed in

the joint to the depth shown on the plans. The silicone joint sealant material shal then be installed in accordance with the manufacturer's recommendations.

2. It shall be the CONTRACTOR's responsibility to prevent spills or excess sealant material from coming in contact with the horizontal surface of the PCCP on either side of the joint. Any such spills or excess material shall be removed to satisfaction of the OWNER at the CONTRACTOR's expense.

3. Silicone joint sealant shall be tested with the field adhesion test as follows:

These simple screening procedures may help detect application problems such as improper cleaning, use of improper primer, poor primer application, or improper joint configuration.

- Make a knife cut horizontally from one side of the joint to the other.
- Make two vertical cuts (from the horizontal cut) approximately 3 inches (76.2 mm) long, at both sides of the joint.
- Place a 1-inch (25.4mm) mark on the sealant tab.
- Grasp the 2-inch (50.8mm) piece of sealant firmly just beyond the 1-inch (25.4mm) mark and pull at a 90° angle.
- If dissimilar substrates are being sealed, check the adhesion of sealant to each substrate separately. This is accomplished by extending the vertical cut along one side of the joint, checking adhesion to the opposite side and then repeating for the other surface.
- The adhesion test is considered passing when 1 inch (25.4mm) of sealant is elongated to 4 inches (101.6mm) without bond loss.

E. TRANSVERSE EXPANSION JOINTS

1. Transverse expansion joints shall be formed at structure approaches as shown on the plans and as specified herein. Transverse expansion joints shall be formed by means of joint filler strips meeting the requirements of Section 2.10, Expansion Joint. The joint strips shall be firmly supported in position by metal holders and end supports. The supports shall be held firmly in position and shall remain in place after completion of the pavement.

2. The metal holders shall be fabricated of sheet steel not less than 16 gage. They shall be in the form of a deep channel, extending down on both sides of the joint strip to a depth of not less than 3/8 inch. They shall be slotted and cut away as necessary, to allow the PCC to make contact with the joint strip at close intervals. The ends of the holders shall be spread to admit the end supports.

3. During placing and compacting the PCC, the joint holder and end supporters shall be secured as to ensure against movement of the joint strip and to keep the top edge of the strip approximately ½ inch below the surface of the finished pavement. After the PCCP has been placed and consolidation completed, the metal holder may be removed and a suitable shallow metal channel substituted which shall fit snugly over the top edge of the joint strip and shall remain there until the joint is edged.

4. Filler shall extend ½ inch less than full width of the PCCP being placed. After the side forms have been removed, any PCC, which has flowed around the ends of the joint filler, shall be removed.

5. The new PCCP shall be saw cut at locations as shown on the plans for placement of the polyethylene joint filler. Forming of the open joint and subsequently removing the forms to allow placement of the joint filler shall not be permitted. The joint shall be saw cut full width of the PCCP and full depth and shall be 1 inch wide with a tolerance of +0 inches and minus 1/8 inch between joint faces. Remove all PCC between the cuts and thoroughly clean the open joint with compressed air.

6. The polyethylene joint filler shall conform to the requirements of 2.10 Expansion Joint. The height of the polyethylene joint filler shall be such that the top surface of the installed filler is ½ inch below the finished surface of the PCCP. The width of the joint filler shall be 1 inch. Prior in inserting the filler into the open joint, all sides of the material which shall be in contact with PCCP shall be coated with a joint lubricant and adhesive recommended by the manufacturer of the filler material.

3.08 TOLERANCES

A. SMOOTHNESS

1. Upon completion of the pavement, any points that are high in excess of the 10' straightedge tolerance of ¼-inch maximum longitudinal and ¼-inch maximum transverse shall be removed by diamond grinding.

2. The profile index requirements herein shall not apply to the pavement within 30 feet of either end of a bridge or intersection. The finished surface of such pavement shall, however meet all other requirements in this section.

3. Equipment that consistently produces a finished surface having a profile index of 12 inches per mile or less shall be used. Should the profile index exceed the rate of 12 inches per mile, for reasons other that to match perimeter condition, the paving operations shall be discontinued until other means and equipment are proposed for trial by the CONTRACTOR and are approved by the OWNER. Such revised methods and equipment shall be discontinued if they do not produce a finished surface having a profile index of 12 inches per mile or less per Colorado Procedure CP 64 (0.2 inches Blanking Band) per Section 12.4.

4. In addition to the requirement for average profile index, all areas representing high points having deviations in excess of 0.5 inch in 25 ft. as defined in the following paragraph, shall be reduced by abrasive means until such deviations as indicated by return of the profilograph do not exceed of 0.5 inch and 25 ft. shall be determined by measurement of the profilograph.

5. After grinding has been completed to reduce all individual deviations in excess of 0.5 inch as provided in the above paragraph, additional grinding shall be

performed if necessary to reduce the profile index, as measured by the profilograph to 12 inches per mile, or less, in any 0.1 mile section along any line parallel to the edge of the pavement.

6. Additional grinding shall be performed as necessary to extend the ground area laterally to the nearest lane line or edge of pavement and longitudinally so that the grinding begins and ends at lines perpendicular to the pavement centerline within any one ground area. It is the intent of this requirement that all ground areas be neat rectangular areas of uniform texture.

7. The CONTRACTOR shall be responsible for providing a CDOT certified profilograph and trained operator.

B. PAVEMENT THICKNESS

1. It is the intent of the specifications that PCCP shall be constructed in accordance with the thickness requirements of the plans and specifications. Thickness measurements shall be made at the rate of not less than one measurement for each 500 linear feet of traffic lane, or fraction thereof, of pavement placed. Insufficient thickness from 0.20 to 0.60 inches shall be price reduced for each lot of 500-foot lane feet, in accordance to Section 3.13, Price Reductions.

2. If the pavement is deficient by more that 0.6 inches, additional measurements shall be made to identify the limits of the deficiency. The area in question shall be entirely bordered by longitudinal or transverse joints and pavement edges. After the deficient area is identified, the OWNER shall determine as to whether the PCCP shall remain in place or be removed in accordance with the following procedures:

   a. The CONTRACTOR shall, at his expense, remove and replace the PCCP in the area determined by the OWNER with new PCCP meeting the thickness requirements. Subgrade shall be lowered as necessary to meet full thickness.

   b. The CONTRACTOR shall leave the deficient section of PCCP in place without payment, if it meets all of the other requirements as determined by the OWNER.

C. COMPRESSIVE STRENGTH

1. When a compressive strength test falls below the specified 28-day compressive strength, a determination shall be made by the OWNER to whether the PCCP shall be removed and replaced or allowed to remain in place. This determination shall be based on ACI 301, sections 1.6.5.3, 1.6.6.3, and 1.6.7.3, except that the Testing Agency shall be mutually agreed upon by the OWNER and the CONTRACTOR. If after review, the OWNER allows the PCCP to remain in place a price reductions for each lot of 100 cubic yards shall be in accordance Section 3.13, Price Reductions.

D. REPAIR OF DEFECTIVE CONCRETE PAVEMENT

1. Defective concrete pavement shall be repaired or replaced at the CONTRACTOR's expense. This corrective work shall be accomplished prior to joint sealing and final smoothness measurement. Defective concrete pavement replaced after smoothness measurement shall be retested for final smoothness. The CONTRACTOR's corrective work plan shall be approved prior to performing the work. When necessary, the extent of defects shall be determined by the inspection of cores at the CONTRACTOR's expense. The following conditions shall require removal and replacement:
   a. Pavement slabs containing multiple cracks through the full depth of the slab that separate the slab into three or more parts.
   b. Pavement slabs having a single diagonal crack that intersects the longitudinal and transverse joint.
   c. Pavement slabs containing honeycombed areas.
   d. Pavement slabs containing more than one void greater in depth than half the pavement thickness.
   e. Pavement slabs containing a cumulative surface area of moderate and severe voids, as defined below, greater than one percent of the slab's total area.
   f. Pavement slabs containing 20 or more severe voids as defined below.

2. A void is a defect in a slab caused by air pockets, clay balls, clumps of sand, cement or reinforcing fibers caused in incomplete mixing or foreign materials such as cans, rags, bottles etc. The defect shall be considered a moderate void when the largest dimension, height, width, or length is at least 13 mm (1/2 inch) but no more than 50 mm (2 inch). The defect shall be considered a severe void when the largest dimension, height, width or length is greater than 50 mm (2inch). The defect shall be considered an extreme void when the largest dimension, height, width or length is greater that 1/3 the thickness of the pavement.

3. Concrete slabs that are determined by the OWNER to be acceptable if repaired shall have voids filled using materials from an Approved Products List and methods approved by the OWNER.

4. Cracks penetrating the full depth of the pavement shall be routed and sealed. The top of the crack shall be routed to a minimum depth of 20 mm (3/4 inch) and a width not less that 10mm (3/8 inch) or more than 15 mm (5/8 inch) by means of a grooving machine. The routing machine shall be capable of following the path of the crack and widening the top of the crack to the required section without spalling or damaging the concrete. Loose and fractured concrete shall be removed and the crack shall be cleaned and sealed in accordance with section 3.07.D, Joint Sealing. Cracks that are tight and do not penetrate the full depth of the pavement shall be left undisturbed.

3.09 CURING

A. GENERAL

1. All PCCP must be cured and protected from weather and temperature extremes for a minimum of 72 hours.

B. Curing Compound Method

1. The curing compound shall not be applied until patching and surface finishing, except grinding, has been completed. When deemed necessary by the OWNER during periods of hot weather, fogging of the PCCP with water shall be continued after curing compound is applied until a cooling effect is no longer required.

2. Refer to section 2.11, Curing Compound Or Combination Cure/Sealer for selecting a curing compound to be used between April 1st and September 14th, and selecting a combination cure-seal to be used between September 15th and March 31st, or when use of de-icing materials may likely be applied to PCCP less than 28 days old.

3. All PCCP, except colored PCCP, shall be cured with a white pigmented curing compound applied uniformly with adequate coverage according to manufacturer's recommendations. A fugitive dye or clear curing compound shall be used on colored PCCP. Surfaces, which are exposed to the air, shall be sprayed uniformly to obtain total coverage of the PCCP surfaces. The rate of application for an approved curing compound shall be 1 gallon per 150 square feet ±50 square feet. Curing compound must be uniformly applied, resulting in a visually well-painted surface. Power operated spraying equipment for application of curing compound shall be equipped with an operational pressure gage and means of controlling the pressure.

4. The compound shall be applied immediately after the moisture sheen begins to disappear from the surface, but before any drying shrinkage or craze cracks begin to appear. In the event of any delay in the application of curing compound, application of water with an atomizing nozzle shall be started immediately and shall be continued until the application of the compound is resumed or started. Evaporation inhibitors shall be a mono-molecular film here in addition to the water, like Dayton J-74 or Master Builders Confilm. These are poly vinyl alcohols that do not change the water cement ratio, but by being hydrophobic they drive the water back into the slab and thus inhibit evaporation. Should the film of compound be damaged from any causes before the expiration of 72 hours after the PCCP is placed, the damaged portion shall be repaired immediately with additional compound.

3.10 PROTECTION OF PAVEMENT DURING INCLEMENT WEATHER

A. The CONTRACTOR shall protect newly placed PCCP from any detrimental precipitation and weather. The CONTRACTOR may choose to proceed with PCCP placement during marginal weather, but they do so at their own risk of later removal. Concrete damaged due to changing weather conditions shall be repaired, or removed and replaced as determined by the OWNER, and at the CONTRACTOR's expense.

B. Placing PCCP shall be stopped and newly placed PCCP to be covered before:

1. The quantity of RAINFALL is sufficient to cause a flow or wash the surface.
2. The quantity of PRECIPITATION or FOREIGN SUBSTANCES might impact the strength or durability.
3. The quantity of wind could cause surface drying leading to cracking

C. The CONTRACTOR must protect pavement from damage due to the above-mentioned causes. Failure to properly protect concrete may constitute cause for removal and replacement of defective pavement. Plastic film is preferred for surface and edge protection.

D. Rain damage varies considerably depending on rainfall intensity, duration, and protective measures taken by the CONTRACTOR. Covering and side forms placed by the CONTRACTOR afford sufficient protection to unhardened concrete in some cases. In other cases, surface and edges may erode to such a degree that removal and replacement of the slab is the only solution. If the concrete is still plastic and is damaged by rain, the damaged surface may be dug out and replaced with fresh concrete. It is not acceptable to refinish the damaged surface without first removing the damaged concrete.

E. All PCCP shall be protected from freezing or frost for a period of 5 days after placing. The temperature of the surface of the concrete shall not be allowed to drop below 40°F during this period. The CONTRACTOR shall show, by evidence of using recording internal or surface mounted external concrete temperature devices, that all slab top, side, and face surfaces were cured at or above the minimum 40°F during the 5-day period. Any evidence of curing below this temperature, such as a "leaf pattern" on surfaces or inside cracks, shall be investigated. The concrete compressive strength must be deemed to be at or above the required minimum 2,500 psi strength prior to opening to traffic, and able to reach the required 28 day strength.

3.11 DAMAGE ANALYSIS

A. All PCCP, which the OWNER suspects might be damaged, shall be assessed for suitability by mutual agreement with the CONTRACTOR. CONTRACTOR shall provide OWNER with approved experts in assessment, and shall replace or repair any damaged PCCP at the CONTRACTOR's expense, as approved by the OWNER.

B. In the event that the CONTRACTOR does not agree with removal and replacement, the CONTRACTOR may elect, at his own expense, to send cores to an experienced petrographer for analysis of the concrete's water-cement ratio, air content, air void spacing factor, and general appearance using ASTM C 856. The number of cores and the core locations should be approved by the OWNER prior to submittal. A petrographic examination of cores taken from the concrete does provide helpful information to determine the extent of the damage. A petrographer can indicate how deep any damage extends and provide recommendations for effective repair, such as diamond grinding. If a general petrographic evaluation does not answer all of the questions on the concrete's durability to the satisfaction of the OWNER, a more detailed analysis of the air-void system using ASTM 0457 should be performed. Surface scaling tests should also be conducted in accordance with ASTM C 672. Rain damaged concrete shall need to be removed if it is determined by the OWNER to be non-durable in terms of abrasion, skid resistance, or freezing & thawing.

3.12 OPEN TO TRAFFIC

A. No traffic or CONTRACTOR's equipment shall be permitted on the pavement until the strength requirement has been reached. Criteria shall be at least 2,500 psi for PCCP over 8" thick, and the design field strength (4000psi) for PCCP less than or equal to 8" thick.

3.13 PRICE REDUCTIONS

A. Insufficient and compressive strength shall be subjected to removal and replacement or price reduction at the following rates:

TABLE 3.13A

Average Thickness Deficiency (inches)	Price Reduction*
0.00 to 0.20	0
0.21 to 0.30	10
0.31 to 0.40	15
0.41 to 0.50	25
0.51 to 0.60	30

* Price reduction is equal to the full price of the entire pavement system. Price reductions reflect the reduction in lifespan of the facility resulting from the deficiency.

TABLE 3.13B

Percent of specified 28 – Day Compressive Strength	PRICE Reduction **
100	0
95	0
94	3
93	6
92	9
91	12
90	15
89	18
88	21
87	24
86	27
85	30

* Strengths less than 85 percent of the design strength shall be removed and replaced.
** Price reduction is equal to the full price of the entire pavement system. Price reductions reflect the reduction in lifespan of the facility resulting from the deficiency.

3.14 MEASUREMENT

A. PCCP

1. The number of square yards of PCCP to be measured for payment shall be determined using filed quantities mutually agreed upon by the CONTRACTOR and the OWNER.

B. Expansion Joints

1. The "expansion joint (saw cut)" shall be measured for payment by the linear foot.

C. Sawed Transverse and Longitudinal Weakened Plane Joints

1. The length of sawed transverse and longitudinal weakened plane joints shall be measured for payment by the linear foot of the transverse and longitudinal weakened plane joints actually sawed. Volunteer or uncontrolled cracks shall not be included in the length of transverse or longitudinal weakened plane joints measured for payment. Joint Sealant and bond breaker placed in sawed transverse and longitudinal weakened plane joints shall not be measured or paid for directly, and shall be included in the contract unit bid per linear foot for the joint.

D. Tie Bars and Dowels

1. Tie bars and dowels shall not be measured or paid for separately.

3.15 TESTING AND INSPECTION

A. Testing of Portland Concrete pavement shall be performed in accordance with Table 3.15 by the CONTRACTOR at no cost to the OWNER.

TABLE 3.15
SCHEDULE FOR MINIMUM MATERIALS SAMPLING AND TESTING
FOR ITEM 11-PORTLAND CEMEMNT CONCRETE PAVEMENT

Test Type	Test Standard	Minimum Frequency of Tests
Compressive Strength	ASTM C 39	One set* per 100 cubic yards (minimum of one/day)
Air Content	ASTM C 231	One test per each first three trucks, then one for every five trucks**
Slump	ASTM C 143	One test per each first three trucks, then one for every five trucks**
Temperature	ASTM C 1064	One test with every air and slump test.
Thickness	ASTM C 174	One core test per 500 linear feet per lane.
Unit Weight	ASTM C 138	One per air content test
Joint Sealant Pull Test	See Section 3.07D, Joint Sealing	One per 1,000 lf of joints

* One set consists of at least 4 cylinders
** If out of specification test all until in specification.

B. Field-testing shall be performed by ACI certified field testers. The OWNER shall determine who is responsible for performing QA testing, and the CONTRACTOR shall be responsible for QC testing.

3.16 PAYMENT

A. The accepted quantities measured in Section 3.14 Measurement shall be paid for at the contact unit price bid as follows:

B. Bid Item for Portland Cement Concrete Pavement (PCCP) shall be full compensation per square yard placed. All PCCP shall include PC, water, mixing, hauling, placing, tie bars, dowels, finishing, curing pavement protection, QC testing, repairs and incidentals necessary for doing all the work shown on the plans or established by the OWNER.

1. There shall be no direct payment for growing the PCCP surface with tines and all work involved shall be considered included in the contract bid prices for other items of work.

2. There shall be no direct payment for Profilograph measurement to the PCCP surface, all work involved shall be included in the contract bid prices as shown on the plans or when established by the OWNER.

3. All grinding and re-grooving of the PCCP that is necessary to meet the requirements of these specifications shall be accomplished solely at the CONTRACTOR's expense and no additional payment shall be allowed.

C. Bid Item for Expansion Joint (Saw Cut) shall be full compensation by the linear foot placed. All expansion Joint shall include sawing, cleaning, furnishing, expansion material, placing the bond breaker, and the joint sealer,

D. Bid Item for Sawed transverse and Longitudinal Weakened Plane Joints shall be full compensation by the linear foot placed. All Expansion Joint shall include sawing, cleaning, furnishing, placing the bond breaker, and the joint sealer,

Item	Description	Payment
1.	Portland Cement Concrete Pavement (PCCP)	$/sq. yard
2.	Expansion Joint (Saw Cut)	$/linear ft.
3.	Sawed Transverse and Longitudinal Weakened Plan Joints	$/linear ft.

3.17 WHITETOPPING

A. General

1. Whitetopping is a concrete overlay of existing asphalt pavement. When loaded by vehicles, the new concrete overlay behaves like a new concrete pavement on a firm base course. All the previous specification shall apply including the following. Concrete for whitetopping (concrete overlay) shall conform to the requirements for Class "P" concrete as specified in subsection 2.01, Materials with the following exception: Class P pavement for whitetopping shall contain a minimum of 55% course aggregate. Course aggregate shall be AASHTO M43 No. 67.

B. Asphalt Surface Preparation

1. Mill existing asphalt surface to the depth indicated on the Project Drawings using self-propelled milling equipment having an effective means for preventing dust from escaping into the air. The recommended minimum depth of asphalt base remaining after milling is 4". All material removed becomes the property of the CONTRACTOR unless otherwise designated in the Contract Documents.

2. Remove all loose foreign material from asphalt surface with compressed air, by brooming or other methods immediately prior to placing concrete.

3. Deposit concrete directly onto the prepared asphalt that has been physically saturated with water prior to placement. Other methods of conveying the concrete may be used when accepted by the OWNER.

4. Do not place concrete when the asphalt surface temperature is less than 10°C (50°F).

C. Whitetopping Placement

1. Whitetopping shall be placed in accordance with the following additional requirements:

a. The CONTRACTOR shall receive delivery tickets, monitor and maintain yield to achieve the minimum thickness. Copy of tickets to be delivered to the ENGINEER at the end of each days paving.

b. Surface Milling: The milled surface shall be full-width. No concrete shall be placed on un-milled surface or on exposed base. If based is exposed, CONTRACTOR shall submit a plan to correct the exposed surface and proceed with corrective action after the ENGINEER approves the plan.

c. Surface Cleaning: The initial cleaning of the milled surface shall be accomplished by brooming the surface using a mechanical broom surface shall be further cleaned by the use of pressure cleaning to insure that all dust, grit, and foreign material is removed from the surface. The area shall be protected as necessary to prevent contamination until placement. When areas are not adequately protected, additional cleaning will be required prior to placement of concrete overlay as directed by the ENGINEER. Milled surface will not be open to traffic.

d. Moisture Coat: After the surface has been milled, cleaned, and accepted, and prior to the concrete overlay placement, a light wetting of the surface will be required. This should be accomplished just ahead of the concrete overlay placement by spraying with a fine mist without allowing ponding on the surface. No asphalt surface shall be allowed to dry prior to concrete overlay placement.

D. Exclusions to Normal PCCP

1. Dowel bars will not be required for whitetopping (concrete overlay)

E. Method of Measurement

1. Asphalt Surface Preparation: The area of asphalt surface preparation shall be computed by the OWNER in square yards from measurements of the locations.

2. PCCP: The area of concrete overlay placement shall be computed by the OWNER in square yards from measurements of the locations.

3.18 BASIS OF PAYMENT

A. Asphalt Surface Preparation

1. Payment shall be made at the contact price per Square yard for asphalt surface preparation. This shall include full compensation for removal and disposal of material.

B. PCCP - Whitetopping

1. For the concrete overlay constructed, payment shall be made at the contract price per square yard. This shall be full compensation for furnishing all materials, and for proportioning, mixing, delivery of concrete to the paving site, all labor, equipment and materials to place, finish, texture, cure the concrete, and to saw or form and seal the joints, in accordance with the Contract Documents.

C.

Item	Description	Payment
1.	Asphalt Surface Preparation	$/sq. yard
2.	Portland Cement Concrete Pavement-Whitetopping	$/sq. yard

END OF SECTION 321313

SECTION 321613 - CONCRETE CURBS, GUTTERS AND VALLEY GUTTERS

PART 1 GENERAL

1.01 DESCRIPTION

A. This section includes concrete curbs, gutters, and valley gutters.

PART 2 MATERIALS

2.01 REFERENCES

A. Materials and construction methods shall meet the requirements of the CDOT Standard Specifications except as noted herein. Concrete shall meet the requirements of the latest edition of MGPEC Item 11 – PCCP.

2.02 MIX DESIGN

A. All concrete shall be ready mixed concrete. Concrete mix designs consist of selecting the aggregates appropriate for the concrete application, and performing a mix design to determine the properties of the proposed mix. The proposed mix shall be tested in accordance with ACI code requirements. The Contractor must submit one (1) mix design per each concrete type intended for use within the public rights-of-way at the start of each calendar year. No concrete shall be placed prior to the submittal of such mix designs.

B. Concrete for bikeways shall be fiber reinforced. The fibers shall be between ¾ - 1 inch in length and shall be used at a rate of 1 ½ lbs. of fibers per cubic yard of concrete.

2.03 ADDITIVES

A. Admixtures shall meet the requirements of MGPEC Item 11 – PCCP.

B. Additives for concrete, other than those specified in the mix design, shall not be used without prior written approval of the Engineer. When approved for use, chemical admixtures or additives shall comply with applicable ASTM or AASHTO standards. Application of admixtures shall be as per manufacturer's specifications.

C. Coloring for colored concrete shall be accomplished by the addition of an approved commercially pure or synthetic mineral pigment as specified in the contract documents.

PART 3 EXECUTION

3.01 BATCHING

A. Batching of concrete shall conform to CDOT Standard Specifications, Section 601.06.

3.02 MIXING

A. Mixing of concrete shall conform to CDOT Standard Specifications, Section 601.07.

B. The Contractor shall submit to the Owner the ready mix delivery ticket for each load upon request by the Owner indicating the following:

1. Supplier's name and date.
2. Truck number.
3. Project number and location.
4. Concrete class designation.
5. Cubic yards batched.
6. Mix design identification.
7. Type, brand, and amount of cement.
8. Brand and amount of all admixtures.
9. Weights of fine and coarse aggregates.
10. Moisture content of fine and coarse aggregates.
11. Gallons of batch water.
12. Time at which water was added.
13. Elapsed time between when water was added and concrete load was in place.
14. Amounts of initial water and the maximum number of gallons of supplemental water allowed to be added for each truck at the job site.
15. Name of individual authorizing supplemental water.
16. Numerical sequence of delivery by indicating cumulative yardage delivered on each ticket.
17. Provide the following titles with blank space to record discharge time, water-cement ratio, air content, slump, and revolutions.

3.03 LIMITATIONS OF PLACING CONCRETE

A. Refer to MGPEC, Item 11 – PCCP.

3.04 PREPARATION OF SUBGRADE

A. No pea gravel, sand, or other material with less than ten (10%) percent passing the #100 sieve will be allowed as bedding beneath any concrete.

B. If in the opinion of the Owner, the foundation soil is of such character that it will swell or shrink with changes in its moisture content to such an extent that the concrete may be damaged; the soil shall be excavated as directed by the Owner refilled, and compacted with material which meets the requirements of these specifications.

C. The subgrade and/or sub-base shall be brought to a firm and unyielding condition with a uniform density. All soft and yielding material and other portions of the subgrade that will not compact readily when rolled or tamped shall be removed and replaced with suitable material. Concrete shall not be placed on a soft, spongy, frozen, or otherwise unsuitable subgrade. Before placing any concrete, the subgrade shall be tested for conformity with the cross-section shown on the plans, using an approved template, or other approved method. The finished subgrade shall be kept smooth, damp, and compacted prior to placing concrete.

D. *Construction Sequence* – All curb, gutter, valley gutters, shall be constructed after installation of sanitary sewer and storm sewer mains, laterals, and service lines have been installed and properly compacted in accordance with these specifications. Water mains and service lines which cross curb, gutter, attached walks, and driveways shall also be installed and properly compacted prior to installation of said curb, gutter, attached walks, and driveway approaches. Tunneling under curb, gutter, and walks will not be allowed. Water valve boxes and manholes should be adjusted to final grade after installation of curb and gutter and pavement. Electrical services should be installed after water services but prior to installation of curb radii except where previous arrangements for use of conduit have been made and approved by the electric company.

E. Prior to constructing the infrastructure, the area in the construction limits of the project should be brought to ±6" of finished subgrade.

3.05 CONSTRUCTION OF CURBS, GUTTERS, VALLEY GUTTERS AND BIKEWAYS

A. Construction requirements for concrete curbs, and bikeways shall conform to MGPEC Item 11 - PCCP except as modified by these specifications.

B. Where a section of concrete curb and gutter, valley gutter, bikeways, driveway, or alley approach has been disturbed, it shall be removed to a joint, if the joint is situated within five feet of the proposed or existing cut, otherwise a full depth straight line shall be saw cut prior to replacement. Where new construction abuts existing, the work shall be accomplished so that the variance in grade between the old and new work does not exceed one quarter (1/4") inch.

3.06 PLACING CONCRETE

A. *Alignment and Grades* – The alignment and grade elevations of the forms shall be checked by the Contractor immediately ahead of concrete placement and necessary corrections will be made. Any forms that have been disturbed or subgrade that has become unsuitable shall be corrected, forms reset, and rechecked. Any variations in grade and alignment shall be subject to approval by the Owner prior to placing the concrete. Forms shall be oiled prior to placement of concrete. The subgrade shall be moist but not wet prior to placing concrete. After the Owner has approved the forms and subgrade, the concrete shall be deposited on the subgrade to the required depth and width in successive batches and in a continuous operation. The concrete shall be placed as uniformly as possible to minimize the amount of spreading necessary. While being placed, the concrete shall be consolidated or vibrated with suitable tools to prevent the formation of voids or honeycomb. All curves with radii less than two-hundred (200) feet shall be constructed with flexible forms. Curb ramps shall not have a lip between the flowline of the gutter and the ramp approach. The flowline of all new valley gutters shall be a straight-line grade between the gutters at each end of the new valley gutter. The Contractor is required to set finishing screed points at minimum intervals of ten (10') feet along the flowline to control the finished elevation. The screed points shall be removed or driven through the plastic concrete into the subgrade after the concrete has been finished to the required elevation. The use of timber screed points is expressly prohibited. No low spots, which cause water to pond, will be acceptable. Curb and gutter, valley gutters, bikeways, driveways, and alley approaches shall be formed and placed true to line, grade, and cross section. The finished surface must be straight and true to within one-quarter (1/4") inch along any ten-foot length of the new concrete in the vertical or horizontal plane.

B. *Construction Stakes* – The Contractor shall provide all construction stakes required for curbs, gutters, walks, and structures and will furnish all necessary information relating to lines and grades. The Contractor shall be held responsible for the reasonable preservation of all such stakes. The Contractor shall not remove stakes until three working days after placement of the concrete unless approved by the Owner.

C. Concrete shall be thoroughly vibrated or properly consolidated. Care shall be taken in vibrating concrete to bring only a continuous film of mortar to the surface. Vibration shall stop before any segregation of the concrete occurs. Any evidence of lack of consolidation or over-consolidation shall be regarded as sufficient reason for requiring the removal of the section involved and its replacement with new concrete at the Contractor's expense.

3.07 TEST REQUIREMENTS

A. The Contractor shall provide quality control testing. Copies of quality control test results shall be kept in a notebook which is kept onsite at all times by the contractor. Formal test results shall be sent to the Owner on a weekly basis. The Owner may provide acceptance testing associated with placing of concrete, base course subgrade preparation, and trench compaction. The Contractor shall coordinate with the Owner as to when he or she is ready for acceptance tests. The Contractor shall furnish the concrete necessary for acceptance testing. The Contractor shall be responsible for the costs associated with re-testing due to failed acceptance tests.

B. Concrete acceptance testing to be completed by the Owner will follow recognized ASTM, AASHTO or ACI procedures. Concrete that does not meet the acceptance criteria shall be removed and replaced.

3.08 FINISHING CONCRETE

A. *Finishing* – After the concrete has been placed and consolidated in the forms, it shall be finished. A wood float shall be used to bring the surface of the concrete to its final form; excessive working of the surface will not be permitted. The Contractor is cautioned against the use of a steel trowel for floating the finished concrete. This practice can excessively seal and trap moisture beneath the surface. Premature surface spalling may occur. Repair of such spalling will be the responsibility of the Contractor/Owner. The final texture of all exposed surfaces shall be obtained by light brooming. For bikeways, and driveway, broom the surface in the direction perpendicular to the main traffic flow. For all concrete surfaces that are designed to carry storm water; such as , curbs and gutters, valley gutters, and crosspans, broom the surface in the direction of flow. After completion of brooming and before the concrete has taken its initial set, all edges in contact with the forms shall be tooled with an edger having a one-quarter (1/4") inch radius. No dusting or topping of the surface or sprinkling with water to facilitate finishing will be permitted.

B. *Remove Forms* – Face forms for concrete curb may be removed for finishing curb face and fillets as soon as the concrete will retain its shape. Do not remove the back forms for concrete curbs until the concrete has been in place for at least six (6) hours.

3.09 JOINTS

A. *Transverse Joints* – Transverse joints (contraction joints) shall be located at intervals of ten (10') feet in curbs, gutters, and crosspans. When combination curb, gutter, and walk are used, the joint shall be continuous through all three elements. Joint depth shall be a minimum of one-third (1/3) the thickness of the concrete. Joint width shall be a maximum of one-quarter (1/4") inch wide except at expansion joints. For tooled joints, the edges adjacent to joint shall be rounded with an edger of one-quarter (1/4") inch radius. Joints for Bikeways shall be removable plastic dummy joint strips or saw-cut at ten (10') feet intervals, tooled joints shall not be allowed on any bikeway except at expansion joints. Saw cutting of joints shall be performed as soon as the concrete surface is hard enough to allow the sawing operation without otherwise marring the concrete surface, prior to any development of shrinkage cracks. Saw-cutting shall proceed around the clock if necessary to meet these requirements.

B. *Expansion Joints* – Expansion joints shall be required as shown in the details. Expansion joints, where required, shall be filled with one-half (1/2") inch thick full depth, preformed expansion joint material as per CDOT Standard Specifications, Subsection 705.01. Expansion joint material must be set vertical and with the top edge 3/4-inch below the finished surface. The joint shall be edged with a one-quarter (1/4") inch radius edging tool. All expansion joints shall be caulked as per detail.

3.10 CURING

A. Curing of the concrete shall meet the requirements of MGPEC, Item 11 –PCCP. Concrete shall be cured by protecting it against loss of moisture, rapid temperature change, freezing, rain or hail, flowing water, and mechanical injury. The edge of concrete exposed by the removal of forms shall be protected immediately to provide these surfaces with continuous curing treatment. The coating shall be protected against marring for a period of at least five (5) days after application. Any coating marred, or otherwise disturbed, shall be given an additional coating. It shall be the Contractor's responsibility to protect the concrete being cured from the elements, traffic, and vandalism. The Contractor shall have the equipment needed for adequate curing available before commencing concrete placement. Inadequate protection by the Contractor shall be cause for suspension of concreting operations and damaged concrete shall be removed and replaced at the Contractor/Owner's expense.

3.11 DRIVEWAY INSTALLATIONS

A. All driveways constructed of concrete within public right-of-way shall conform to the requirements of these specifications. When constructed of concrete, the thickness shall not be less than six (6") inches. Driveways abutting a curb, gutter, and drive-over curb and gutter or driveway gutter pan or any cross gutter shall have a strip of non-extruding expansion joint material one-half (1/2") inch thick, conforming to the cross-sections of the driveway, placed between the driveway and the curb or cross gutter so as to provide for the expansion of the concrete driveway.

3.12 PROTECTION/DEFACED/DAMAGED OR DEFECTIVE CONCRETE

A. The Contractor shall be responsible for taking adequate steps to protect concrete placed during precipitous, hot, or cold weather. Limitations on the placing and protecting of concrete shall conform to MGPEC Item 11 – PCCP. Any concrete damaged by precipitation or extreme temperatures or otherwise defective shall be removed and replaced at the Contractor expense.

B. *Cold Weather Concreting* – Cold weather concreting requirements shall start when the ambient temperature is less than 40 degrees F. Concrete shall not be placed when the ambient temperature falls below 35 degrees F. When the ambient temperature drops to or is below 35 degrees F, the contractor shall maintain the concrete surface temperature above 50 degrees F until the concrete reaches a field compressive strength of 4000 psi. It shall be the Contractor's responsibility to determine for himself the necessity of undertaking protective measures and the type of protective measures to be used. Once the concrete reaches a field comprehensive strength of 4000 psi, the cold weather protection can be discontinued.

C. *Potential Frost Damage* – The following procedures shall be followed if the temperature of the surface of the concrete falls below 32 degrees F before the concrete reaches the required compressive field strength of 4000 psi.

1. The Contractor will take cores at locations designated by the Owner.
2. The Owner will take immediate possession of the cores and submit the cores to a petrographer for examination in accordance with ASTM C 856.
3. All costs associated with coring, transmittal of cores, petrographic examination, and reports shall be born by the Contractor regardless of the outcome of the petrographic examination.
4. Concrete damaged by the frost as determined by the petrographic examination shall be removed and replaced at the Contractor's expense.

3.13 BACKFILLING

A. Once the concrete has gained sufficient strength to not be damaged, the space adjoining the concrete shall be backfilled with suitable material, properly compacted, and brought flush with the surface of the concrete and adjoining ground surface. In embankments, the backfill shall be level with the top of the concrete and the maximum slope shall be 4:1. Where detached walks occur, the space between the curb and walk shall be backfilled on a straight line from the top of walk to the top of curb not to exceed 4:1 slope.

3.14 OPENING TO TRAFFIC

A. Walks and bikeways shall not be opened to pedestrian or bicycle traffic for at least twenty-four (24) hours after placement; driveways, curb, gutter, and valley gutters shall not be opened to vehicular traffic for at least seven (7) days after placement or until field-cured test cylinder breaks show an average compressive strength of 80% of the 28th day mix design strength psi. Exception will be made where high-early quick strength cement is used. In cases where high-early strength cements are used, the Contractor shall request a variance from the Engineer for opening to traffic sooner. The Contractor shall maintain suitable barricades to comply with the foregoing requirements.

END OF SECTION 321613

Appendix G
Subsurface Soils Investigation - Sample

NOTE: The subsurface soils investigation in this book is strictly to be used as an example as a working copy of a soils report that may be found on real life projects. The report included is provided as a learning tool and not intended to be evaluated, reviewed, or used in any other manner.

REFERENCE: Aims Community College
North Campus Refresh
Greeley, Colorado
Project Manual – Volume 2 or 2
July 29, 2016
Prepared by DLR Group

GEOTECHNICAL EXPLORATION REPORT
AIMS COMMUNITY COLLEGE
NORTH CAMPUS REFRESH
GREELEY, COLORADO
EEC PROJECT NO. 1162022

Prepared for:

Facilities and Operations Department
Aims Community College
Greeley, Colorado 80632

Attn: Mr. Michael Millsapps (michael.millsapps@aims.edu)

Prepared by:

Earth Engineering Consultants, LLC
4396 Greenfield Drive
Windsor, Colorado 80550

EARTH ENGINEERING CONSULTANTS, LLC

April 14, 2016

Facilities and Operations Department
Aims Community College
Greeley, Colorado 80632

Attn: Mr. Michael Millsapps (michael.millsapps@aims.edu)

Re: Geotechnical Exploration Report
Aims Community College
North Campus Refresh
Greeley, Colorado
EEC Project No. 1162022

Mr. Millsapps:

Enclosed, herewith, are the results of the geotechnical subsurface exploration completed by Earth Engineering Consultants, LLC (EEC) personnel for the referenced project. For this exploration, a total of twenty-four (24) soil borings were drilled at the approximate locations indicated on the enclosed *Test Boring Location Diagram*. The borings were extended to depths of approximately 10 to 35 feet below existing site grades. Individual boring logs, including results of laboratory testing, are included as a part of the attached report. This exploration was completed in general accordance with our proposal dated February 24, 2016.

We understand this project involves the construction of new buildings and building additions at the Aims Community College North Campus. Foundation loads for the structures are generally expected to be light to moderate in the range of 100 to 260 kips for column loads and 2 klf for wall loads and floor loads are expected to be light. Maximum column loads for the proposed three-story structure could reach into the 450 kip range depending on final column configuration.

Based on the materials observed within the test borings, expected site grading and anticipated foundation loads in the 250 kip column range, we believe the new buildings could be supported on conventional type spread footings bearing on native materials or a zone of engineered fill material placed and compacted as described within the enclosed report. Heavier column loads would require alternate foundation approaches. In general, it appears the in-situ site materials or similar import fill materials could be used for support of interior slab-on-grades, exterior flatwork, and site pavements. Post-construction movement cannot be eliminated. Additionally

4396 GREENFIELD DRIVE
WINDSOR, COLORADO 80550
(970) 545-3908 FAX (970) 663-0282

the essentially cohesive subsoil materials may be subject to strength loss and instability when wetted. Close monitoring and evaluation during the construction phase should be performed to reduce the potential for post-construction movement.

Geotechnical recommendations concerning design and construction of foundation systems and support of floor slabs and pavements, are provided within the text of the enclosed report. We appreciate the opportunity to be of service to you on this project. If you have any questions concerning the enclosed report, or if we can be of further service to you in any other way, please do not hesitate to contact us.

Very truly yours,
Earth Engineering Consultants, LLC

COLORADO LICENSED
ETHAN PATRICK WIECHERT
40838
4-15-16
PROFESSIONAL ENGINEER

Ethan P. Wiechert, P.E.
Senior Project Engineer

Reviewed by: Lester L. Litton, P.E.
Principal Engineer

EPW/LLL/dla

GEOTECHNICAL EXPLORATION REPORT
AIMS COMMUNITY COLLEGE
NORTH CAMPUS REFRESH
GREELEY, COLORADO
EEC PROJECT NO. 1162022

April 14, 2016

INTRODUCTION

The geotechnical subsurface exploration for the proposed North Campus Refresh Project at Aims Community College in Greeley, Colorado, has been completed. For this exploration, twenty-four (24) soil borings were drilled within the proposed improvement areas at the North Campus site extending to approximate depths of 10 to 35 feet below existing site grades. This exploration was completed in general accordance with our proposal dated February 24, 2016.

Based on information provided to us from Aims Community College (Aims), the proposed development will generally include the construction of new buildings and building additions at the existing Aims North Campus generally at Academy Road and Westview Lane in Greeley. The approximate layout of the existing and new buildings is indicated on the attached boring location diagram. The new buildings will generally be single-story, slab-on-grade metal buildings and will include a new fleet storage building, a new facilities building, and new office and mailroom building along with a new road salt storage facility. The proposed building additions will include a single-story, slab-on-grade welding addition and a three-story, steel frame, slab-on-grade Advanced Engineering Technology (AET) addition.

Foundation loads for the single-story structures and single-story addition are expected to be light with continuous wall loads less than 2 kips per lineal foot and individual column loads less than 100 kips. Foundation column loads in the AET addition may be on the order of 260 kips. Those loads could increase to be in the range of 450 kips depending on the column location/spacing. Floor loads are expected to be light. Paved drive and parking for the development is expected to carry light to moderate traffic consisting of private autos and light trucks with designated areas carrying heavier maintenance vehicles.

Select existing structures on the site will be razed prior to development of the new structures and pavements. Site grading will include removal of existing berms adjacent to the razed building and other areas of the site with maximum cuts and fills on the order of 5 feet. A portion of the existing pavements will also be removed.

The purpose of this report is to describe the subsurface conditions encountered in the test borings, analyze and evaluate the test data and provide geotechnical recommendations concerning design and construction of foundations and support of floor slabs, exterior flatwork, and pavements. Preliminary pavement sections are also provided based on presumed subgrade conditions and assumed vehicle loads.

EXPLORATION AND TESTING PROCEDURES

The test boring locations were established in the field by estimating locations relative to existing site features/structures. The boring locations were adjusted as needed to avoid existing facilities and limit access routes needed by the drilling equipment. A diagram indicating the approximate drilling locations is included with this report. The locations of the borings should be considered accurate only to the degree implied by field measurement methods.

The test borings were completed using a truck mounted, CME-55 drill rig equipped with a hydraulic head employed in drilling and sampling operations. The boreholes were advanced using 4-inch nominal diameter continuous flight augers. Samples of the subsurface materials encountered were obtained using split barrel and California barrel sampling procedures in general accordance with ASTM Specifications D1586 and D3550, respectively.

In the split barrel and California barrel sampling procedures, standard sampling spoons are advanced into the ground by means of a 140-pound hammer falling a distance of 30 inches. The number of blows required to advance the split barrel and California barrel samplers is recorded and is used to estimate the in-situ relative density of cohesionless soils and, to a lesser degree of accuracy, the consistency of cohesive soils and hardness of weathered bedrock. In the California barrel sampling procedure, relatively intact samples are obtained in removable brass liners. All samples obtained in the field were sealed and returned to our laboratory for further examination, classification and testing.

Moisture content tests were completed on each of the recovered samples. Atterberg limits and washed sieve analysis tests were completed on selected samples to evaluate the quantity and plasticity of fines in the subgrades. Washed sieve analysis tests were utilized to develop information on grain size distribution. Swell/consolidation tests were completed to evaluate the potential for the subgrade materials to change volume with variation in moisture and load. Water soluble sulfate tests were completed on selected samples. Results of the outlined tests are indicated on the attached boring logs and summary sheets.

As part of the testing program, all samples were examined in the laboratory and classified in general accordance with the attached General Notes and the Unified Soil Classification System, based on the soil's texture and plasticity. The estimated group symbol for the Unified Soil Classification System is indicated on the boring logs and a brief description of that classification system is included with this report. Classification of the bedrock was based on visual and tactual observation of disturbed samples and auger cuttings. Coring and/or petrographic analysis may reveal other rock types.

SITE AND SUBSURFACE CONDITIONS

The attached boring location diagram shows the existing building layout along with the proposed building additions and new building locations. The existing welding technology building and Mildred E. Hanson building will remain while the other site buildings will be razed as a part of the North Campus Refresh project. Existing pavements, sidewalks and landscaping will also be removed from the proposed improvement areas. Existing site berms will also be regraded to meet new subgrade and grade requirements.

Based on results of the field borings and laboratory testing, subsurface conditions can be generalized as follows. A surficial layer of asphalt pavement with aggregate base or surficial topsoil and vegetation were encountered at the surface at the boring locations. Slightly cohesive to cohesive clayey sand and/or sandy clay subsoils were generally encountered beneath the surfacing and commonly extended to sands and gravels at approximate depths of approximately 5 to 15 feet below existing site grades. Sand and gravel zones were not encountered at all boring locations. Sandstone/siltstone/claystone bedrock was commonly encountered beneath the overburden soils at approximate depths generally in the range of 8 to 20 feet below existing site grades. Bedrock was encountered deeper at a depth of approximately 36 feet in test boring B-4. Bedrock was also not observed at all of the boring locations. The stratification boundaries indicated on the boring logs represent the approximate locations of changes in soil types. In-situ, the transition of materials may be gradual and indistinct.

The upper essentially cohesive soils encountered beneath the surfacing materials varied from medium stiff to very stiff in consistency and exhibited low swell potential and typically low bearing capacity characteristics. The swell potential of these soils are shown on the enclosed swell-consolidation curves included with this report. The underlying sands and gravels were commonly medium dense to dense. The underlying sandstone/siltstone/claystone was brown/grey/rust in color, moderately hard and exhibited moderate to high bearing characteristics.

GROUNDWATER CONDITIONS

Observations were made while drilling and after completion of the borings to detect the presence and depth to hydrostatic groundwater. At the time of drilling, free water was observed at depths of approximately 8 to 19 feet below site grades. The water depths observed at the time of drilling are indicated in the upper right hand corner of the boring logs.

The borings were backfilled immediately after completion; therefore subsequent groundwater measurements were not obtained. Fluctuations in groundwater levels can occur over time depending on variations in hydrologic conditions and other conditions not apparent at the time of this report. Longer term monitoring of water levels in cased wells, which are sealed from the influence of surface water would be required to evaluate fluctuations in groundwater levels at the site. We have typically noted deepest groundwater levels in late winter and shallowest groundwater levels in mid to late summer

ANALYSIS AND RECOMMENDATIONS

Swell – Consolidation Test Results

Swell/consolidation testing is performed to evaluate the swell or collapse potential of soil or bedrock to assist in determining/evaluating foundation, floor slab and pavement design criteria. In the swell/consolidation test, samples obtained directly from the California barrel sampler are placed in a laboratory apparatus and inundated with water under a pre-established load. The swell-index is the resulting amount of swell or collapse under the initial loading condition expressed as a percent of the sample's initial thickness. After the initial monitoring period, additional incremental loads are applied to evaluate swell pressure and/or consolidation response.

For this exploration, we carried out twenty-six (26) swell-consolidation tests throughout the site. The swell index values for the samples analyzed revealed low swell characteristics.

Colorado Association of Geotechnical Engineers (CAGE) uses the following information to provide uniformity in terminology between geotechnical engineers to provide a relative correlation of performance risk to measured swell. "The representative percent swell values are not necessarily measured values; rather, they are a judgment of the swell of the soil and/or bedrock profile likely to influence slab performance." Geotechnical engineers use this information to also evaluate the swell potential risks for foundation performance based on the risk categories.

TABLE I: Recommended Representative Swell Potential Descriptions and Corresponding Slab Performance Risk Categories

Slab Performance Risk Category	Representative Percent Swell (500 psf Surcharge)	Representative Percent Swell (1000 psf Surcharge)
Low	0 to < 3	0 < 2
Moderate	3 to < 5	2 to < 4
High	5 to < 8	4 to < 6
Very High	> 8	> 6

Site Preparation

Based on our understanding of the proposed development, we expect cuts and fills less than 5 feet may be completed to achieve design grades. In addition, two (2) existing buildings will be razed. All existing improvements including foundations, floor slabs, pavements, landscaping (including tree root systems and any dry and desiccated soils associated with the root systems), curb-and-gutter, flatwork/sidewalks and any undocumented fill materials should be removed from the site improvement areas.

Isolated areas of apparent fill materials were identified during our field borings; close observation during construction is advised to identify and evaluate any in-place fills. Fill materials were likely placed to develop site grades and backfill materials would have been placed in prior building areas. After removing all improvements including any undocumented fill materials, and prior to placement of any fill and/or site improvements, we recommend the exposed soils be scarified to a depth of at least 9 inches, adjusted in moisture content to within ±2% of standard Proctor optimum moisture content and compacted to at least 95% of the material's standard Proctor maximum dry density as determined in accordance with ASTM Specification D698.

Fill soils required to develop the building, pavement/flatwork and site subgrades should consist of approved, low-volume-change materials, which are free from organic matter and debris. It is our opinion the on-site sandy lean clay/clayey sand soils or similar import soils could be used as fill in these areas, provided adequate moisture treatment and compaction procedures are followed. Granular structural fill materials could also be used to develop the site grades; recommendations outlined in this report assume subgrades consisting of materials similar to the site sandy lean clay/clayey sand. Placement of defined depths of granular structural fill could result in alternative recommendations such as reduced pavement thickness or increased bearing capacity. We recommend the fill soils be placed in loose lifts not to exceed 9 inches thick and adjusted in moisture content and compacted as recommended for the scarified soils. If the site soils or similar import materials are used as fill material, care will be needed to maintain the recommended moisture content prior to and during construction of overlying improvements.

Care should be exercised after preparation of the subgrades to avoid disturbing the subgrade materials. Positive drainage should be developed away from the structures and pavements to avoid wetting of subgrade materials. Subgrade materials becoming wet subsequent to construction of the site improvements can result in unacceptable performance.

As indicated on the attached boring logs and laboratory summary sheets, low swelling soils are present on this site. However, swell potential can change with changes in moisture content and density of the subgrade materials. This report provides recommendations to help mitigate the effects of soil shrinkage or expansion for cohesive subgrades. Even if these procedures are followed, some movement and at least minor cracking in the site improvements should be anticipated. The severity of cracking and other cosmetic damage such as uneven floor slabs/exterior flatwork will probably increase if any modification of the site results in excessive drying and/or wetting of the site soils. Eliminating the risk of movement and cosmetic distress may not be feasible, but it may be possible to further reduce the risk of movement if significantly more extensive measures are used during construction. We would be pleased to discuss other construction alternatives with you upon request.

In areas where excavations will extend below existing groundwater table such as utility excavation, placement of cleaner granular fill material may be desirable. Those materials should be placed in lifts and compacted to at least 70% relative density.

Areas of deeper fills may experience settlement from within the placed fill materials. Settlement on the order of 1 to 1.5% of the fill depth would be estimated. The rate of settlement will be dependent on the type of fill material placed and construction methods. Granular soils will consolidate essentially immediately upon placement of overlying loads. Cohesive soils will consolidate at a significantly slower rate.

Foundation Systems – General Considerations

The site appears suitable for support of lightly loaded shallow foundations based on the results of our field exploration and review of the proposed development plans. Consideration should be given to support of heavier loads (greater than 250 kips) on a deep foundation system or use of an over excavation and backfill procedure to support of heavier loads.

Conventional Spread Footing Foundations

The native undisturbed lean clay to clayey sand generally exhibited low swell potential and low to moderate bearing characteristics. Placed fill materials are present on the site; either documentation of the fill placement densities should be reviewed to confirm suitability of those fill materials to support the proposed improvements or those materials should be removed as recommended in the section *Site Preparation*. Subgrade and bearing soils should also be closely observed at the time of construction to evaluate in-place materials and assess the suitability of those soils for support.

Based on the materials observed in the test borings, we expect conventional type spread footings could be used to support the proposed slab on grade buildings provided the footings are placed on approved native subgrade material or moisture/density controlled fill material and the maximum anticipated wall and column loads do not exceed 4 klf and 260 kips, respectively. If actual design loads exceed the assumed values as previously presented, we should be consulted to provide supplemental design criteria for alternative foundations.

Footings bearing on approved native subsoils or moisture/density conditioned soils could be designed for a maximum net allowable total load bearing pressure of 2,000 psf. The net bearing pressure refers to the pressure at foundation bearing level in excess of the minimum surrounding overburden pressure. Total load should include full dead and live loads.

Footings should be proportioned to reduce differential foundation movement. We estimate the total long term settlement of footings designed as outlined above would be less than one inch. Greater settlements will occur in foundations with the higher column loads.

The backfill soils adjacent to the foundations should be placed in loose lifts not to exceed 9 inches thick, moisture conditioned to within ±2% of the material's standard Proctor optimum moisture content, and compacted to be at least 95% of standard Proctor maximum dry density, ASTM D698.

After placement of the fill materials, care should be taken to avoid wetting or drying of those materials. Bearing materials, which are loosened or disturbed by the construction activities, or materials which become dry and desiccated or wet and softened, should be removed and replaced or reworked in place prior to construction of the overlying improvements.

Exterior foundations and foundations in unheated areas should be located at least 30 inches below adjacent exterior grade to provide frost protection. We recommend formed continuous footings have a minimum width of 16 inches and isolated column foundations have a minimum width of 30 inches.

Seismic

The site soil conditions consist of approximately 10 to greater than 25 feet of overburden soils overlying moderately hard bedrock. For those site conditions, the 2012 International Building Code indicates a Seismic Site Classification of D.

Lateral Earth Pressures

For any area of the proposed development having below grade construction, such as basement or below building walls, those portions will be subject to lateral earth pressures. Passive lateral earth pressures may help resist the driving forces for retaining wall or other similar site structures. Active lateral earth pressures could be used for design of structures where some movement of the structure is anticipated, such as retaining walls. The total deflection of structures for design with active earth pressure is estimated to be on the order of one half of one percent of the height of the down slope side of the structure. We recommend at-rest pressures be used for design of structures where rotation of the walls is restrained such as below grade building walls. Passive pressures and friction between the footing and bearing soils could be used for design of resistance to movement.

Coefficient values for backfill with anticipated types of soils for calculation of active, at-rest and passive earth pressures are provided in the table below. Equivalent fluid pressure is equal to the coefficient times the appropriate soil unit weight. Those coefficient values are based on horizontal backfill with backfill soils consisting of on-site essentially cohesive subsoils, or approved imported granular materials with friction angles of 25 and 35 degrees respectively. For the at-rest and active earth pressures, slopes down and away from the structure would result in reduced driving forces with slopes up and away from the structures resulting in greater forces on the walls. The passive resistance would be reduced with slopes away from the wall. The top 30 inches of soil on the passive resistance side of walls could be used as a surcharge load; however, should not be used as a part of the passive resistance value. Frictional resistance is equal to the tangent of the friction angle times the normal force. Surcharge loads or point loads placed in the backfill can also create additional loads on below grade walls. Those situations should be designed on an individual basis.

Table II – Lateral Earth Pressure Design Values

Soil Type	On-Site Low Plasticity Cohesive	Imported Medium Dense Granular
Wet Unit Weight	120	135
Saturated Unit Weight	135	140
Friction Angle (ϕ) (assumed)	25°	35°
Active Pressure Coefficient	0.40	0.27
At-rest Pressure Coefficient	0.58	0.43
Passive Pressure Coefficient	2.46	3.70

The outlined values do not include factors of safety nor allowances for hydrostatic loads and are based on assumed friction angles, which should be verified prior to construction. Care should be taken to develop appropriate drainage systems behind below grade walls to eliminate potential for hydrostatic loads developing on the walls. Those systems would likely include perimeter drain systems extending to sump areas or free outfall where reverse flow cannot occur into the system. Where necessary, appropriate hydrostatic load values should be used for design.

Floor Slabs

Conventional slab-on-grade construction could be considered for the site provided certain precautions are adhered to. To reduce floor slab movement, we recommend floor slabs bear on reconditioned site soils or properly placed and compacted engineered fill material as outlined under *Site Preparation*. It is our opinion the on-site essentially cohesive soils could be used as fill in these areas, provided adequate moisture treatment and compaction procedures are followed. If the site lean clay soils or similar import materials are used as fill material, care will be needed to maintain the recommended moisture content prior to and during construction of overlying improvements. This procedure will not fully eliminate the possibilities of slab movement; but movements should be reduced and tend to be more uniform.

For structural design of concrete slabs-on-grade, a modulus of subgrade reaction (k-value) of 100 pounds per cubic inch (pci) may be used for floors supported on compacted on-site or similar import soils.

Positive separations and/or isolation joints should be provided between slabs and all foundations, columns or utility lines to allow independent movement. Control joints should be provided in slabs to control the location and extent of cracking. Interior trench backfill placed beneath slabs should be compacted in a similar manner as previously described for imported structural fill material.

Pavements

We expect the site pavements will include areas designated for low volume automobile traffic/parking and limited areas of heavier/higher volume traffic. For heavier traffic areas an assumed equivalent daily load axle (EDLA) of 25 was selected and for automobile/parking areas an EDLA of 10 was selected.

Proof rolling and recompacting the subgrade is recommended immediately prior to placement of the aggregate road base section. Soft or weak areas delineated by the proof rolling operations should be undercut or stabilized in-place to achieve the appropriate subgrade support. Based on the subsurface conditions encountered at the site and the laboratory test results, it is recommended the on-site private drives and parking areas be designed using an R-value of 10.

Three to five feet of moisture conditioned/engineered fill material may be placed in areas to develop the subgrades. Existing improvements/fill materials should be stripped and site fill materials should be placed and compacted as recommended in the *Site Preparation* section of this report. Pumping conditions could develop within a moisture treatment process of on-site or imported essentially cohesive soils. Subgrade stabilization may be needed to develop a stable subgrade for paving. A stabilized subgrade could also reduce the thickness of the overlying pavement structure. Stabilization could include incorporating at least 12% (by weight) Class C fly ash into the upper 12 inches of subgrade.

Hot bituminous pavement (HBP) underlain by crushed aggregate base course with or without a fly ash treated subgrade, and non-reinforced concrete pavement could be considered as alternatives for the proposed on-site paved sections. Eliminating the risk of movement within the proposed pavement section may not be feasible due to the characteristics of the subsurface materials; but it may be possible to further reduce the risk of movement if significantly more extensive subgrade stabilization measures are used during construction. We would be pleased to discuss other construction alternatives with you upon request.

Pavement design methods are intended to provide structural sections with adequate thickness over a particular subgrade such that wheel loads are reduced to a level the subgrade can support. The support characteristics of the subgrade for pavement design do not account for shrink/swell movements of an expansive clay subgrade or consolidation of a wetted subgrade. Thus, the pavement may be adequate from a structural standpoint, yet still experience cracking and deformation due to shrink/swell related movement of the subgrade. Therefore it's important to minimize moisture changes in the subgrade to reduce shrink/swell movements.

Recommended pavement sections are provided below in Table III. The HBP could be grading SX (75) or S (75) with PG 58-28 oil. The aggregate base should be Class 5 or Class 6 base. Portland cement concrete should be an exterior pavement mix with a minimum 28-day compressive strength of 4,500 psi and should be air entrained. HBP pavements may show rutting and distress in truck loading and turning areas. Concrete pavements should be considered in those areas. The recommended pavement sections are minimums and periodic maintenance should be expected.

TABLE III – RECOMMENDED PAVEMENT SECTIONS		
	Local Residential Roadways/Parking	**Minor Collectors / Heavy Duty Roadways**
EDLA Reliability Resilient Modulus PSI Loss	10 75% 3562 psi 2.5	25 85% 3562 psi 2.2
Design Structure Number	2.60	3.20
Composite Section without Fly Ash – Option A Hot Mix Asphalt (HMA) Aggregate Base Course Design Structure Number	 4" 8" (2.64)	 5-1/2" 8" (3.30)
Composite Section with Fly Ash – Option B Hot Mix Asphalt (HMA) Aggregate Base Course ABC Fly Ash Treated Subgrade Design Structure Number	 4" 6" 12" (3.02)	 4" 8" 12" (3.24)
PCC (Non-reinforced) – placed on an approved subgrade	5-1/2"	7"

Longitudinal and transverse joints should be provided as needed in concrete pavements for expansion/contraction and isolation. The location and extent of joints should be based upon the final pavement geometry. Sawed joints should be cut in accordance with current ACI criteria. All joints should be sealed to prevent entry of foreign material and dowelled where necessary for load transfer.

The collection and diversion of surface drainage away from paved areas is critical to the satisfactory performance of the pavement. Drainage design should provide for the removal of water from paved areas in order to reduce the potential for wetting of the subgrade soils.

Long-term pavement performance will be dependent upon several factors, including maintaining subgrade moisture levels and providing for preventive maintenance. The following recommendations should be considered the minimum:

- The subgrade and the pavement surface should be adequately sloped to promote proper surface drainage
- Install pavement drainage surrounding areas anticipated for frequent wetting (e.g. garden centers and wash racks)
- Install joint sealant and seal cracks immediately

- Seal all landscaped areas in, or adjacent to pavements to minimize or prevent moisture migration to subgrade soils
- Placing compacted, low permeability backfill against the exterior side of curb and gutter
- Placing curb, gutter, and/or sidewalk directly on approved proof rolled subgrade soils

Preventive maintenance should be planned and provided for through an on-going pavement management program. Preventive maintenance activities are intended to slow the rate of pavement deterioration, and to preserve the pavement investment. Preventive maintenance consists of both localized maintenance (e.g. crack and joint sealing and patching) and global maintenance (e.g. surface sealing). Preventive maintenance is usually the first priority when implementing a planned pavement maintenance program and provides the highest return on investment for pavements. Prior to implementing any maintenance, additional engineering observation is recommended to determine the type and extent of preventive maintenance.

Site grading is generally accomplished early in the construction phase. However as construction proceeds, the subgrade may be disturbed due to utility excavations, construction traffic, desiccation, or rainfall. As a result, the pavement subgrade may not be suitable for pavement construction and corrective action will be required. The subgrade should be carefully evaluated at the time of pavement construction for signs of disturbance, such as but not limited to drying, or rutting. If disturbance has occurred, pavement subgrade areas should be reworked, moisture conditioned, and properly compacted to the recommendations in this report immediately prior to paving.

If during or after placement of the stabilization or initial lift of pavement, the area is observed to be yielding under vehicle traffic or construction equipment, it is recommended that EEC be contacted for additional alternative methods of stabilization, or a change in the pavement section.

Other Considerations

Positive drainage should be developed away from the structures and pavement areas with a minimum slope of 1 inch per foot for the first 10 feet away from the improvements in landscape areas. Care should be taken in planning of landscaping adjacent to the building to avoid features which would pond water adjacent to the foundations or stemwalls. Placement of plants which require irrigation systems or could result in fluctuations of the moisture content of the subgrade material should be avoided adjacent to site improvements. Irrigation systems should not be placed within 5 feet of the perimeter of the buildings and parking areas. Spray heads should be designed not to spray water on or immediately adjacent to the structures or site pavements. Roof drains should be designed to discharge at least 5 feet away from the structures and away from the pavement areas.

Excavations into the on-site clays and proposed import materials can be expected to stand on relatively steep temporary slopes during construction. However, if excavations extend into the underlying sand and gravel layers and/or underlying groundwater, caving soils may be encountered. The individual contractor(s) should be made responsible for designing and constructing stable, temporary excavations as required to maintain stability of both the excavation sides and bottom. All excavations should be sloped or shored in the interest of safety following local and federal regulations, including current OSHA excavation and trench safety standards.

Water Soluble Sulfates

The water soluble sulfate (SO_4) testing of the on-site overburden materials taken during our subsurface exploration are provided in Table IV below.

TABLE IV - Water Soluble Sulfate Test Results

Sample Location	Description	Soluble Sulfate Content (mg/kg)	Soluble Sulfate Content (%)
B-3, S-2 at 4'	Sandy Lean Clay/Clayey Sand	70	0.01
B-16, S-5 at 24'	Sandstone	160	0.02
B-22, S-4 at 19'	Claystone/Siltstone/Sandstone	130	0.01

Based on the results as presented in Table IV above, ACI 318, Section 4.2 indicates the site overburden soils have a low risk of sulfate attack on Portland cement concrete. Therefore **Class 0** sulfate exposure could be used for concrete on and below site grade within the on-site overburden soils. Import materials should be evaluated to determine sulfate levels.

GENERAL COMMENTS

The analysis and recommendations presented in this report are based upon the data obtained from the soil borings performed at the indicated locations and from any other information discussed in this report. This report does not reflect any variations, which may occur between borings or across the site. The nature and extent of such variations may not become evident until construction. If variations appear evident, it will be necessary to re-evaluate the recommendations of this report.

It is recommended that the geotechnical engineer be retained to review the plans and specifications so comments can be made regarding the interpretation and implementation of our geotechnical recommendations in the design and specifications. It is further recommended that the geotechnical engineer be retained for testing and observations during earthwork phases to help determine that the design requirements are fulfilled.

This report has been prepared for the exclusive use of Aims Community College for specific application to the project discussed and has been prepared in accordance with generally accepted geotechnical engineering practices. No warranty, expressed or implied, is made. In the event that any changes in the nature, design, or location of the project as outlined in this report are planned, the conclusions and recommendations contained in this report shall not be considered valid unless the changes are reviewed and the conclusions of this report are modified or verified in writing by the geotechnical engineer.

DRILLING AND EXPLORATION

DRILLING & SAMPLING SYMBOLS:

SS: Split Spoon - 13/8" I.D., 2" O.D., unless otherwise noted
ST: Thin-Walled Tube - 2" O.D., unless otherwise noted
R: Ring Barrel Sampler - 2.42" I.D., 3" O.D. unless otherwise noted
PA: Power Auger
HA: Hand Auger
DB: Diamond Bit = 4", N, B
AS: Auger Sample
HS: Hollow Stem Auger
PS: Piston Sample
WS: Wash Sample
FT: Fish Tail Bit
RB: Rock Bit
BS: Bulk Sample
PM: Pressure Meter
WB: Wash Bore

Standard "N" Penetration: Blows per foot of a 140 pound hammer falling 30 inches on a 2-inch O.D. split spoon, except where noted.

WATER LEVEL MEASUREMENT SYMBOLS:

WL : Water Level
WCI: Wet Cave in
DCI: Dry Cave in
AB : After Boring
WS : While Sampling
WD : While Drilling
BCR: Before Casing Removal
ACR: After Casting Removal

Water levels indicated on the boring logs are the levels measured in the borings at the time indicated. In pervious soils, the indicated levels may reflect the location of ground water. In low permeability soils, the accurate determination of ground water levels is not possible with only short term observations.

DESCRIPTIVE SOIL CLASSIFICATION

Soil Classification is based on the Unified Soil Classification system and the ASTM Designations D-2488. Coarse Grained Soils have move than 50% of their dry weight retained on a #200 sieve; they are described as: boulders, cobbles, gravel or sand. Fine Grained Soils have less than 50% of their dry weight retained on a #200 sieve; they are described as : clays, if they are plastic, and silts if they are slightly plastic or non-plastic. Major constituents may be added as modifiers and minor constituents may be added according to the relative proportions based on grain size. In addition to gradation, coarse grained soils are defined on the basis of their relative in-place density and fine grained soils on the basis of their consistency. Example: Lean clay with sand, trace gravel, stiff (CL); silty sand, trace gravel, medium dense (SM).

CONSISTENCY OF FINE-GRAINED SOILS

Unconfined Compressive Strength, Qu, psf	Consistency
< 500	Very Soft
500 - 1,000	Soft
1,001 - 2,000	Medium
2,001 - 4,000	Stiff
4,001 - 8,000	Very Stiff
8,001 - 16,000	Very Hard

RELATIVE DENSITY OF COARSE-GRAINED SOILS:

N-Blows/ft	Relative Density
0-3	Very Loose
4-9	Loose
10-29	Medium Dense
30-49	Dense
50-80	Very Dense
80 +	Extremely Dense

PHYSICAL PROPERTIES OF BEDROCK

DEGREE OF WEATHERING:

Slight	Slight decomposition of parent material on joints. May be color change.
Moderate	Some decomposition and color change throughout.
High	Rock highly decomposed, may be extremely broken.

HARDNESS AND DEGREE OF CEMENTATION:

Limestone and Dolomite:

Hard	Difficult to scratch with knife.
Moderately Hard	Can be scratched easily with knife. Cannot be scratched with fingernail.
Soft	Can be scratched with fingernail.

Shale, Siltstone and Claystone:

Hard	Can be scratched easily with knife, cannot be scratched with fingernail.
Moderately Hard	Can be scratched with fingernail.
Soft	Can be easily dented but not molded with fingers.

Sandstone and Conglomerate:

Well Cemented	Capable of scratching a knife blade.
Cemented	Can be scratched with knife.
Poorly Cemented	Can be broken apart easily with fingers.

UNIFIED SOIL CLASSIFICATION SYSTEM

Criteria for Assigning Group Symbols and Group Names Using Laboratory Tests				Soil Classification: Group Symbol	Soil Classification: Group Name
Coarse - Grained Soils more than 50% retained on No. 200 sieve	Gravels more than 50% of coarse fraction retained on No. 4 sieve	Clean Gravels Less than 5% fines	Cu≥4 and 1<Cc≤3[E]	GW	Well-graded gravel[F]
			Cu<4 and/or 1>Cc>3[E]	GP	Poorly-graded gravel[F]
		Gravels with Fines more than 12% fines	Fines classify as ML or MH	GM	Silty gravel[G,H]
			Fines Classify as CL or CH	GC	Clayey Gravel[F,G,H]
	Sands 50% or more coarse fraction passes No. 4 sieve	Clean Sands Less than 5% fines	Cu≥6 and 1<Cc≤3[E]	SW	Well-graded sand[I]
			Cu<6 and/or 1>Cc>3[E]	SP	Poorly-graded sand[I]
		Sands with Fines more than 12% fines	Fines classify as ML or MH	SM	Silty sand[G,H,I]
			Fines classify as CL or CH	SC	Clayey sand[G,H,I]
Fine-Grained Soils 50% or more passes the No. 200 sieve	Silts and Clays Liquid Limit less than 50	inorganic	PI>7 and plots on or above "A" Line	CL	Lean clay[K,L,M]
			PI<4 or plots below "A" Line	ML	Silt[K,L,M]
		organic	Liquid Limit - oven dried / Liquid Limit - not dried <0.75	OL	Organic clay[K,L,M,N] / Organic silt[K,L,M,O]
	Silts and Clays Liquid Limit 50 or more	inorganic	PI plots on or above "A" Line	CH	Fat clay[K,L,M]
			PI plots below "A" Line	MH	Elastic Silt[K,L,M]
		organic	Liquid Limit - oven dried / Liquid Limit - not dried <0.75	OH	Organic clay[K,L,M,P] / Organic silt[K,L,M,O]
Highly organic soils		Primarily organic matter, dark in color, and organic odor		PT	Peat

[A]Based on the material passing the 3-in. (75-mm) sieve

[B]If field sample contained cobbles or boulders, or both, add "with cobbles or boulders, or both" to group name.

[C]Gravels with 5 to 12% fines required dual symbols:
GW-GM well graded gravel with silt
GW-GC well-graded gravel with clay
GP-GM poorly-graded gravel with silt
GP-GC poorly-graded gravel with clay

[D]Sands with 5 to 12% fines require dual symbols:
SW-SM well-graded sand with silt
SW-SC well-graded sand with clay
SP-SM poorly graded sand with silt
SP-SC poorly graded sand with clay

[E] $Cu = D_{60}/D_{10}$ $Cc = \frac{(D_{30})^2}{D_{10} \times D_{60}}$

[F]If soil contains ≥15% sand, add "with sand" to

[G]If fines classify as CL-ML, use dual symbol GC-CM, or SC-SM.

[H]If fines are organic, add "with organic fines" to group name

[I]If soil contains >15% gravel, add "with gravel" to group name

[J]If Atterberg limits plots shaded area, soil is a CL-ML, Silty clay

[K]If soil contains 15 to 29% plus No. 200, add "with sand" or "with gravel", whichever is predominant.

[L]If soil contains ≥ 30% plus No. 200 predominantly sand, add "sandy" to group name.

[M]If soil contains ≥30% plus No. 200 predominantly gravel, add "gravelly" to group name.

[N]PI≥4 and plots on or above "A" line.

[O]PI≤4 or plots below "A" line.

[P]PI plots on or above "A" line.

[Q]PI plots below "A" line.

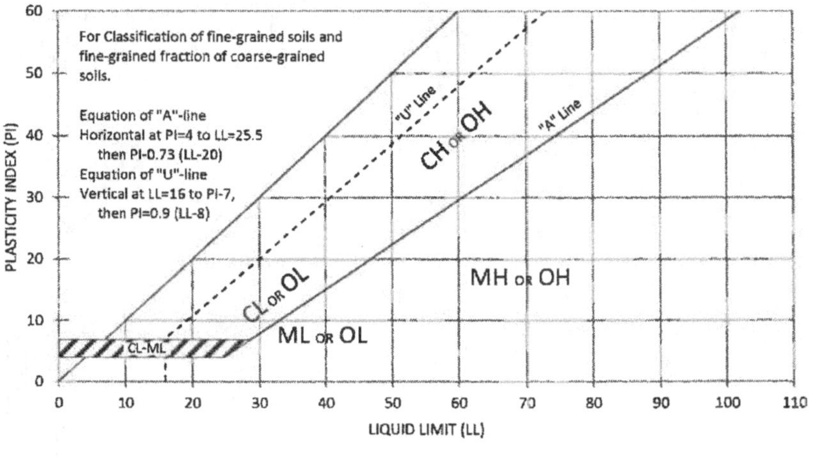

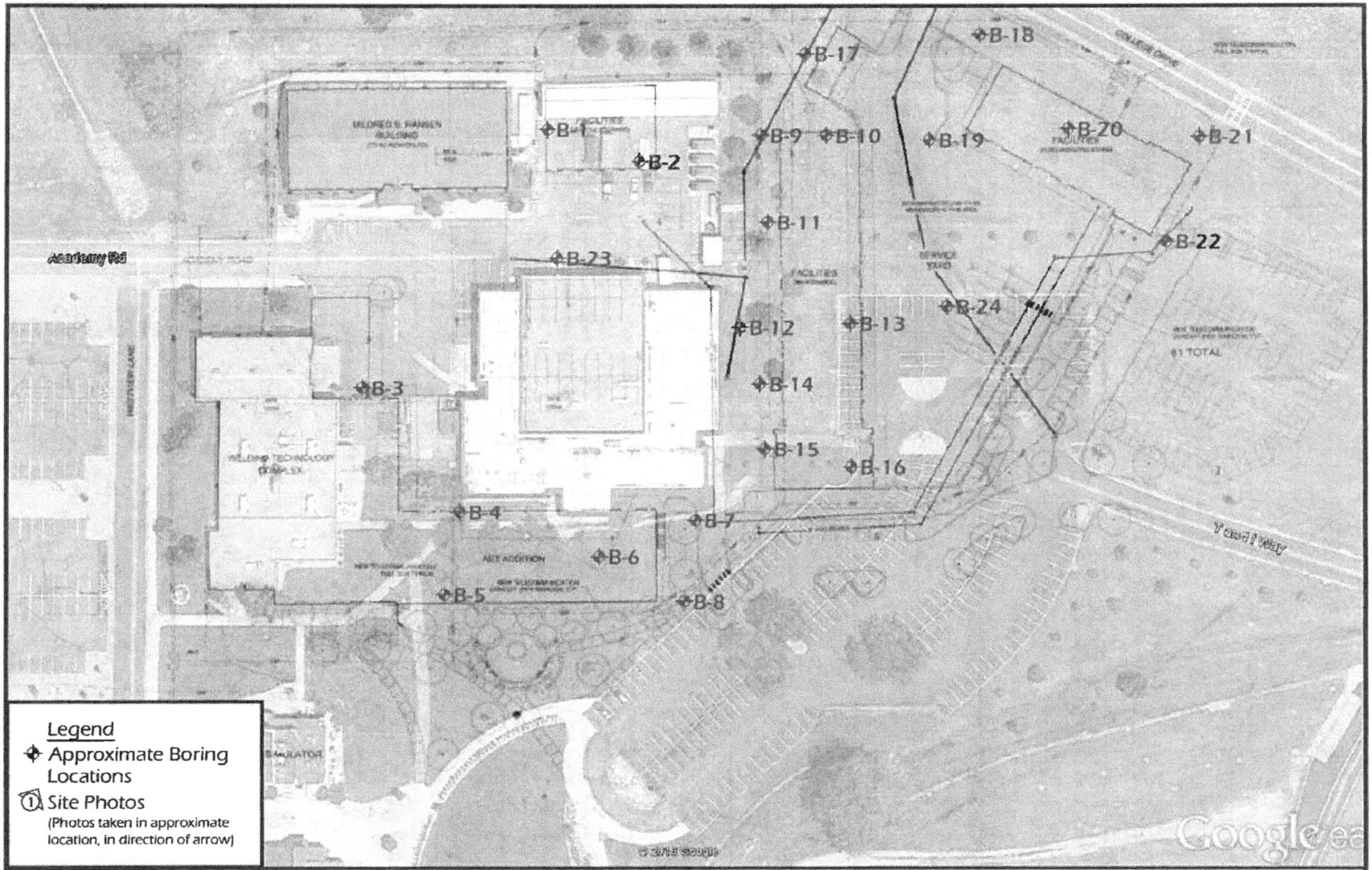

Boring Location Diagram
Aims Community College - North Campus Refresh
Greeley, Colorado
EEC Project Number: 1162022 Date: March 2016

AIMS COMMUNITY COLLEGE NORTH CAMPUS REFRESH
GREELEY, COLORADO
EEC PROJECT NO. 1162022
MARCH 2016

AIMS COMMUNITY COLLEGE - NORTH CAMPUS REFRESH
GREELEY, COLORADO

PROJECT NO: 1162022	LOG OF BORING B-1		DATE: MARCH 2016		
RIG TYPE: CME55	*SHEET 1 OF 2*		WATER DEPTH		
FOREMAN: DG	START DATE	3/25/2016	WHILE DRILLING	10'	
AUGER TYPE: 4" CFA	FINISH DATE	3/25/2016	AFTER DRILLING	N/A	
SPT HAMMER: AUTOMATIC	SURFACE ELEV	N/A	24 HOUR	N/A	

SOIL DESCRIPTION	TYPE	D (FEET)	N (BLOWS/FT)	QU (PSF)	MC (%)	DD (PCF)	A-LIMITS LL	A-LIMITS PI	-200 (%)	SWELL PRESSURE	SWELL % @ 500 PSF
3.5" ASPHALT											
4" AGGREGATE BASE		1									
SILTY / CLAYEY SAND (SM/SC)		2									
light brown / tan / red											
very stiff		3									
		4									
	CS	5	9	5000	18.7	103.7	32	17	42.8	<500 psf	None
		6									
		7									
		8									
SAND & GRAVEL (SP/GP)		9									
brown											
medium dense	SS	10	14	--	4.0				5.3		
		11									
		12									
SANDSTONE		13									
grey / rust / brown											
		14									
with intermittent claystone zones											
	CS	15	50/6"	3000	19.1	105.0					
		16									
		17									
		18									
		19									
	SS	20	50/8"	--	28.6						
		21									
		22									
		23									
		24									
	CS	25	50/2"	9000+	18.3						

Continued on Sheet 2 of 2

AIMS COMMUNITY COLLEGE - NORTH CAMPUS REFRESH

GREELEY, COLORADO

PROJECT NO: 1162022	LOG OF BORING B-1		DATE: MARCH 2016	
RIG TYPE: CME55	*SHEET 2 OF 2*		WATER DEPTH	
FOREMAN: DG	START DATE	3/25/2016	WHILE DRILLING	18'
AUGER TYPE: 4" CFA	FINISH DATE	3/25/2016	AFTER DRILLING	N/A
SPT HAMMER: AUTOMATIC	SURFACE ELEV	N/A	24 HOUR	N/A

SOIL DESCRIPTION	TYPE	D (FEET)	N (BLOWS/FT)	QU (PSF)	MC (%)	DD (PCF)	A-LIMITS LL	A-LIMITS PI	-200 (%)	SWELL PRESSURE	SWELL % @ 500 PSF
Continued from Sheet 1 of 2		26									
SANDSTONE		27									
grey / rust / brown		28									
		29									
	SS	30	50/6"	--	24.9						
		31									
		32									
		33									
		34									
	CS	35	50/7"	5000	23.6	99.1					
BOTTOM OF BORING DEPTH 35.0'		36									
		37									
		38									
		39									
		40									
		41									
		42									
		43									
		44									
		45									
		46									
		47									
		48									
		49									
		50									

AIMS COMMUNITY COLLEGE - NORTH CAMPUS REFRESH
GREELEY, COLORADO

PROJECT NO: 1162022	LOG OF BORING B-2		DATE: MARCH 2016	
RIG TYPE: CME55	*SHEET 1 OF 1*		WATER DEPTH	
FOREMAN: DG	START DATE	3/25/2016	WHILE DRILLING	10'
AUGER TYPE: 4" CFA	FINISH DATE	3/25/2016	AFTER DRILLING	N/A
SPT HAMMER: AUTOMATIC	SURFACE ELEV	N/A	24 HOUR	N/A

SOIL DESCRIPTION	TYPE	D (FEET)	N (BLOWS/FT)	QU (PSF)	MC (%)	DD (PCF)	A-LIMITS LL	A-LIMITS PI	-200 (%)	SWELL PRESSURE	SWELL % @ 500 PSF
3.5" ASPHALT											
5" AGGREGATE BASE		1									
CLAYEY SAND & GRAVEL (SC/GC) (Possible Fill)		2									
brown											
medium dense	CS	3	13	1000	12.4	115.9				<500 psf	None
		4									
SANDY LEAN CLAY (CL)	SS	5	9	8000	17.0						
brown / black, mottled											
very stiff to medium stiff		6									
with gravel											
		7									
		8									
		9									
brown / tan	CS	10	8	1000	21.4	108.2					
		11									
		12									
		13									
SANDSTONE		14									
brown / grey / rust											
	SS	15	50/11"	–	23.2						
		16									
		17									
		18									
		19									
	CS	20	--	--	20.2						
BOTTOM OF BORING DEPTH 20.0'		21									
		22									
		23									
		24									
		25									

AIMS COMMUNITY COLLEGE - NORTH CAMPUS REFRESH
GREELEY, COLORADO

PROJECT NO: 1162022	LOG OF BORING B-3	DATE: MARCH 2016	
RIG TYPE: CME55	*SHEET 1 OF 1*	WATER DEPTH	
FOREMAN: DG	START DATE 3/18/2016	WHILE DRILLING	9.5'
AUGER TYPE: 4" CFA	FINISH DATE 3/18/2016	AFTER DRILLING	N/A
SPT HAMMER: AUTOMATIC	SURFACE ELEV N/A	24 HOUR	N/A

SOIL DESCRIPTION	TYPE	D (FEET)	N (BLOWS/FT)	QU (PSF)	MC (%)	DD (PCF)	A-LIMITS LL	A-LIMITS PI	-200 (%)	SWELL PRESSURE	SWELL % @ 500 PSF
2.5" ASPHALT											
3" AGGREGATE BASE		1									
SANDY LEAN CLAY (CL)		2									
brown / tan											% @ 150 psf
stiff to very stiff	CS	3	13	3000	17.5	110.1	28	13	57.9	<150 psf	None
with calcareous deposits											
		4									
with traces of gravel	SS	5	18	3000	14.8						
		6									
		7									
		8									
		9									
SAND & GRAVEL (SP/GP)	CS	10	42	--	4.2	117.6					
brown											
dense		11									
		12									
		13									
		14									
SANDSTONE	SS	15	50	--	24.0						
brown / grey / rust											
		16									
		17									
		18									
		19									
	CS	20	50/9"	--	21.9						
BOTTOM OF BORING DEPTH 20.0'											
		21									
		22									
		23									
		24									
		25									

AIMS COMMUNITY COLLEGE - NORTH CAMPUS REFRESH
GREELEY, COLORADO

PROJECT NO: 1162022	LOG OF BORING B-4 *SHEET 1 OF 2*	DATE: MARCH 2016
RIG TYPE: CME55		WATER DEPTH
FOREMAN: DG	START DATE 3/16/2016	WHILE DRILLING 19'
AUGER TYPE: 4" CFA	FINISH DATE 3/16/2016	AFTER DRILLING N/A
SPT HAMMER: AUTOMATIC	SURFACE ELEV N/A	24 HOUR N/A

SOIL DESCRIPTION	TYPE	D (FEET)	N (BLOWS/FT)	QU (PSF)	MC (%)	DD (PCF)	A-LIMITS LL	A-LIMITS PI	-200 (%)	SWELL PRESSURE	SWELL % @ 500 PSF
TOPSOIL & VEGETATION											
		1									
SANDY LEAN CLAY (CL) brown / tan, mottled very stiff with traces of gravel		2									
	CS	3	13	8000	14.0	118.4					
		4									
	SS	5	14	6500	16.1						
		6									
		7									
		8									
		9									
	CS	10	16	5500	14.5	105.8	31	16	40.8	<500 psf	None
SILTY / CLAYEY SAND (SM/SC) brown / tan medium dense		11									
		12									
POORLY GRADED GRAVEL (GP) brown dense		13									
		14									
	SS	15	41/11"	--	1.8						
		16									
		17									
		18									
		19									
SAND & GRAVEL (SP/GP) brown with apparent cobbles	SS	20	Cuttings	--	9.9						
		21									
		22									
		23									
		24									
	SS	25	Cuttings	--	5.3						

Continued on Sheet 2 of 2

Earth Engineering Consultants, LLC

AIMS COMMUNITY COLLEGE - NORTH CAMPUS REFRESH
GREELEY, COLORADO

PROJECT NO: 1162022	LOG OF BORING B-4		DATE: MARCH 2016		
RIG TYPE: CME55	SHEET 2 OF 2		WATER DEPTH		
FOREMAN: DG	START DATE	3/16/2016	WHILE DRILLING	19'	
AUGER TYPE: 4" CFA	FINISH DATE	3/16/2016	AFTER DRILLING	N/A	
SPT HAMMER: AUTOMATIC	SURFACE ELEV	N/A	24 HOUR	N/A	

SOIL DESCRIPTION	TYPE	D (FEET)	N (BLOWS/FT)	QU (PSF)	MC (%)	DD (PCF)	A-LIMITS LL	A-LIMITS PI	-200 (%)	SWELL PRESSURE	SWELL % @ 500 PSF
Continued from Sheet 1 of 2		26									
SAND & GRAVEL (SP/GP)		27									
brown											
with apparent cobbles		28									
		29									
	SS	30	Cuttings	–	29.4						
		31									
		32									
		33									
		34									
	SS	35	Cuttings								
		36									
SANDSTONE		37									
brown / rust											
		38									
		39									
	SS	40	50/6"	1500	24.4						
BOTTOM OF BORING DEPTH 40.5'		41									
		42									
		43									
		44									
		45									
		46									
		47									
		48									
		49									
		50									

AIMS COMMUNITY COLLEGE - NORTH CAMPUS REFRESH

GREELEY, COLORADO

PROJECT NO: 1162022	LOG OF BORING B-5	DATE:	MARCH 2016
RIG TYPE: CME55	*SHEET 1 OF 1*	WATER DEPTH	
FOREMAN: DG	START DATE: 3/16/2016	WHILE DRILLING	16'
AUGER TYPE: 4" CFA	FINISH DATE: 3/16/2016	AFTER DRILLING	N/A
SPT HAMMER: AUTOMATIC	SURFACE ELEV: N/A	24 HOUR	N/A

SOIL DESCRIPTION	TYPE	D (FEET)	N (BLOWS/FT)	QU (PSF)	MC (%)	DD (PCF)	A-LIMITS LL	A-LIMITS PI	-200 (%)	SWELL PRESSURE	SWELL % @ 500 PSF
TOPSOIL & VEGETATION											
		1									
SANDY LEAN CLAY (CL) brown stiff		2									
		3									
		4									
brown / tan mottled	CS	5	7	4000	17.7	106.9	27	6	49.8	<500 psf	None
		6									
		7									
		8									
		9									
SANDY LEAN CLAY / CLAYEY SAND (CL/SC)	SS	10	9	1500	13.8						
brown medium stiff with traces of gravel		11									
		12									
SAND & GRAVEL (SP/GP) brown dense		13									
		14									
	CS	15	50	--	1.9	137.2					
		16									
		17									
CLAYSTONE grey / rust with intermittent sandstone		18									
		19									
	SS	20	42	6000	22.4						
BOTTOM OF BORING DEPTH 20.5'		21									
		22									
		23									
		24									
		25									

Earth Engineering Consultants, LLC

AIMS COMMUNITY COLLEGE - NORTH CAMPUS REFRESH
GREELEY, COLORADO

PROJECT NO: 1162022	LOG OF BORING B-6		DATE: MARCH 2016		
RIG TYPE: CME55	*SHEET 1 OF 1*		WATER DEPTH		
FOREMAN: DG	START DATE	3/16/2016	WHILE DRILLING	17'	
AUGER TYPE: 4" CFA	FINISH DATE	3/16/2016	AFTER DRILLING	N/A	
SPT HAMMER: AUTOMATIC	SURFACE ELEV	N/A	24 HOUR	N/A	

SOIL DESCRIPTION	TYPE	D (FEET)	N (BLOWS/FT)	QU (PSF)	MC (%)	DD (PCF)	A-LIMITS LL	A-LIMITS PI	-200 (%)	SWELL PRESSURE	SWELL % @ 500 PSF
TOPSOIL & VEGETATION											
		1									
SANDY LEAN CLAY (CL) brown / tan, mottled very stiff to stiff with traces of gravel		2									
	CS	3	20	9000+	12.2	121.2				<500 psf	None
		4									
	SS	5	10	3000	14.5						
		6									
		7									
		8									
		9									
	CS	10	28	2500	15.7	119.5					
SAND & GRAVEL (SP/GP) brown medium dense to dense with apparent cobbles		11									
		12									
		13									
		14									
	SS	15	53	--	1.7						
		16									
		17									
		18									
		19									
	SS	20	44	--	14.1						
BOTTOM OF BORING DEPTH 20.5'		21									
		22									
		23									
		24									
		25									

AIMS COMMUNITY COLLEGE - NORTH CAMPUS REFRESH

GREELEY, COLORADO

PROJECT NO: 1162022	LOG OF BORING B-7		DATE: MARCH 2016		
RIG TYPE: CME55	*SHEET 1 OF 1*		WATER DEPTH		
FOREMAN: DG	START DATE	3/16/2016	WHILE DRILLING	None	
AUGER TYPE: 4" CFA	FINISH DATE	3/16/2016	AFTER DRILLING	N/A	
SPT HAMMER: AUTOMATIC	SURFACE ELEV	N/A	24 HOUR	N/A	

SOIL DESCRIPTION	TYPE	D (FEET)	N (BLOWS/FT)	QU (PSF)	MC (%)	DD (PCF)	A-LIMITS LL	A-LIMITS PI	-200 (%)	SWELL PRESSURE	SWELL % @ 500 PSF
TOPSOIL & VEGETATION											
		1									
SILTY / CLAYEY SAND (SM/SC)											
brown		2									
medium dense											
		3									
		4									
	CS	5	18	2500	8.0	113.2				<500 psf	None
		6									
		7									
		8									
		9									
CLAYEY SAND / SANDY LEAN CLAY (SC/CL)	SS	10	10	2500	16.0						
brown / tan											
medium dense to dense / stiff		11									
with traces of gravel											
		12									
		13									
		14									
	CS	15	42	–	12.1						
gravel seam 15-16'											
		16									
		17									
		18									
CLAYSTONE / SILTSTONE / SANDSTONE											
brown / grey / rust		19									
highly weathered											
	SS	20	35	9000+	19.3						
BOTTOM OF BORING DEPTH 20.5'		21									
		22									
		23									
		24									
		25									

AIMS COMMUNITY COLLEGE - NORTH CAMPUS REFRESH

GREELEY, COLORADO

PROJECT NO: 1162022	LOG OF BORING B-8		DATE: MARCH 2016		
RIG TYPE: CME55	SHEET 1 OF 2		WATER DEPTH		
FOREMAN: DG	START DATE	3/16/2016	WHILE DRILLING	16'	
AUGER TYPE: 4" CFA	FINISH DATE	3/16/2016	AFTER DRILLING	N/A	
SPT HAMMER: AUTOMATIC	SURFACE ELEV	N/A	24 HOUR	N/A	

SOIL DESCRIPTION	TYPE	D (FEET)	N (BLOWS/FT)	QU (PSF)	MC (%)	DD (PCF)	A-LIMITS LL	A-LIMITS PI	-200 (%)	SWELL PRESSURE	SWELL % @ 500 PSF
TOPSOIL & VEGETATION											
		1									
SILTY CLAYEY SAND (SM/SC)											
brown / tan, mottled		2									
medium dense											
		3									
		4									
	CS	5	9	9000	12.9	116.2	29	9	36.2	<500 psf	None
		6									
		7									
		8									
		9									
brown											
with traces of gravel	SS	10	12	2500	17.7						
		11									
POORLY GRADED GRAVEL (GP)		12									
brown											
dense		13									
		14									
	SS	15	32/9"	--	2.0						
		16									
		17									
		18									
		19									
	SS	20	40	7500	21.4						
CLAYSTONE / SANDSTONE / SILTSTONE											
brown / grey / rust		21									
highly weathered											
		22									
		23									
		24									
SANDSTONE											
brown / rust	CS	25	50/6"	--	23.4	95.1					

Continued on Sheet 2 of 2

AIMS COMMUNITY COLLEGE - NORTH CAMPUS REFRESH

GREELEY, COLORADO

PROJECT NO: 1162022	LOG OF BORING B-8		DATE: MARCH 2016	
RIG TYPE: CME55	*SHEET 2 OF 2*		WATER DEPTH	
FOREMAN: DG	START DATE	3/16/2016	WHILE DRILLING	16'
AUGER TYPE: 4" CFA	FINISH DATE	3/16/2016	AFTER DRILLING	N/A
SPT HAMMER: AUTOMATIC	SURFACE ELEV	N/A	24 HOUR	N/A

SOIL DESCRIPTION	TYPE	D (FEET)	N (BLOWS/FT)	QU (PSF)	MC (%)	DD (PCF)	A-LIMITS LL	A-LIMITS PI	-200 (%)	SWELL PRESSURE	SWELL % @ 500 PSF
Continued from Sheet 1 of 2		26									
SANDSTONE		27									
brown / grey / rust		28									
		29									
with claystone & siltstone zones	CS	30	50/6"	9000+	20.3						
		31									
		32									
		33									
		34									
	CS	35	50/4"	--	18.9	109.0					
BOTTOM OF BORING DEPTH 35.0'		36									
		37									
		38									
		39									
		40									
		41									
		42									
		43									
		44									
		45									
		46									
		47									
		48									
		49									
		50									

AIMS COMMUNITY COLLEGE - NORTH CAMPUS REFRESH

GREELEY, COLORADO

PROJECT NO: 1162022	LOG OF BORING B-9		DATE: MARCH 2016	
RIG TYPE: CME55	*SHEET 1 OF 1*		WATER DEPTH	
FOREMAN: DG	START DATE	3/28/2016	WHILE DRILLING	16'
AUGER TYPE: 4" CFA	FINISH DATE	3/28/2016	AFTER DRILLING	N/A
SPT HAMMER: AUTOMATIC	SURFACE ELEV	N/A	24 HOUR	N/A

SOIL DESCRIPTION	TYPE	D (FEET)	N (BLOWS/FT)	QU (PSF)	MC (%)	DD (PCF)	A-LIMITS LL	A-LIMITS PI	-200 (%)	SWELL PRESSURE	SWELL % @ 500 PSF
TOPSOIL & VEGETATION											
		1									
SANDY LEAN CLAY/CLAYEY SAND (CL/SC) (Possible brown Berm Fill) stiff / medium dense		2									
		3									
		4									
	CS	5	19	3000	10.5	110.0				<500 psf	None
		6									
		7									
		8									
		9									
SANDY LEAN CLAY (CL) brown / tan, mottled very stiff with traces of gravel	SS	10	10	8000	11.9						
		11									
		12									
		13									
		14									
	CS	15	6	3000	23.4	102.9					
LEAN CLAY (CL) with Organics grey / rust stiff		16									
		17									
		18									
		19									
SANDY LEAN CLAY (CL) brown / tan, soft mottled, with traces of gravel	SS	20	7	1000	17.2						
BOTTOM OF BORING DEPTH 20.5'		21									
		22									
		23									
		24									
		25									

AIMS COMMUNITY COLLEGE - NORTH CAMPUS REFRESH
GREELEY, COLORADO

PROJECT NO: 1162022	LOG OF BORING B-10		DATE: MARCH 2016		
RIG TYPE: CME55	*SHEET 1 OF 1*		WATER DEPTH		
FOREMAN: DG	START DATE	3/28/2016	WHILE DRILLING	12.5'	
AUGER TYPE: 4" CFA	FINISH DATE	3/28/2016	AFTER DRILLING	N/A	
SPT HAMMER: AUTOMATIC	SURFACE ELEV	N/A	24 HOUR	N/A	

SOIL DESCRIPTION	TYPE	D (FEET)	N (BLOWS/FT)	QU (PSF)	MC (%)	DD (PCF)	A-LIMITS LL	A-LIMITS PI	-200 (%)	SWELL PRESSURE	SWELL % @ 500 PSF
TOPSOIL & VEGETATION											
		1									
CLAYEY GRAVEL (GP) brown / tan medium dense		2									
		3									
		4									
	CS	5	18	8000	7.2	108.2					
		6									
		7									
		8									
		9									
SILTY SANDY LEAN CLAY (CL) brown / tan medium stiff with gravel	SS	10	6	1000	20.1						
		11									
		12									
		13									
		14									
	CS	15	5	1000	30.0	105.1					
		16									
SANDSTONE / SILTSTONE brown / rust		17									
		18									
		19									
	SS	20	50/9"	3000	25.1						
BOTTOM OF BORING DEPTH 20.5'		21									
		22									
		23									
		24									
		25									

AIMS COMMUNITY COLLEGE - NORTH CAMPUS REFRESH
GREELEY, COLORADO

PROJECT NO: 1162022	LOG OF BORING B-11		DATE: MARCH 2016	
RIG TYPE: CME55	*SHEET 1 OF 1*		WATER DEPTH	
FOREMAN: DG	START DATE	3/28/2016	WHILE DRILLING	14
AUGER TYPE: 4" CFA	FINISH DATE	3/28/2016	AFTER DRILLING	N/A
SPT HAMMER: AUTOMATIC	SURFACE ELEV	N/A	24 HOUR	N/A

SOIL DESCRIPTION	TYPE	D (FEET)	N (BLOWS/FT)	QU (PSF)	MC (%)	DD (PCF)	A-LIMITS LL	A-LIMITS PI	-200 (%)	SWELL PRESSURE	SWELL % @ 500 PSF
TOPSOIL & VEGETATION											
		1									
SILTY / CLAYEY SAND (SM/SC) (Possible Berm Fill)											
brown		2									
medium dense to loose											
	CS	3	14	3500	10.1	110.8	NL	NP	28.1	<500 psf	None
gravel zones		4									
	SS	5	37	2000	8.8						
		6									
		7									
		8									
		9									
SILTY / CLAYEY SAND (SM/SC)	CS	10	14	9000+	12.7	114.7	29	13	35.9	<500 psf	None
brown / rust, mottled											
loose to medium dense		11									
		12									
		13									
		14									
brown / tan											
with traces of gravel	SS	15	8	500	23.8						
		16									
		17									
		18									
CLAYSTONE / SILTSTONE		19									
brown / rust / grey											
highly weathered	CS	20	15	7000	23.9	100.8					
BOTTOM OF BORING DEPTH 20.0'											
		21									
		22									
		23									
		24									
		25									

Earth Engineering Consultants, LLC

AIMS COMMUNITY COLLEGE - NORTH CAMPUS REFRESH
GREELEY, COLORADO

PROJECT NO: 1162022	LOG OF BORING B-12		DATE: MARCH 2016		
RIG TYPE: CME55	*SHEET 1 OF 1*		WATER DEPTH		
FOREMAN: DG	START DATE	3/28/2016	WHILE DRILLING	16'	
AUGER TYPE: 4" CFA	FINISH DATE	3/28/2016	AFTER DRILLING	N/A	
SPT HAMMER: AUTOMATIC	SURFACE ELEV	N/A	24 HOUR	N/A	

SOIL DESCRIPTION	TYPE	D (FEET)	N (BLOWS/FT)	QU (PSF)	MC (%)	DD (PCF)	A-LIMITS LL	A-LIMITS PI	-200 (%)	SWELL PRESSURE	SWELL % @ 500 PSF
TOPSOIL & VEGETATION											
		1									
SANDY LEAN CLAY / CLAYEY SAND (CL/SC) (Possible											
stiff / medium dense Berm Fil		2									
with traces of gravel											
		3									
		4									
	CS	5	11	3000	14.7	113.9				<500 psf	None
		6									
		7									
		8									
		9									
SANDY LEAN CLAY (CL)	SS	10	15	1500	18.5						
brown / rust / grey, mottled											
medium stiff		11									
mottled											
		12									
		13									
		14									
GRAVEL with SAND (GP)	CS	15	37	--	12.8	108.5					
brown											
dense		16									
		17									
		18									
		19									
CLAYSTONE / SILTSTONE / SANDSTONE											
grey / rust / brown	SS	20	33	6000	23.0						
BOTTOM OF BORING DEPTH 20.5'		21									
		22									
		23									
		24									
		25									

AIMS COMMUNITY COLLEGE - NORTH CAMPUS REFRESH

GREELEY, COLORADO

PROJECT NO: 1162022	LOG OF BORING B-13	DATE: MARCH 2016
RIG TYPE: CME55	*SHEET 1 OF 2*	WATER DEPTH
FOREMAN: DG	START DATE 3/18/2016	WHILE DRILLING 9'
AUGER TYPE: 4" CFA	FINISH DATE 3/18/2016	AFTER DRILLING N/A
SPT HAMMER: AUTOMATIC	SURFACE ELEV N/A	24 HOUR N/A

SOIL DESCRIPTION	TYPE	D (FEET)	N (BLOWS/FT)	QU (PSF)	MC (%)	DD (PCF)	A-LIMITS LL	A-LIMITS PI	-200 (%)	SWELL PRESSURE	SWELL % @ 500 PSF
3" ASPHALT											
2" AGGREGATE BASE		1									
SILTY / CLAYEY SAND (SM/SC)		2									
brown											
loose to medium dense	CS	3	7	4000	17.8	113.0	27	13	42.4	<500 psf	None
		4									
SILTY SANDY LEAN CLAY (CL)	SS	5	10	3000	18.7						
brown / tan											
stiff		6									
with traces of gravel		7									
		8									
		9									
	CS	10	21	--	14.8						
		11									
		12									
CLAYSTONE / SILTSTONE / SANDSTONE		13									
brown / grey / rust											
highly weathered		14									
	SS	15	36/10"	5000	23.7						
		16									
		17									
		18									
		19									
SANDSTONE	CS	20	50/6.5"	2500	25.0	98.4					
brown / rust											
		21									
		22									
		23									
		24									
SANDSTONE / CLAYSTONE / SILTSTONE											
rust / brown / grey	SS	25	50/6"	4500	25.0						
Continued on Sheet 2 of 2											

Earth Engineering Consultants, LLC

AIMS COMMUNITY COLLEGE - NORTH CAMPUS REFRESH
GREELEY, COLORADO

PROJECT NO: 1162022	LOG OF BORING B-13	DATE: MARCH 2016	
RIG TYPE: CME55	*SHEET 2 OF 2*	WATER DEPTH	
FOREMAN: DG	START DATE 3/18/2016	WHILE DRILLING	9'
AUGER TYPE: 4" CFA	FINISH DATE 3/18/2016	AFTER DRILLING	N/A
SPT HAMMER: AUTOMATIC	SURFACE ELEV N/A	24 HOUR	N/A

SOIL DESCRIPTION	TYPE	D (FEET)	N (BLOWS/FT)	QU (PSF)	MC (%)	DD (PCF)	A-LIMITS LL	A-LIMITS PI	-200 (%)	SWELL PRESSURE	SWELL % @ 500 PSF
Continued from Sheet 1 of 2		26									
SANDSTONE / CLAYSTONE / SILTSTONE		27									
rust / brown / grey		28									
		29									% @ 1000 psf
	CS	30	50/6"	3500	26.4	97.4	NL	NP	16.5	<1000 psf	None
		31									
		32									
		33									
		34									
	SS	35	50/7"	2000	25.7						
BOTTOM OF BORING DEPTH 35.5'		36									
		37									
		38									
		39									
		40									
		41									
		42									
		43									
		44									
		45									
		46									
		47									
		48									
		49									
		50									

AIMS COMMUNITY COLLEGE - NORTH CAMPUS REFRESH
GREELEY, COLORADO

PROJECT NO: 1162022	LOG OF BORING B-14		DATE: MARCH 2016	
RIG TYPE: CME55	*SHEET 1 OF 1*		WATER DEPTH	
FOREMAN: DG	START DATE	3/28/2016	WHILE DRILLING	16.5'
AUGER TYPE: 4" CFA	FINISH DATE	3/28/2016	AFTER DRILLING	N/A
SPT HAMMER: AUTOMATIC	SURFACE ELEV	N/A	24 HOUR	N/A

SOIL DESCRIPTION	TYPE	D (FEET)	N (BLOWS/FT)	QU (PSF)	MC (%)	DD (PCF)	A-LIMITS LL	A-LIMITS PI	-200 (%)	SWELL PRESSURE	SWELL % @ 500 PSF
TOPSOIL & VEGETATION											
		1									
CLAYEY SAND to SANDY LEAN CLAY (SC/CL) (Possible											
brown / tan Berm F		2									
dense											
	CS	3	50	9000	6.1	117.7				<500 psf	None
		4									
	SS	5	33/8"	1000	6.0						
		6									
		7									
		8									
		9									
SANDY LEAN CLAY (CL)	CS	10	23	9000+	12.6	121.5					
tan / brown / rust, mottled											
very stiff		11									
		12									
		13									
		14									
	SS	15	40	3500	9.5				20.6		
CLAYEY SAND & GRAVEL (SC/GP)											
brown		16									
medium dense											
		17									
		18									
		19									
	CS	20	34	--	12.3						
BOTTOM OF BORING DEPTH 20.0'											
		21									
		22									
		23									
		24									
		25									

AIMS COMMUNITY COLLEGE - NORTH CAMPUS REFRESH

GREELEY, COLORADO

PROJECT NO: 1162022	LOG OF BORING B-15		DATE: MARCH 2016	
RIG TYPE: CME55	*SHEET 1 OF 1*		WATER DEPTH	
FOREMAN: DG	START DATE	3/28/2016	WHILE DRILLING	14'
AUGER TYPE: 4" CFA	FINISH DATE	3/28/2016	AFTER DRILLING	N/A
SPT HAMMER: AUTOMATIC	SURFACE ELEV	N/A	24 HOUR	N/A

SOIL DESCRIPTION	TYPE	D (FEET)	N (BLOWS/FT)	QU (PSF)	MC (%)	DD (PCF)	A-LIMITS LL	A-LIMITS PI	-200 (%)	SWELL PRESSURE	SWELL % @ 500 PSF
CLAYEY SAND (SC) (Possible Berm Fill)		1									
brown / rust		2									
		3									
		4									
SILTY / CLAYEY SAND (SM/SC)	CS	5	15	9000+	12.1	118.6	24	9	35.4	<500 psf	None
brown / rust		6									
medium dense		7									
		8									
SAND & GRAVEL (SP/GP)		9									
brown											
dense	SS	10	33	--	3.5						
		11									
		12									
		13									
		14									
	CS	15	30	9000+	18.3	110.5					
CLAYSTONE		16									
grey / brown / rust											
highly weathered		17									
		18									
		19									
SANDSTONE											
brown / rust	SS	20	50/8"	--	27.7						
BOTTOM OF BORING DEPTH 20.5'		21									
		22									
		23									
		24									
		25									

AIMS COMMUNITY COLLEGE - NORTH CAMPUS REFRESH

GREELEY, COLORADO

PROJECT NO: 1162022	LOG OF BORING B-16		DATE: MARCH 2016	
RIG TYPE: CME55	*SHEET 1 OF 1*		WATER DEPTH	
FOREMAN: DG	START DATE	3/18/2016	WHILE DRILLING	10'
AUGER TYPE: 4" CFA	FINISH DATE	3/18/2016	AFTER DRILLING	N/A
SPT HAMMER: AUTOMATIC	SURFACE ELEV	N/A	24 HOUR	N/A

SOIL DESCRIPTION	TYPE	D (FEET)	N (BLOWS/FT)	QU (PSF)	MC (%)	DD (PCF)	A-LIMITS LL	A-LIMITS PI	-200 (%)	SWELL PRESSURE	SWELL % @ 500 PSF
TOPSOIL & VEGETATION											
		1									
SANDY LEAN CLAY to CLAYEY SAND (CL/SC)											
brown		2									
very stiff / medium dense											
		3									
		4									
	CS	5	10	6000	16.2	109.8	25	10	46.9	<500 psf	None
		6									
		7									
		8									
		9									
with intermittent gravel seams to 14.5'											
	SS	10	50	--	4.6						
		11									
		12									
		13									
		14									% @ 1000 psf
	CS	15	30	9000+	21.1		48	20	86.9	1200 psf	0.1%
CLAYSTONE / SILTSTONE / SANDSTONE											
brown / grey / rust		16									
highly weathered											
		17									
		18									
		19									
SANDSTONE	SS	20	50/9.5"	1000	27.4						
brown / rust											
		21									
		22									
		23									
		24									
	CS	25	50/5.5"	5000	24.0	99.9					

Continued on Sheet 2 of 2

AIMS COMMUNITY COLLEGE - NORTH CAMPUS REFRESH

GREELEY, COLORADO

PROJECT NO: 1162022	LOG OF BORING B-16		DATE: MARCH 2016	
RIG TYPE: CME55	*SHEET 2 OF 2*		WATER DEPTH	
FOREMAN: DG	START DATE	3/18/2016	WHILE DRILLING	10'
AUGER TYPE: 4" CFA	FINISH DATE	3/18/2016	AFTER DRILLING	N/A
SPT HAMMER: AUTOMATIC	SURFACE ELEV	N/A	24 HOUR	N/A

SOIL DESCRIPTION	TYPE	D (FEET)	N (BLOWS/FT)	QU (PSF)	MC (%)	DD (PCF)	A-LIMITS LL	A-LIMITS PI	-200 (%)	SWELL PRESSURE	SWELL % @ 500 PSF
Continued from Sheet 1 of 2		26									
SANDSTONE		27									
brown / rust											
		28									
		29									
claystone / siltstone zones	SS	30	50/6"	500	29.2						
		31									
		32									
		33									
		34									
	CS	35	50/5"	9000+	19.8	112.6					
BOTTOM OF BORING DEPTH 35.0'											
		36									
		37									
		38									
		39									
		40									
		41									
		42									
		43									
		44									
		45									
		46									
		47									
		48									
		49									
		50									

AIMS COMMUNITY COLLEGE - NORTH CAMPUS REFRESH

GREELEY, COLORADO

PROJECT NO: 1162022	LOG OF BORING B-17		DATE: MARCH 2016		
RIG TYPE: CME55	*SHEET 1 OF 1*		WATER DEPTH		
FOREMAN: DG	START DATE	3/28/2016	WHILE DRILLING	19'	
AUGER TYPE: 4" CFA	FINISH DATE	3/28/2016	AFTER DRILLING	N/A	
SPT HAMMER: AUTOMATIC	SURFACE ELEV	N/A	24 HOUR	N/A	

SOIL DESCRIPTION	TYPE	D (FEET)	N (BLOWS/FT)	QU (PSF)	MC (%)	DD (PCF)	A-LIMITS LL	A-LIMITS PI	-200 (%)	SWELL PRESSURE	SWELL %@ 500 PSF
TOPSOIL & VEGETATION											
		1									
CLAYEY SAND / SANDY LEAN CLAY (SC/CL) (Possible Fill) brown medium dense / very stiff		2									% @ 150 psf
	CS	3	24	9000+	6.0	112.5				180 psf	0.1%
		4									
CLAYEY SAND (SC) (Possible Fill) brown dense with calcareous deposits & traces of gravel	SS	5	38	6000	5.5						
		6									
		7									
		8									
		9									
CLAYEY SAND / SANDY LEAN CLAY (SC/CL) brown medium dense / very stiff with calcareous deposits	CS	10	16	9000+	8.6	116.9					
		11									
		12									
		13									
		14									
CLAYEY GRAVEL (GC) brown loose	SS	15	8	3000	9.9						
		16									
		17									
		18									
		19									
SILTY SANDY LEAN CLAY (CL) brown / tan, medium stiff	CS	20	7	1500	19.9	110.6					
BOTTOM OF BORING DEPTH 20.0'		21									
		22									
		23									
		24									
		25									

Earth Engineering Consultants, LLC

AIMS COMMUNITY COLLEGE - NORTH CAMPUS REFRESH

GREELEY, COLORADO

PROJECT NO: 1162022	LOG OF BORING B-18	DATE: MARCH 2016
RIG TYPE: CME55	*SHEET 1 OF 2*	WATER DEPTH

FOREMAN: DG	START DATE	3/18/2016	WHILE DRILLING	16'
AUGER TYPE: 4" CFA	FINISH DATE	3/18/2016	AFTER DRILLING	N/A
SPT HAMMER: AUTOMATIC	SURFACE ELEV	N/A	24 HOUR	N/A

SOIL DESCRIPTION	TYPE	D (FEET)	N (BLOWS/FT)	QU (PSF)	MC (%)	DD (PCF)	A-LIMITS LL	A-LIMITS PI	-200 (%)	SWELL PRESSURE	SWELL % @ 500 PSF
TOPSOIL & VEGETATION											
		1									
SANDY LEAN CLAY (CL)											
brown / tan, mottled		2									
very stiff to stiff											
with traces of gravel	CS	3	15	9000+	10.7	118.3					
		4									
	SS	5	40	4000	7.3						
		6									
GRAVEL (GP)											
brown		7									
dense											
		8									
		9									
	SS	10	50	--	2.1				10.7		
		11									
		12									
		13									
CLAYSTONE / SILTSTONE / SANDSTONE		14									
grey / brown / rust											
	SS	15	43	9000+	18.0						
		16									
		17									
		18									
		19									% @ 1000 psf
SANDSTONE	CS	20	50/9"	3000	26.0	96.0	NL	NP	28.8	<1000 psf	None
brown / rust											
with intermittent claystone seams		21									
		22									
		23									
increase in claystone with depth		24									
	SS	25	50/9"	9000	22.9						
Continued on Sheet 2 of 2											

AIMS COMMUNITY COLLEGE - NORTH CAMPUS REFRESH

GREELEY, COLORADO

PROJECT NO: 1162022	LOG OF BORING B-18	DATE: MARCH 2016
RIG TYPE: CME55	*SHEET 2 OF 2*	WATER DEPTH
FOREMAN: DG	START DATE 3/18/2016	WHILE DRILLING 16'
AUGER TYPE: 4" CFA	FINISH DATE 3/18/2016	AFTER DRILLING N/A
SPT HAMMER: AUTOMATIC	SURFACE ELEV N/A	24 HOUR N/A

SOIL DESCRIPTION	TYPE	D (FEET)	N (BLOWS/FT)	QU (PSF)	MC (%)	DD (PCF)	A-LIMITS LL	A-LIMITS PI	-200 (%)	SWELL PRESSURE	SWELL % @ 500 PSF
Continued from Sheet 1 of 2		26									
SANDSTONE		27									
brown / rust											
with claystone seams		28									
		29									
	CS	30	50/5"	3500	21.7	98.8					
		31									
		32									
		33									
		34									
	SS	35	50/6"	3000	26.0						
BOTTOM OF BORING DEPTH 35.5'		36									
		37									
		38									
		39									
		40									
		41									
		42									
		43									
		44									
		45									
		46									
		47									
		48									
		49									
		50									

AIMS COMMUNITY COLLEGE - NORTH CAMPUS REFRESH
GREELEY, COLORADO

PROJECT NO: 1162022	LOG OF BORING B-19		DATE: MARCH 2016	
RIG TYPE: CME55	*SHEET 1 OF 1*		WATER DEPTH	
FOREMAN: DG	START DATE	3/28/2016	WHILE DRILLING	8'
AUGER TYPE: 4" CFA	FINISH DATE	3/28/2016	AFTER DRILLING	N/A
SPT HAMMER: AUTOMATIC	SURFACE ELEV	N/A	24 HOUR	N/A

SOIL DESCRIPTION	TYPE	D (FEET)	N (BLOWS/FT)	QU (PSF)	MC (%)	DD (PCF)	A-LIMITS LL	A-LIMITS PI	-200 (%)	SWELL PRESSURE	SWELL % @ 500 PSF
TOPSOIL & VEGETATION											
		1									
LEAN CLAY (CL)											
brown		2									
with traces of gravel											
		3									
POORLY GRADED GRAVEL (GP)		4									
brown											
dense	CS	5	44	--	5.3	124.2					
		6									
		7									
		8									
CLAYSTONE		9									
grey / brown / rust											
	SS	10	42	9000+	17.4						
		11									
		12									
		13									
		14									% @ 1000 psf
SANDSTONE	CS	15	50/10"	8500	23.5	101.3	26	6	48.0	<1000 psf	None
brown / rust											
		16									
		17									
		18									
		19									
	SS	20	50/6"	1500	23.6						
BOTTOM OF BORING DEPTH 20.5'		21									
		22									
		23									
		24									
		25									

Earth Engineering Consultants, LLC

AIMS COMMUNITY COLLEGE - NORTH CAMPUS REFRESH
GREELEY, COLORADO

PROJECT NO: 1162022	LOG OF BORING B-20		DATE: MARCH 2016
RIG TYPE: CME55	*SHEET 1 OF 1*		WATER DEPTH
FOREMAN: DG	START DATE	3/28/2016	WHILE DRILLING: 9'
AUGER TYPE: 4" CFA	FINISH DATE	3/28/2016	AFTER DRILLING: N/A
SPT HAMMER: AUTOMATIC	SURFACE ELEV	N/A	24 HOUR: N/A

SOIL DESCRIPTION	TYPE	D (FEET)	N (BLOWS/FT)	QU (PSF)	MC (%)	DD (PCF)	A-LIMITS LL	A-LIMITS PI	-200 (%)	SWELL PRESSURE	SWELL % @ 500 PSF
TOPSOIL & VEGETATION											
		1									
SANDY LEAN CLAY (CL)											
brown / tan		2									
with calcareous deposits & traces of gravel	CS	3	11	6000	10.8	115.0					
		4									
SAND & GRAVEL (SP/GP)	SS	5	40	--	4.3						
brown											
dense		6									
		7									
		8									
		9									
	CS	10	41	--	7.3						
		11									
SANDSTONE		12									
brown / rust											
		13									
		14									
	SS	15	50/10"	--	18.7						
		16									
		17									
		18									
		19									
	SS	20	50/9"	1500-3000	25.8						
BOTTOM OF BORING DEPTH 20.5'		21									
		22									
		23									
		24									
		25									

AIMS COMMUNITY COLLEGE - NORTH CAMPUS REFRESH

GREELEY, COLORADO

PROJECT NO: 1162022	LOG OF BORING B-21		DATE:	MARCH 2016	
RIG TYPE: CME55	*SHEET 1 OF 1*		WATER DEPTH		
FOREMAN: DG	START DATE	3/18/2016	WHILE DRILLING		15'
AUGER TYPE: 4" CFA	FINISH DATE	3/18/2016	AFTER DRILLING		N/A
SPT HAMMER: AUTOMATIC	SURFACE ELEV	N/A	24 HOUR		N/A

SOIL DESCRIPTION	TYPE	D (FEET)	N (BLOWS/FT)	QU (PSF)	MC (%)	DD (PCF)	A-LIMITS LL	A-LIMITS PI	-200 (%)	SWELL PRESSURE	SWELL % @ 500 PSF
TOPSOIL & VEGETATION											
		1									
SANDY LEAN CLAY to CLAYEY SAND (CL/SC)											
brown		2									
very stiff / medium dense											
with traces of gravel		3									
		4									
	CS	5	14	9000+	8.3	116.5	26	11	37.8	<500 psf	None
		6									
		7									
		8									
		9									
brown / tan											
	SS	10	10	3500	13.1						
		11									
		12									
SAND & GRAVEL (SP/GP)		13									
brown											
dense		14									
	SS	15	50	--	2.7						
		16									
		17									
		18									
		19									
CLAYSTONE											
brown / grey / rust, highly weathered	CS	20	39	--	39.5						
BOTTOM OF BORING DEPTH 20.0'											
		21									
		22									
		23									
		24									
		25									

Earth Engineering Consultants, LLC

AIMS COMMUNITY COLLEGE - NORTH CAMPUS REFRESH

GREELEY, COLORADO

PROJECT NO: 1162022	LOG OF BORING B-22 *SHEET 1 OF 2*		DATE: MARCH 2016		
RIG TYPE: CME55			WATER DEPTH		
FOREMAN: DG	START DATE	3/18/2016	WHILE DRILLING		12'
AUGER TYPE: 4" CFA	FINISH DATE	3/18/2016	AFTER DRILLING		N/A
SPT HAMMER: AUTOMATIC	SURFACE ELEV	N/A	24 HOUR		N/A

SOIL DESCRIPTION	TYPE	D (FEET)	N (BLOWS/FT)	QU (PSF)	MC (%)	DD (PCF)	A-LIMITS LL	A-LIMITS PI	-200 (%)	SWELL PRESSURE	SWELL % @ 500 PSF
RECLAIMED ASPHALT PAVEMENT SURFACING											
SANDY LEAN CLAY / CLAYEY SAND (CL/SC)		1									
brown / tan											
very stiff / medium dense		2									
with traces of gravel											
		3									
		4									
	CS	5	13	9000	14.5	116.9				<500 psf	None
		6									
		7									
		8									
SAND & GRAVEL (SP/GP)		9									
brown / tan											
dense	SS	10	50	--	2.8						
		11									
		12									
		13									
		14									
	CS	15		--	10.9						
		16									
CLAYSTONE / SILTSTONE / SANDSTONE		17									
grey / brown / rust											
		18									
		19									
	SS	20	55	9000+	21.8						
		21									
		22									
		23									
		24									
SANDSTONE											% @ 1000 psf
with intermittent claystone seams	CS	25	50/6"	4000	25.0	98.6				<1000 psf	None
Continued on Sheet 2 of 2											

Earth Engineering Consultants, LLC

AIMS COMMUNITY COLLEGE - NORTH CAMPUS REFRESH
GREELEY, COLORADO

PROJECT NO: 1162022	LOG OF BORING B-22		DATE: MARCH 2016	
RIG TYPE: CME55	SHEET 2 OF 2		WATER DEPTH	
FOREMAN: DG	START DATE	3/18/2016	WHILE DRILLING	12'
AUGER TYPE: 4" CFA	FINISH DATE	3/18/2016	AFTER DRILLING	N/A
SPT HAMMER: AUTOMATIC	SURFACE ELEV	N/A	24 HOUR	N/A

SOIL DESCRIPTION	TYPE	D (FEET)	N (BLOWS/FT)	QU (PSF)	MC (%)	DD (PCF)	A-LIMITS LL	A-LIMITS PI	-200 (%)	SWELL PRESSURE	SWELL % @ 500 PSF
Continued from Sheet 1 of 2		26									
SANDSTONE		27									
with intermittent claystone seams		28									
		29									
	SS	30	50/5"	3000	22.0						
		31									
claystone / siltstone zones		32									
		33									
		34									
	CS	35	50/5"	–	20.9						
BOTTOM OF BORING DEPTH 35.0'		36									
		37									
		38									
		39									
		40									
		41									
		42									
		43									
		44									
		45									
		46									
		47									
		48									
		49									
		50									

AIMS COMMUNITY COLLEGE - NORTH CAMPUS REFRESH
GREELEY, COLORADO

PROJECT NO: 1162022	LOG OF BORING B-23		DATE: MARCH 2016		
RIG TYPE: CME55	SHEET 1 OF 1		WATER DEPTH		
FOREMAN: DG	START DATE	3/25/2016	WHILE DRILLING	None	
AUGER TYPE: 4" CFA	FINISH DATE	3/25/2016	AFTER DRILLING	N/A	
SPT HAMMER: AUTOMATIC	SURFACE ELEV	N/A	24 HOUR	N/A	

SOIL DESCRIPTION	TYPE	D (FEET)	N (BLOWS/FT)	QU (PSF)	MC (%)	DD (PCF)	A-LIMITS LL	A-LIMITS PI	-200 (%)	SWELL PRESSURE	SWELL % @ 500 PSF
4" ASPHALT											
3.5" AGGREGATE BASE		1									
SANDY LEAN CLAY (CL)		2									
brown / tan											% @ 150 psf
	CS	3	12	5000	17.1	111.9				200 psf	0.1%
		4									
increase in sand with depth	SS	5	11	3500	15.3						
		6									
		7									
		8									
SAND & GRAVEL (SP/GP)		9									
brown											
dense	SS	10	34	--	3.5						
BOTTOM OF BORING DEPTH 10.5'		11									
		12									
		13									
		14									
		15									
		16									
		17									
		18									
		19									
		20									
		21									
		22									
		23									
		24									
		25									

AIMS COMMUNITY COLLEGE - NORTH CAMPUS REFRESH

GREELEY, COLORADO

PROJECT NO: 1162022	LOG OF BORING B-24		DATE: MARCH 2016	
RIG TYPE: CME55	*SHEET 1 OF 1*		WATER DEPTH	
FOREMAN: DG	START DATE	3/18/2016	WHILE DRILLING	None
AUGER TYPE: 4" CFA	FINISH DATE	3/18/2016	AFTER DRILLING	N/A
SPT HAMMER: AUTOMATIC	SURFACE ELEV	N/A	24 HOUR	N/A

SOIL DESCRIPTION	TYPE	D (FEET)	N (BLOWS/FT)	QU (PSF)	MC (%)	DD (PCF)	A-LIMITS LL	A-LIMITS PI	-200 (%)	SWELL PRESSURE	SWELL % @ 500 PSF
2" ASPHALT											
3" AGGREGATE BASE		1									
SANDY LEAN CLAY (CL)		2									
brown / rust, mottled											% @ 150 psf
	CS	3	6	3000	20.3	105.0	33	20	42.0	<150 psf	None
stiff											
		4									
brown / red / tan	SS	5	6	1500	22.2						
		6									
		7									
		8									
gravel seam 8-9.5'											
		9									
CLAYSTONE	SS	10	9	1000	26.8						
grey / brown, highly weathered											
BOTTOM OF BORING DEPTH 10.5'		11									
		12									
		13									
		14									
		15									
		16									
		17									
		18									
		19									
		20									
		21									
		22									
		23									
		24									
		25									

SWELL / CONSOLIDATION TEST RESULTS

Material Description:	Light Brown / Tan / Red Silty / Clayey Sand (SM/SC)		
Sample Location:	Boring 1, Sample 1, Depth 4'		
Liquid Limit: 32	Plasticity Index: 17		% Passing #200: 42.8%
Beginning Moisture: 18.7%	Dry Density: 110.7 pcf		Ending Moisture: 21.1%
Swell Pressure: <500 psf		% Swell @ 500:	None

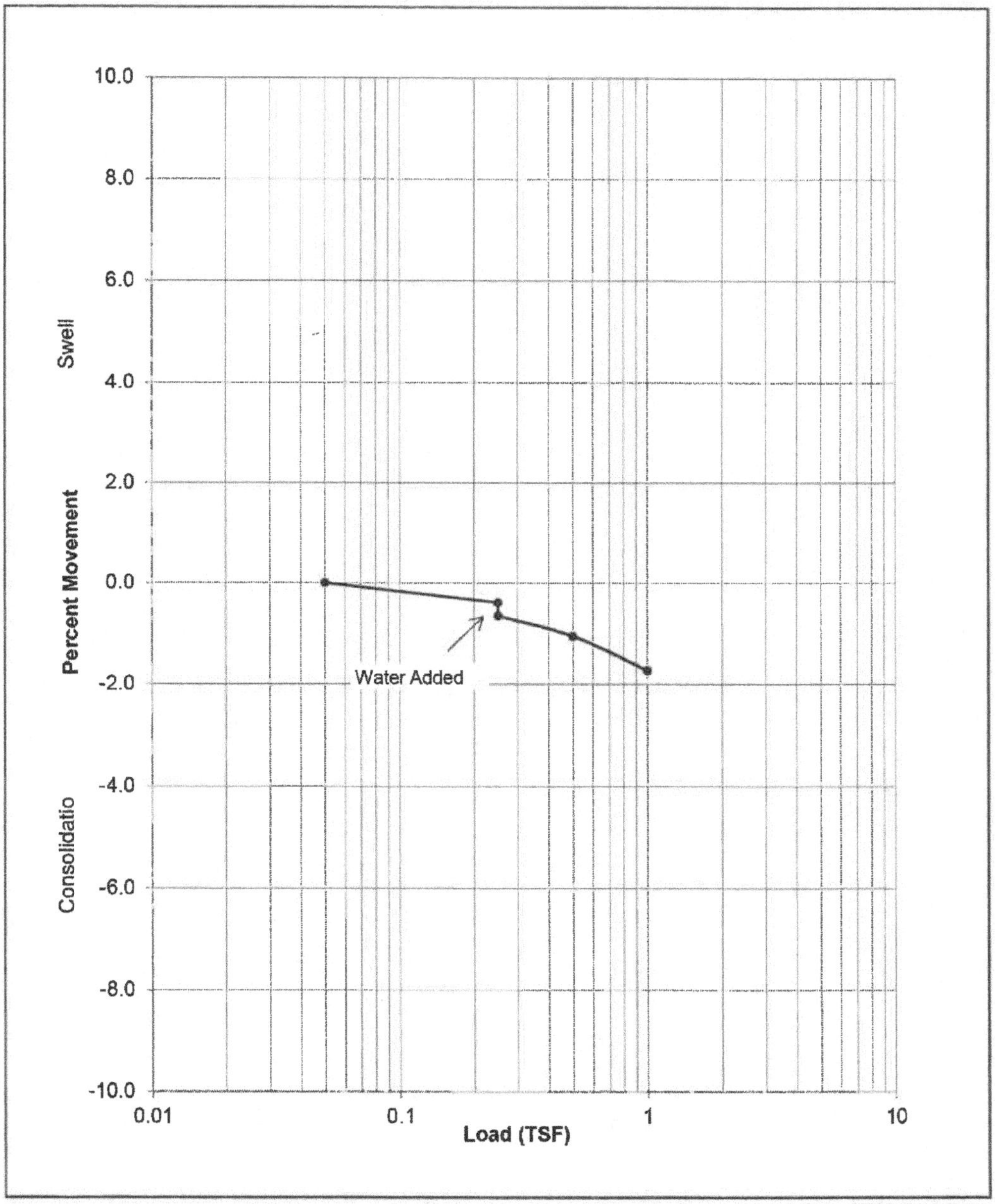

Project: Aims Community College - North Campus Refresh
Location: Greeley, Colorado
Project #: 1162022
Date: March 2016

SWELL / CONSOLIDATION TEST RESULTS

Material Description: Clayey Sand & Gravel (SC/GC)		
Sample Location: Boring 2, Sample 1, Depth 2'		
Liquid Limit:	Plasticity Index:	% Passing #200:
Beginning Moisture: 12.4%	Dry Density: 127.7 pcf	Ending Moisture: 14.2%
Swell Pressure: <500 psf	% Swell @ 500: None	

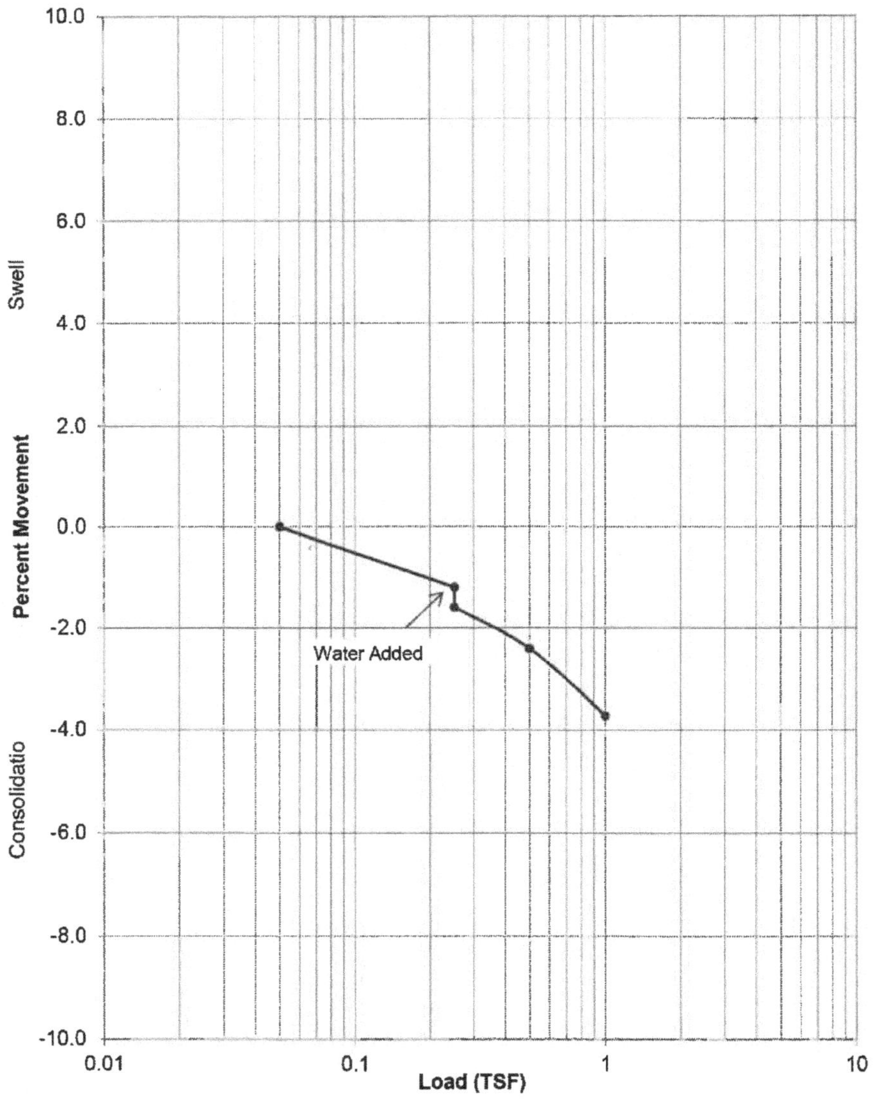

Project: Aims Community College - North Campus Refresh
Location: Greeley, Colorado
Project #: 1162022
Date: March 2016

EEC

SWELL / CONSOLIDATION TEST RESULTS

Material Description:	Brown / Tan Sandy Lean Clay (CL)			
Sample Location:	Boring 3, Sample 1, Depth 2'			
Liquid Limit: 28		Plasticity Index: 13		% Passing #200: 57.9%
Beginning Moisture: 17.5%		Dry Density: 108.2 pcf		Ending Moisture: 17.3%
Swell Pressure: <150 psf			% Swell @ 150: None	

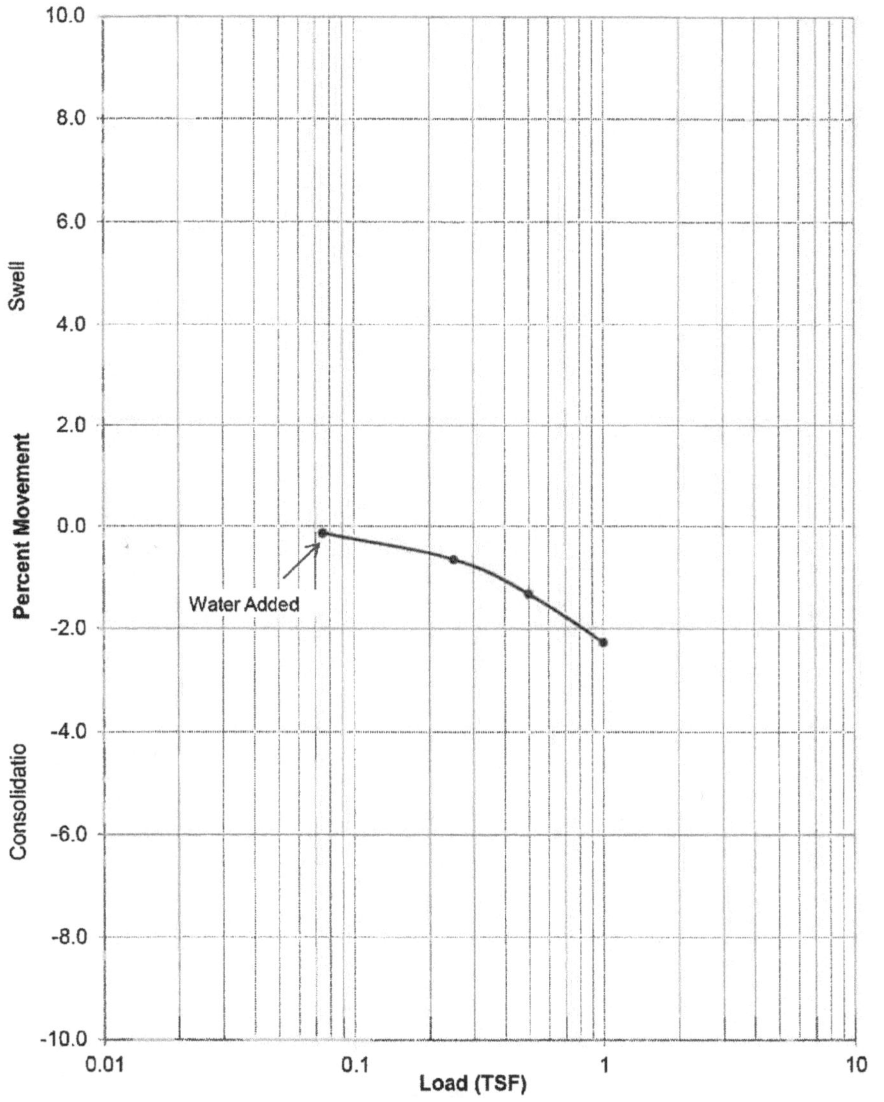

Project: Aims Community College - North Campus Refresh
Location: Greeley, Colorado
Project #: 1162022
Date: March 2016

EEC

SWELL / CONSOLIDATION TEST RESULTS

Material Description:	Brown / Tan Silty Clayey Sand (SM/SC)		
Sample Location:	Boring 4, Sample 3, Depth 9'		
Liquid Limit: 31	Plasticity Index: 16		% Passing #200: 40.8%
Beginning Moisture: 14.5%	Dry Density: 106.6 pcf		Ending Moisture: 27.1%
Swell Pressure: <500 psf		% Swell @ 500: None	

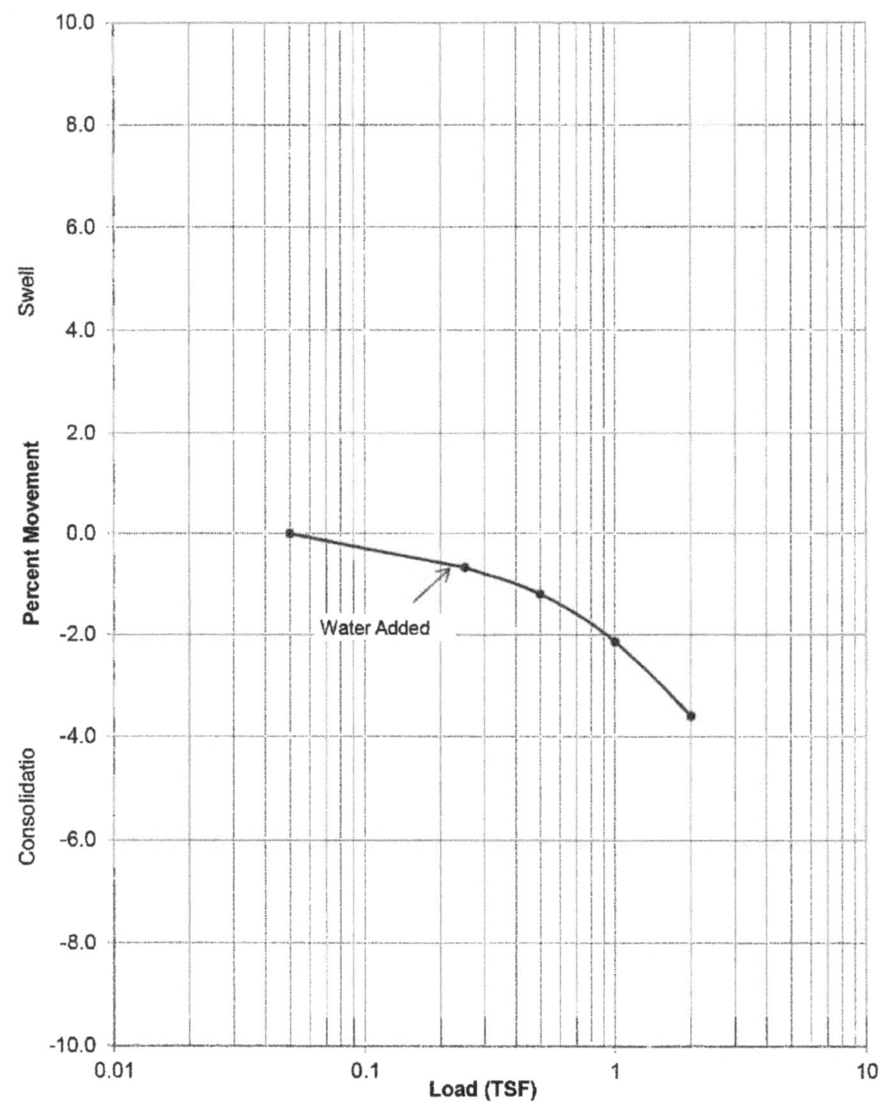

Project: Aims Community College - North Campus Refresh
Location: Greeley, Colorado
Project #: 1162022
Date: March 2016

SWELL / CONSOLIDATION TEST RESULTS

Material Description:	Sandy Lean Clay (CL)		
Sample Location:	Boring 5, Sample 1, Depth 4'		
Liquid Limit: 27	Plasticity Index: 6		% Passing #200: 49.8%
Beginning Moisture: 17.7%	Dry Density: 105.9 pcf		Ending Moisture: 19.9%
Swell Pressure: <500 psf		% Swell @ 500:	None

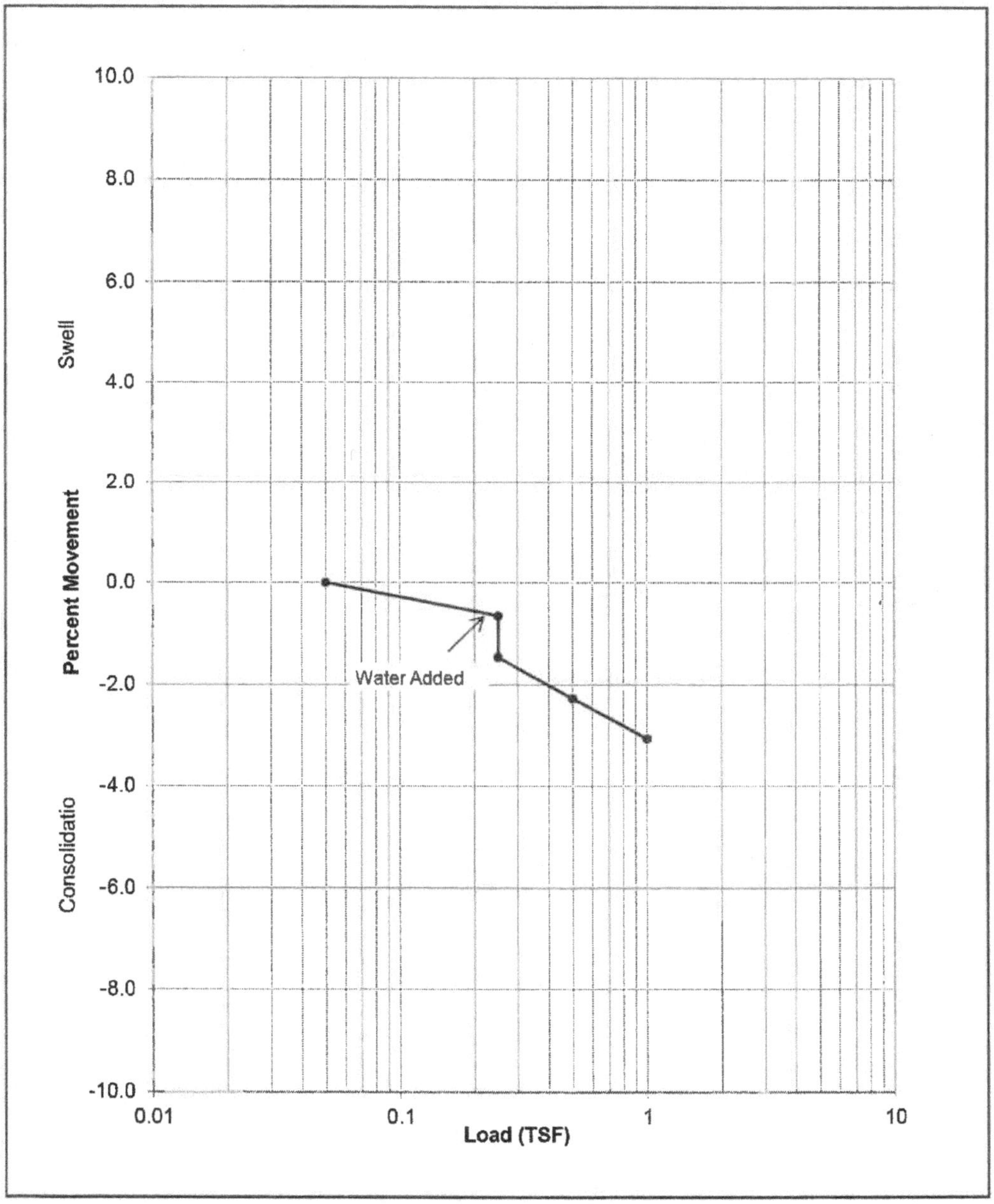

Project: Aims Community College - North Campus Refresh
Location: Greeley, Colorado
Project #: 1162022
Date: March 2016

SWELL / CONSOLIDATION TEST RESULTS

Material Description:	Brown / Tan Sandy Lean Clay (CL)		
Sample Location:	Boring 6, Sample 1, Depth 2'		
Liquid Limit:	Plasticity Index:		% Passing #200:
Beginning Moisture: 12.2%	Dry Density: 122.3 pcf		Ending Moisture: 13.7%
Swell Pressure: <500 psf		% Swell @ 500:	None

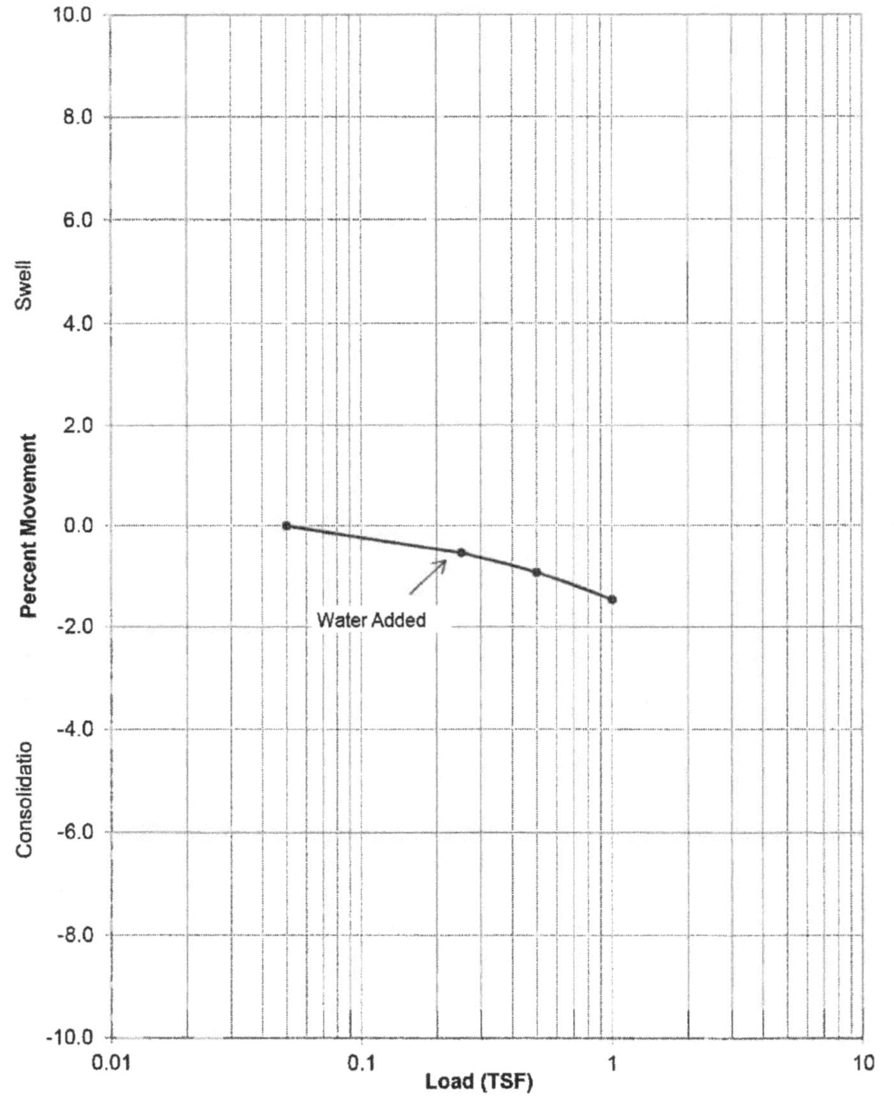

Project: Aims Community College - North Campus Refresh
Location: Greeley, Colorado
Project #: 1162022
Date: March 2016

SWELL / CONSOLIDATION TEST RESULTS

Material Description:	Brown Silty / Clayey Sand (SM/SC)		
Sample Location:	Boring 7, Sample 1, Depth 4'		
Liquid Limit:	Plasticity Index:	% Passing #200:	
Beginning Moisture: 8.0%	Dry Density: 113.8 pcf	Ending Moisture: 16.0%	
Swell Pressure: <500 psf		% Swell @ 500:	None

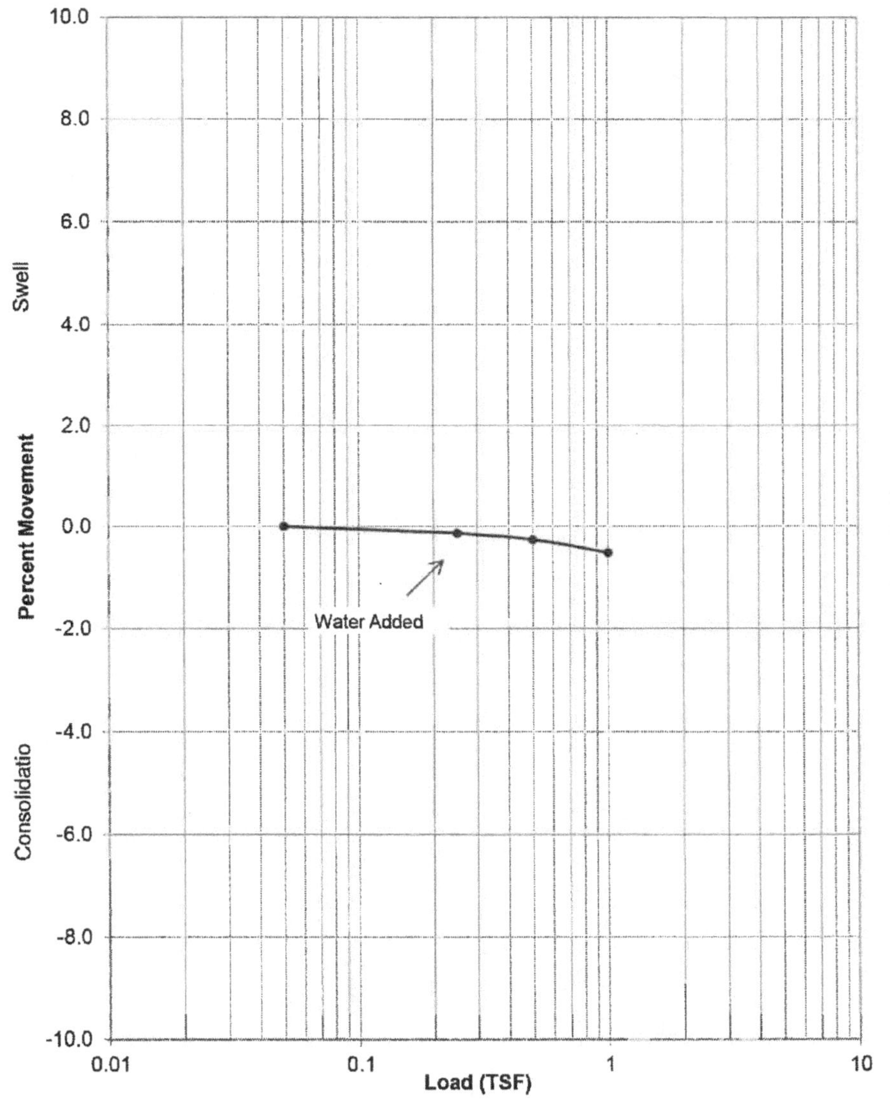

Project: Aims Community College - North Campus Refresh
Location: Greeley, Colorado
Project #: 1162022
Date: March 2016

EEC

SWELL / CONSOLIDATION TEST RESULTS

Material Description:	Brown / Tan Silty / Clayey Sand (SM/SC)		
Sample Location:	Boring 8, Sample 1, Depth 4'		
Liquid Limit: 29	Plasticity Index: 9		% Passing #200: 36.2%
Beginning Moisture: 12.9%	Dry Density: 116.8 pcf		Ending Moisture: 18.0%
Swell Pressure: <500 psf		% Swell @ 500: None	

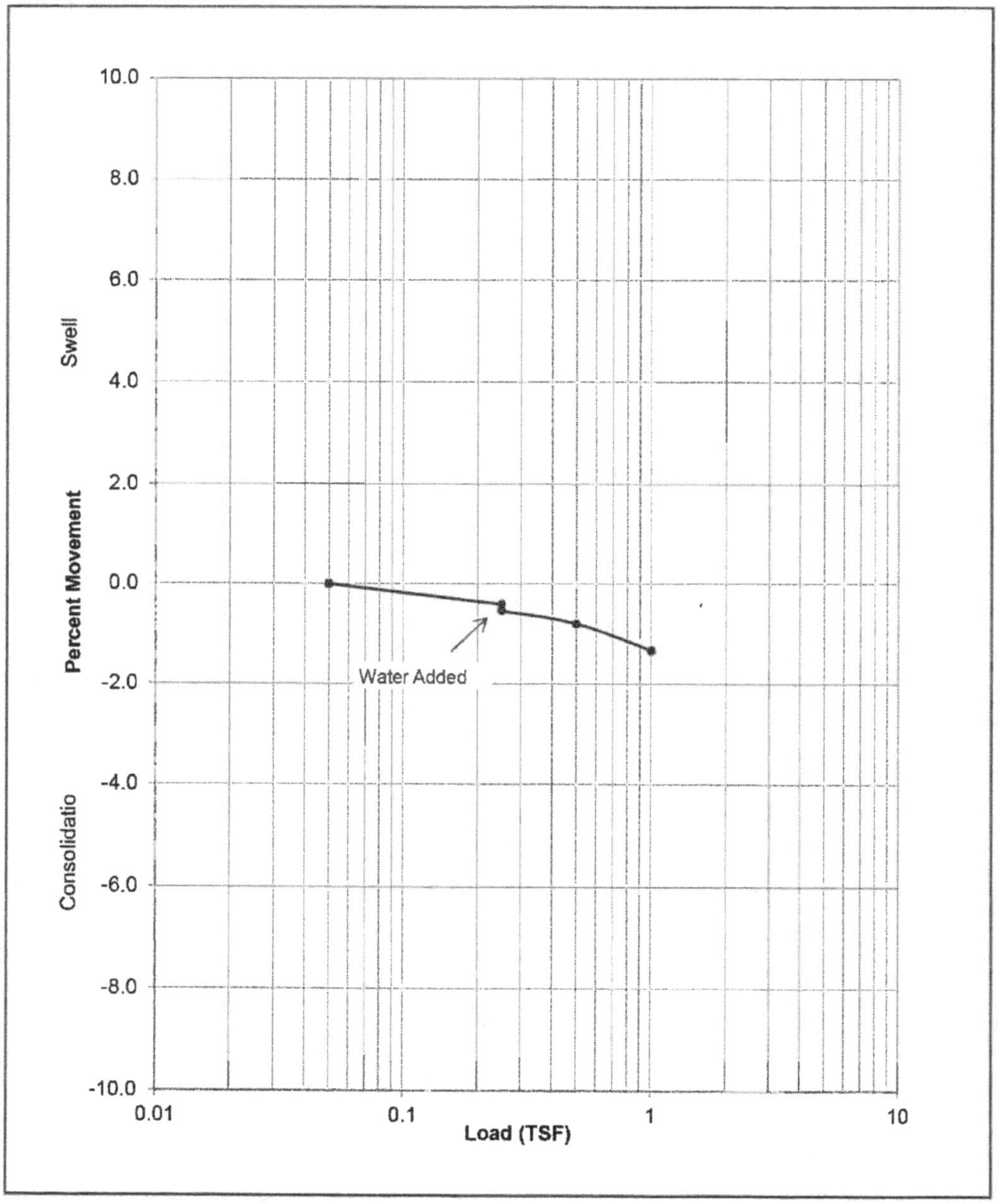

Project: Aims Community College - North Campus Refresh
Location: Greeley, Colorado
Project #: 1162022
Date: March 2016

SWELL / CONSOLIDATION TEST RESULTS

Material Description:	Brown Sandy Lean Clay / Clayey Sand (CL/SC)		
Sample Location:	Boring 9, Sample 1, Depth 4'		
Liquid Limit:	Plasticity Index:	% Passing #200:	
Beginning Moisture: 10.5%	Dry Density: 114.3 pcf	Ending Moisture: 16.0%	
Swell Pressure: <500 psf	% Swell @ 500:	None	

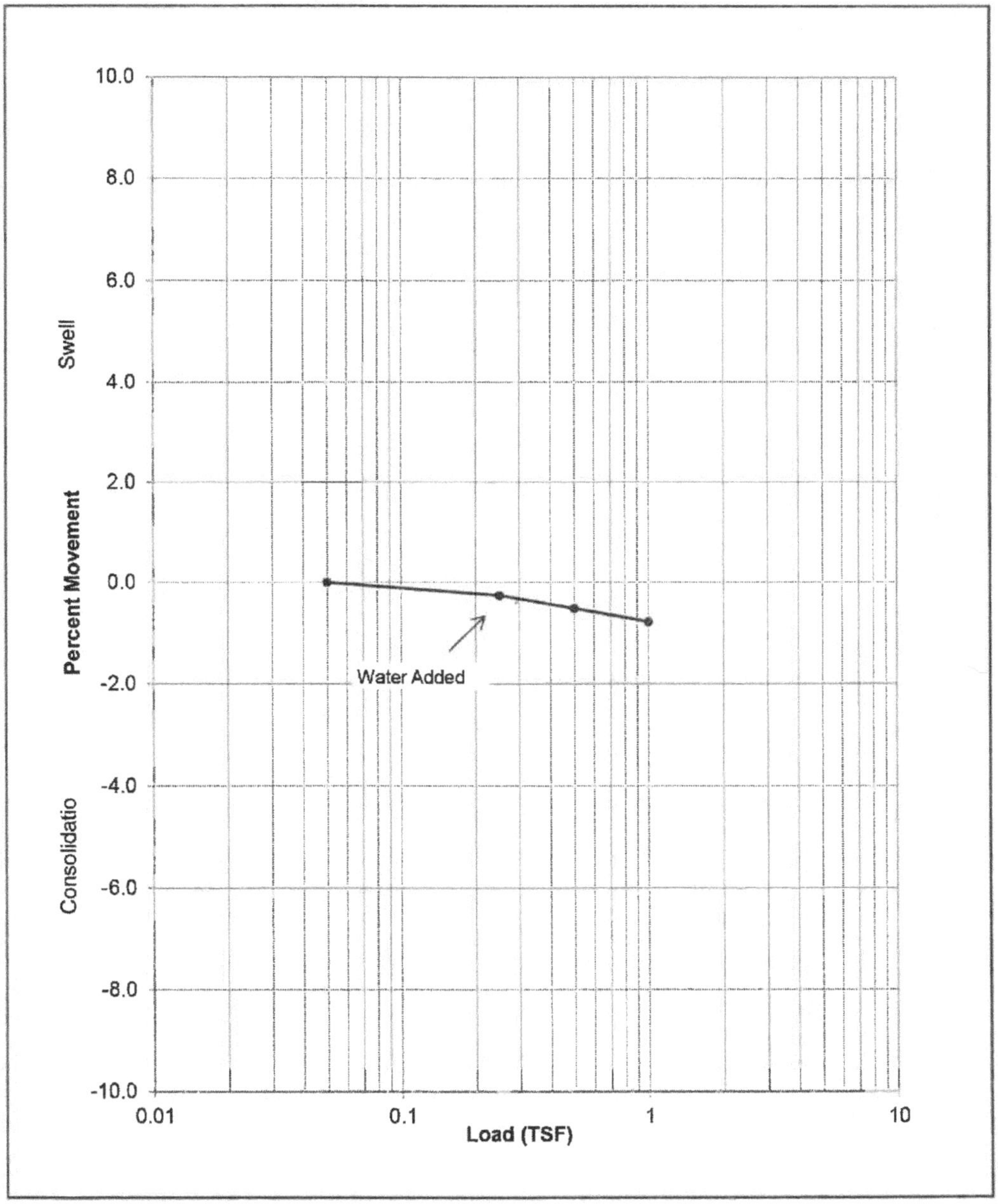

Project: Aims Community College - North Campus Refresh
Location: Greeley, Colorado
Project #: 1162022
Date: March 2016

EEC

SWELL / CONSOLIDATION TEST RESULTS

Material Description:	Brown Silty / Clayey Sand (SM/SC)			
Sample Location:	Boring 11, Sample 1, Depth 2'			
Liquid Limit: NL		Plasticity Index: NP		% Passing #200: 28.1%
Beginning Moisture: 10.1%		Dry Density: 112.5 pcf		Ending Moisture: 17.1%
Swell Pressure: <500 psf			% Swell @ 500: None	

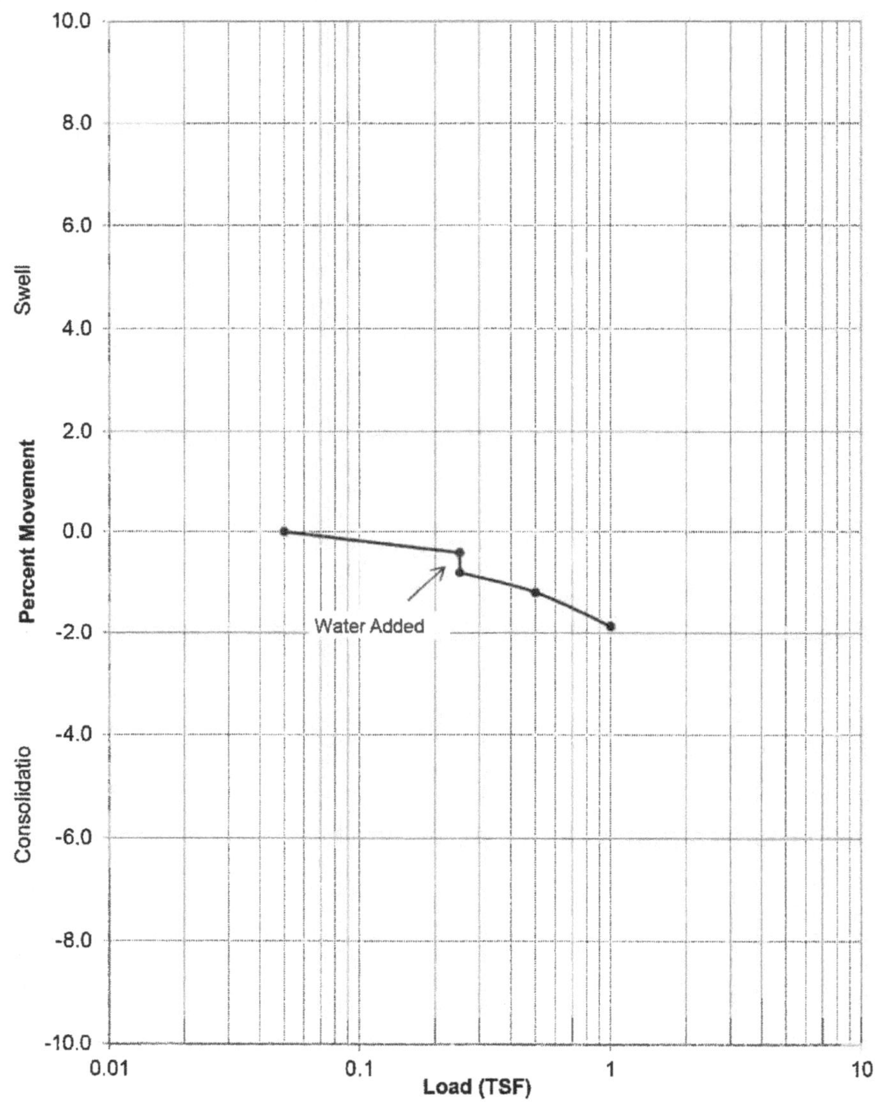

Project: Aims Community College - North Campus Refresh
Location: Greeley, Colorado
Project #: 1162022
Date: March 2016

EEC

SWELL / CONSOLIDATION TEST RESULTS

Material Description:	Brown / Rust Silty / Clayey Sand (SM/SC)		
Sample Location:	Boring 11, Sample 3, Depth 9'		
Liquid Limit: 29	Plasticity Index: 13	% Passing #200: 35.9%	
Beginning Moisture: 12.7%	Dry Density: 119.5 pcf	Ending Moisture: 15.9%	
Swell Pressure: <500 psf	% Swell @ 500: None		

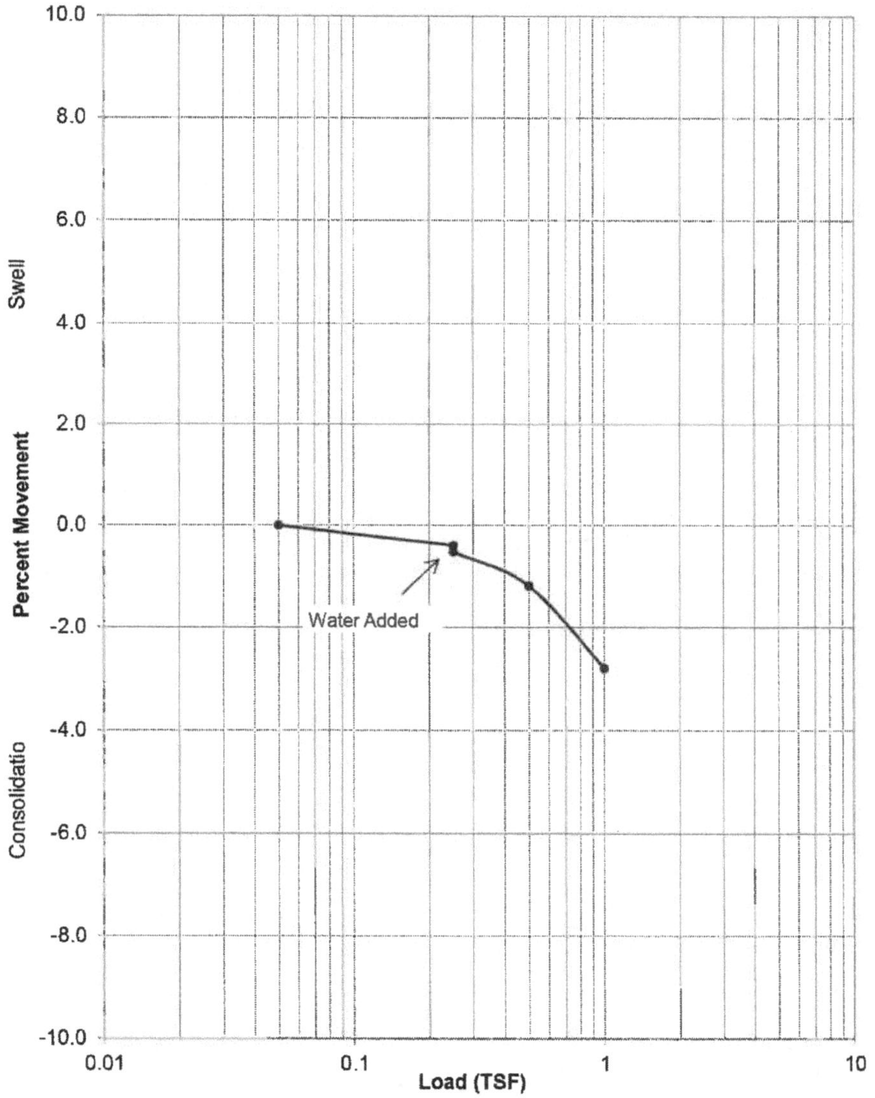

Project: Aims Community College - North Campus Refresh
Location: Greeley, Colorado
Project #: 1162022
Date: March 2016

EEC

SWELL / CONSOLIDATION TEST RESULTS

Material Description:	Sandy Lean Clay / Clayey Sand (CL/SC)		
Sample Location:	Boring 12, Sample 1, Depth 4'		
Liquid Limit:	Plasticity Index:	% Passing #200:	
Beginning Moisture: 14.7%	Dry Density: 112.3 pcf	Ending Moisture: 13.9%	
Swell Pressure: <500 psf	% Swell @ 500:	None	

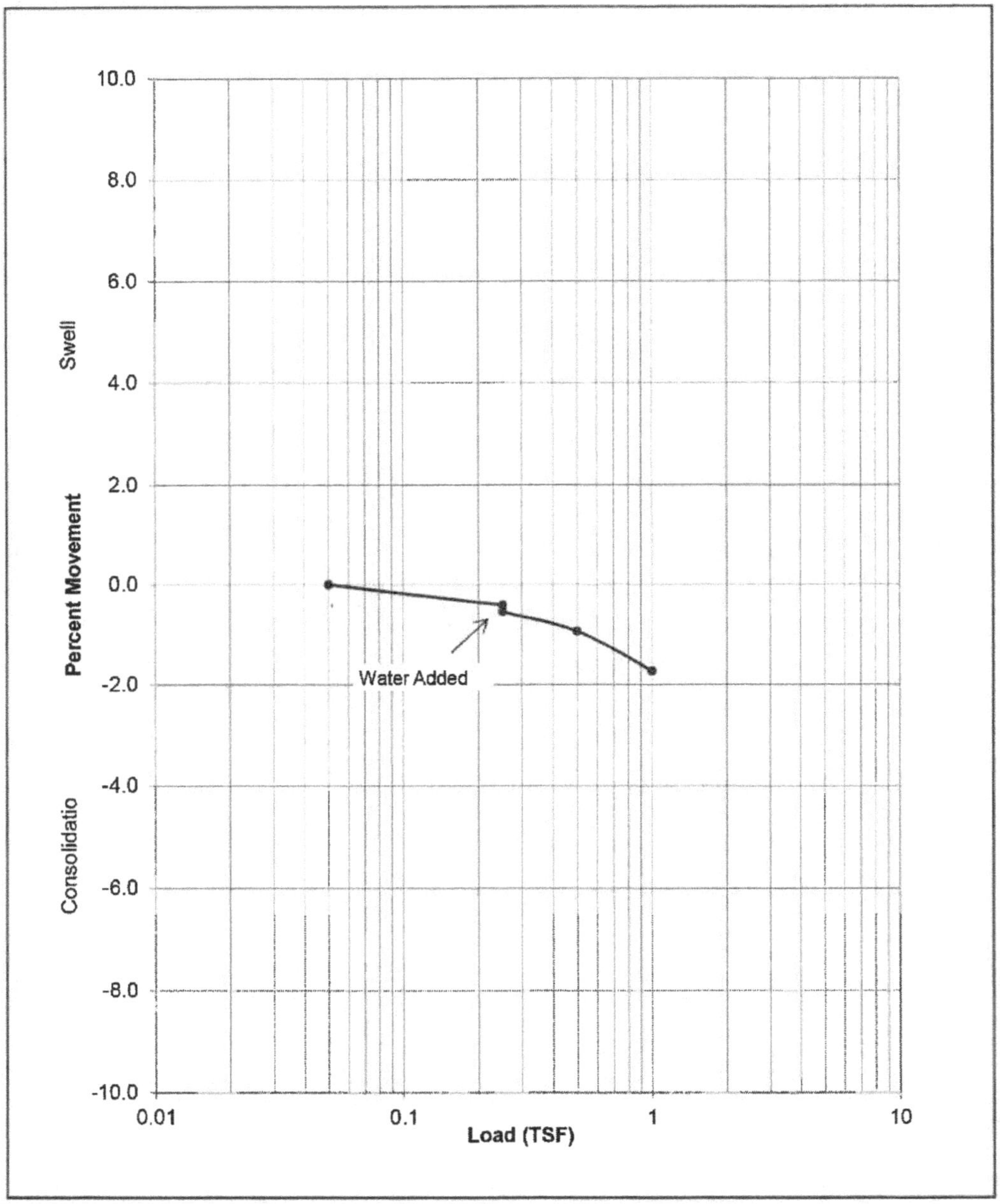

Project: Aims Community College - North Campus Refresh
Location: Greeley, Colorado
Project #: 1162022
Date: March 2016

SWELL / CONSOLIDATION TEST RESULTS

Material Description:	Brown Silty / Clayey Sand (SM/SC)		
Sample Location:	Boring 13, Sample 1, Depth 2'		
Liquid Limit: 27	Plasticity Index: 13	% Passing #200: 42.4%	
Beginning Moisture: 17.8%	Dry Density: 108.5 pcf	Ending Moisture: 18.4%	
Swell Pressure: <500 psf		% Swell @ 500:	None

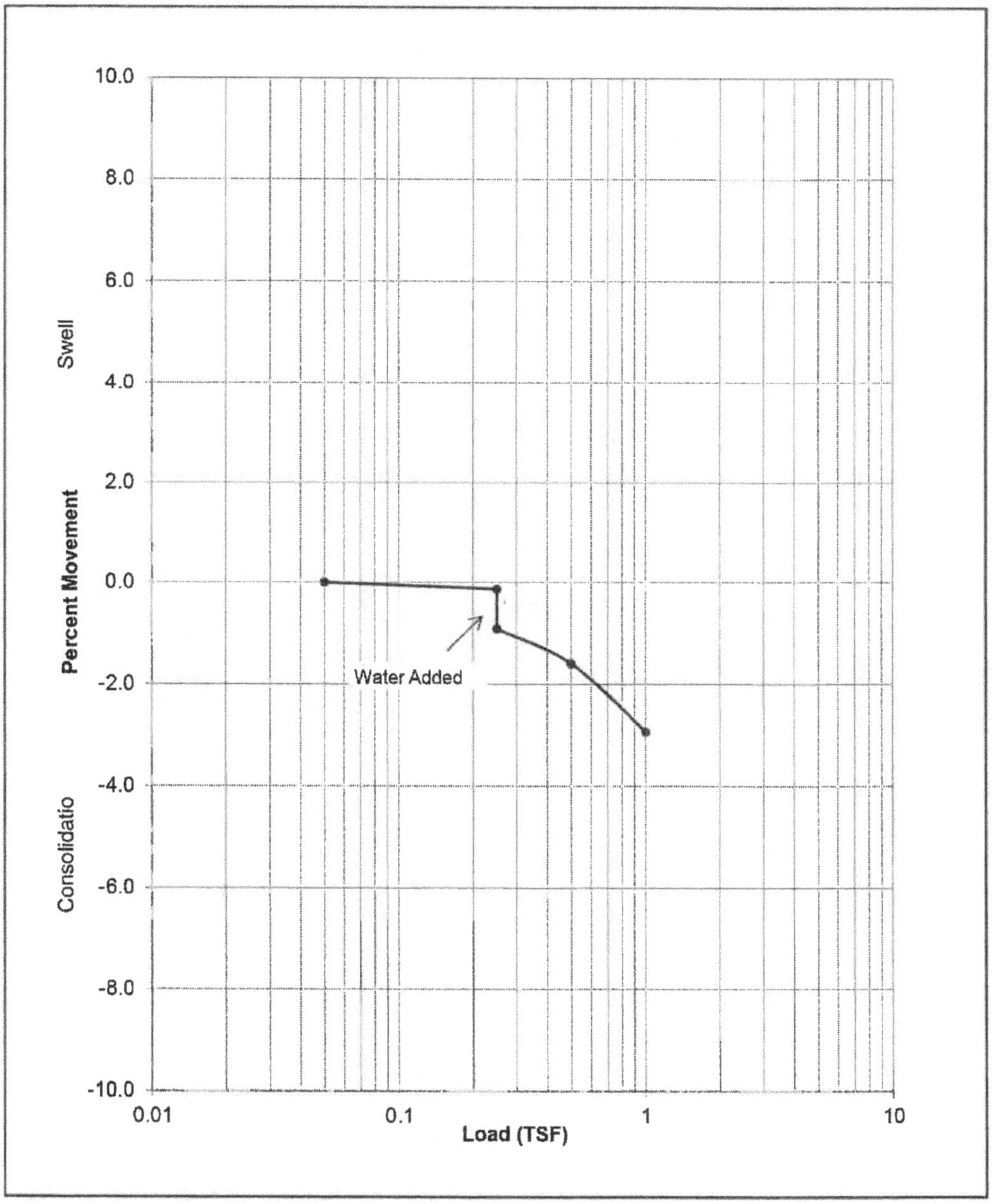

Project: Aims Community College - North Campus Refresh
Location: Greeley, Colorado
Project #: 1162022
Date: March 2016

EEC

SWELL / CONSOLIDATION TEST RESULTS

Material Description: Rust / Brown / Grey Sandstone / Claystone / Siltstone		
Sample Location: Boring 13, Sample 7, Depth 29'		
Liquid Limit: NL	Plasticity Index: NP	% Passing #200: 16.5%
Beginning Moisture: 26.4%	Dry Density: 97.2 pcf	Ending Moisture: 26.0%
Swell Pressure: <1000 psf	% Swell @ 1000: None	

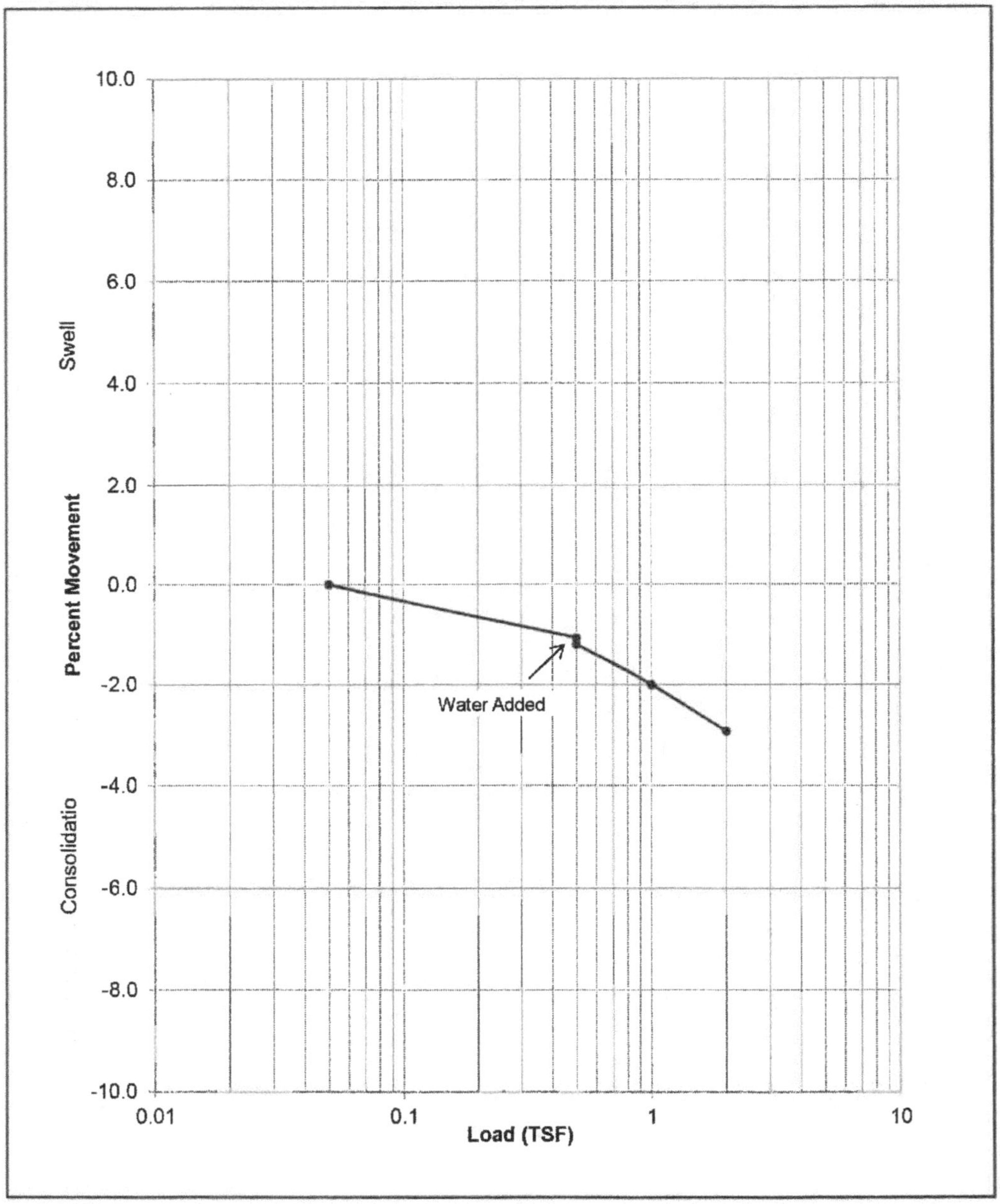

Project: Aims Community College - North Campus Refresh
Location: Greeley, Colorado
Project #: 1162022
Date: March 2016

EEC

SWELL / CONSOLIDATION TEST RESULTS

Material Description:	Clayey Sand to Sandy Lean Clay (SC/CL)		
Sample Location:	Boring 14, Sample 1, Depth 2'		
Liquid Limit:	Plasticity Index:	% Passing #200:	
Beginning Moisture: 6.1%	Dry Density: 113.9 pcf	Ending Moisture: 16.7%	
Swell Pressure: <500 psf		% Swell @ 500:	None

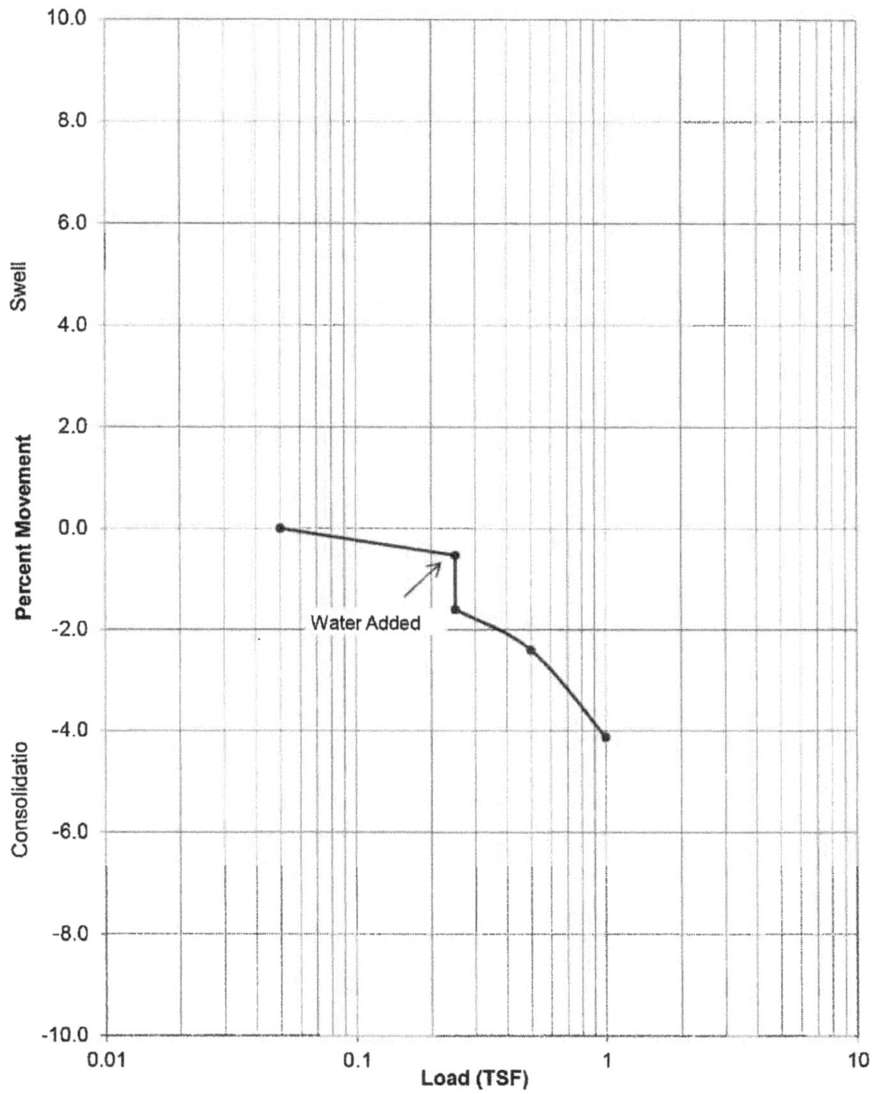

Project: Aims Community College - North Campus Refresh
Location: Greeley, Colorado
Project #: 1162022
Date: March 2016

SWELL / CONSOLIDATION TEST RESULTS

Material Description:	Brown / Rust Clayey Sand (SC)		
Sample Location:	Boring 15, Sample 1, Depth 4'		
Liquid Limit: 24	Plasticity Index: 9		% Passing #200: 35.4%
Beginning Moisture: 12.1%	Dry Density: 117.5 pcf		Ending Moisture: 14.1%
Swell Pressure: <500 psf		% Swell @ 500: None	

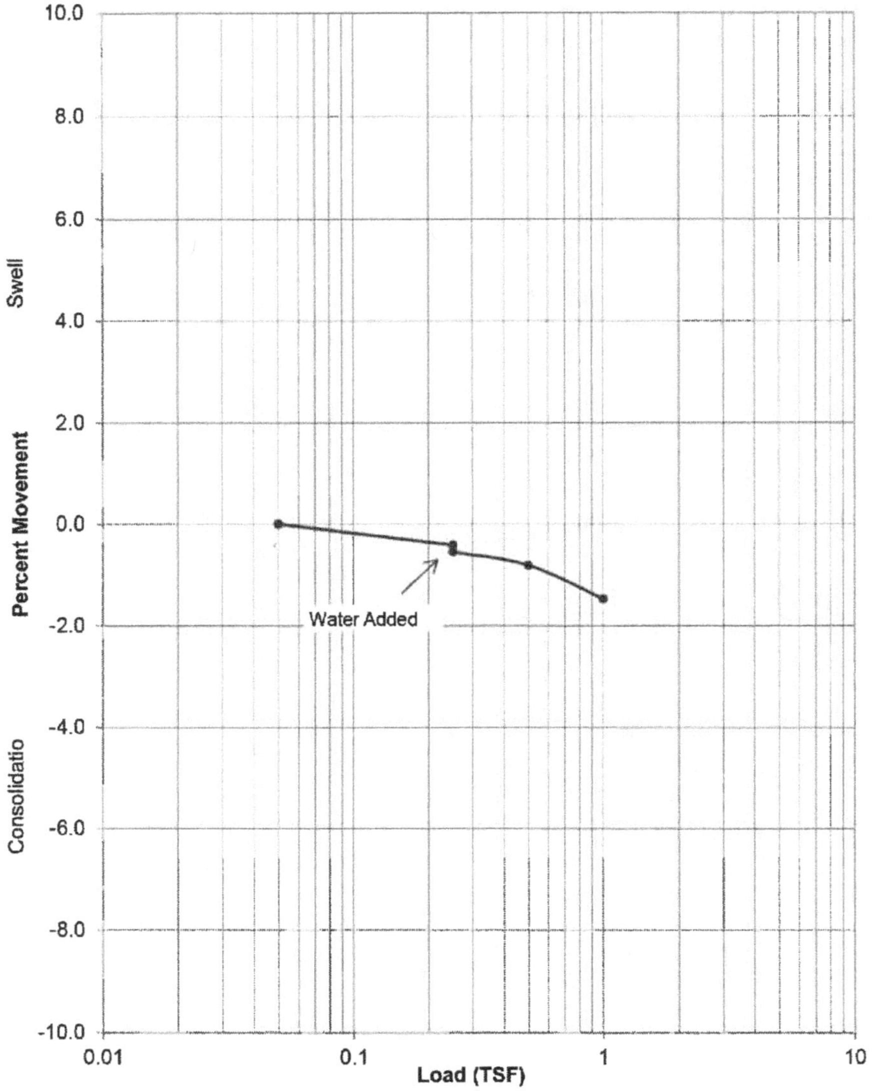

Project: Aims Community College - North Campus Refresh
Location: Greeley, Colorado
Project #: 1162022
Date: March 2016

EEC

SWELL / CONSOLIDATION TEST RESULTS

Material Description:	Brown Sandy Lean Clay to Clayey Sand (CL/SC)		
Sample Location:	Boring 16, Sample 1, Depth 4'		
Liquid Limit: 25	Plasticity Index: 10		% Passing #200: 46.9%
Beginning Moisture: 16.2%	Dry Density: 112.4 pcf		Ending Moisture: 16.3%
Swell Pressure: <500 psf		% Swell @ 500: None	

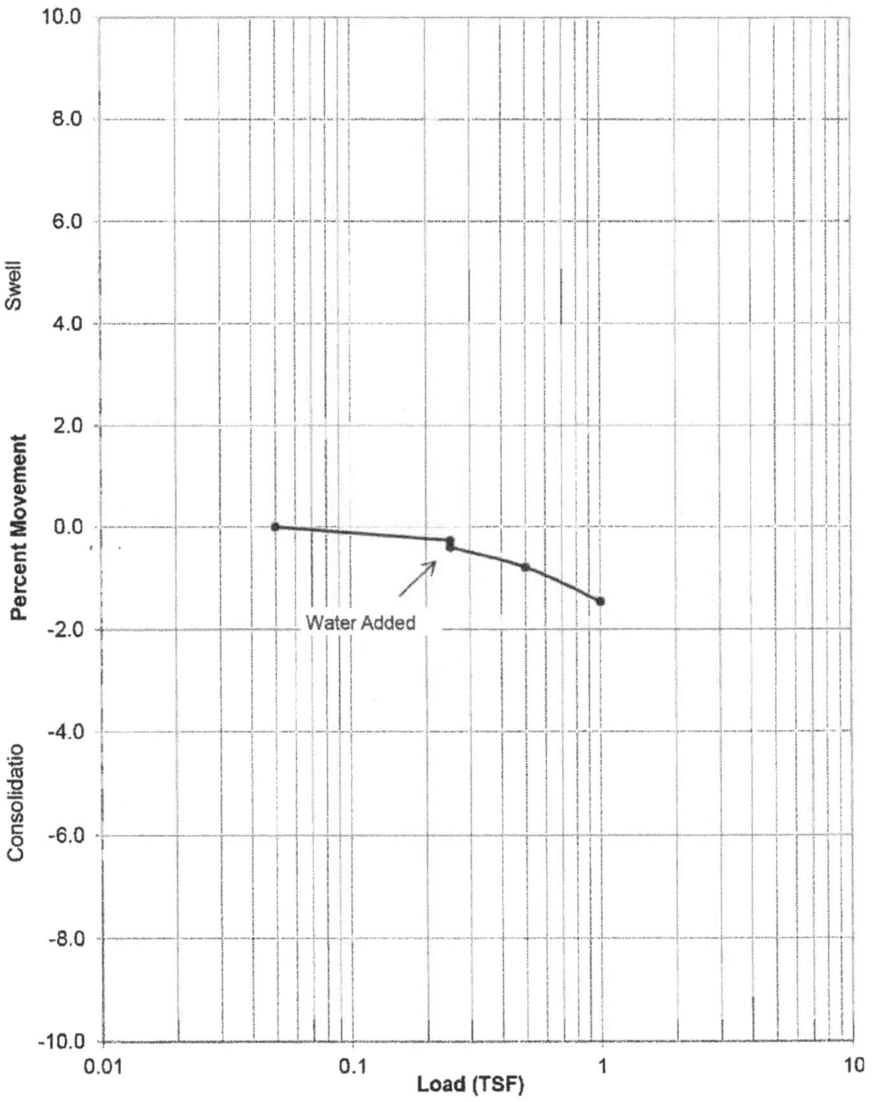

Project: Aims Community College - North Campus Refresh
Location: Greeley, Colorado
Project #: 1162022
Date: March 2016

EEC

SWELL / CONSOLIDATION TEST RESULTS

Material Description: Brown / Grey / Rust Claystone / Siltstone / Sandstone		
Sample Location: Boring 16, Sample 3, Depth 14'		
Liquid Limit: 48	Plasticity Index: 20	% Passing #200: 86.9%
Beginning Moisture: 21.1%	Dry Density: 109 pcf	Ending Moisture: 20.4%
Swell Pressure: 1200 psf	% Swell @ 1000: 0.1%	

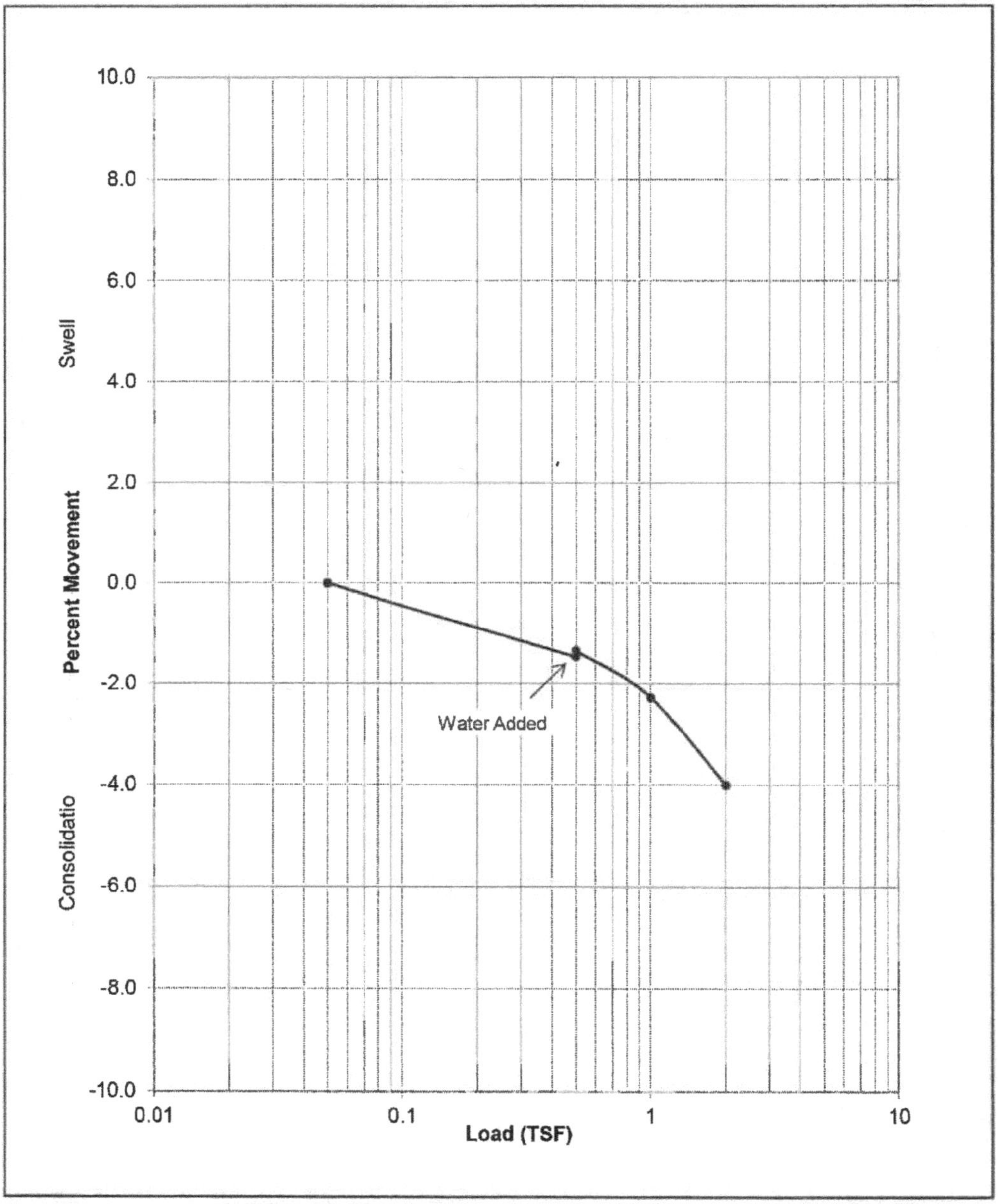

Project: Aims Community College - North Campus Refresh
Location: Greeley, Colorado
Project #: 1162022
Date: March 2016

EEC

SWELL / CONSOLIDATION TEST RESULTS

Material Description: Brown Clayey Sand / Sandy Lean Clay (SC/CL)		
Sample Location: Boring 17, Sample 1, Depth 2'		
Liquid Limit:	Plasticity Index:	% Passing #200:
Beginning Moisture: 6.0%	Dry Density: 109.1 pcf	Ending Moisture: 17.8%
Swell Pressure: 180 psf	% Swell @ 150: 0.1%	

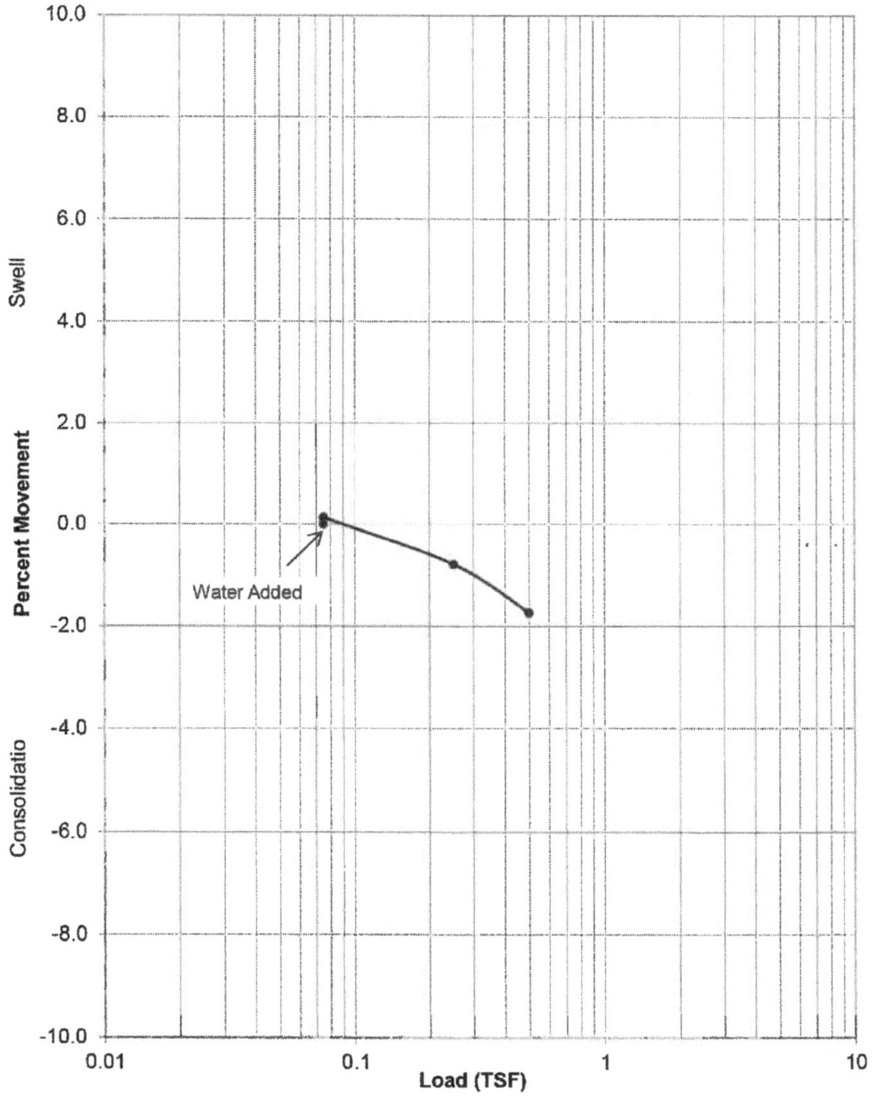

Project: Aims Community College - North Campus Refresh
Location: Greeley, Colorado
Project #: 1162022
Date: March 2016

EEC

SWELL / CONSOLIDATION TEST RESULTS

Material Description: Brown / Rust Sandstone		
Sample Location: Boring 18, Sample 5, Depth 19'		
Liquid Limit: NL	Plasticity Index: NP	% Passing #200: 28.8%
Beginning Moisture: 26.0%	Dry Density: 95.9 pcf	Ending Moisture: 27.0%
Swell Pressure: <1000 psf	% Swell @ 1000: None	

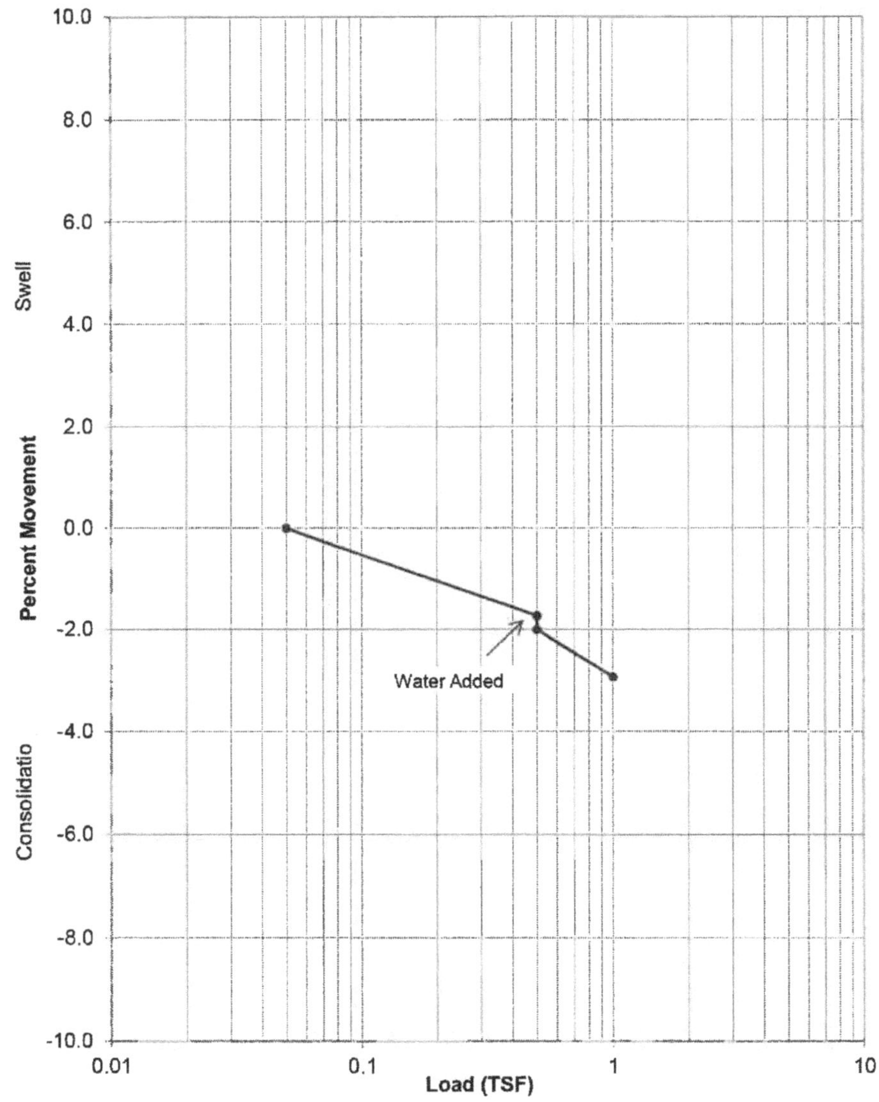

Project: Aims Community College - North Campus Refresh
Location: Greeley, Colorado
Project #: 1162022
Date: March 2016

SWELL / CONSOLIDATION TEST RESULTS

Material Description:	Brown / Rust Sandstone		
Sample Location:	Boring 19, Sample 3, Depth 14'		
Liquid Limit: 26	Plasticity Index: 6		% Passing #200: 48.0%
Beginning Moisture: 23.5%	Dry Density: 98.9 pcf		Ending Moisture: 25.5%
Swell Pressure: <1000 psf		% Swell @ 1000: None	

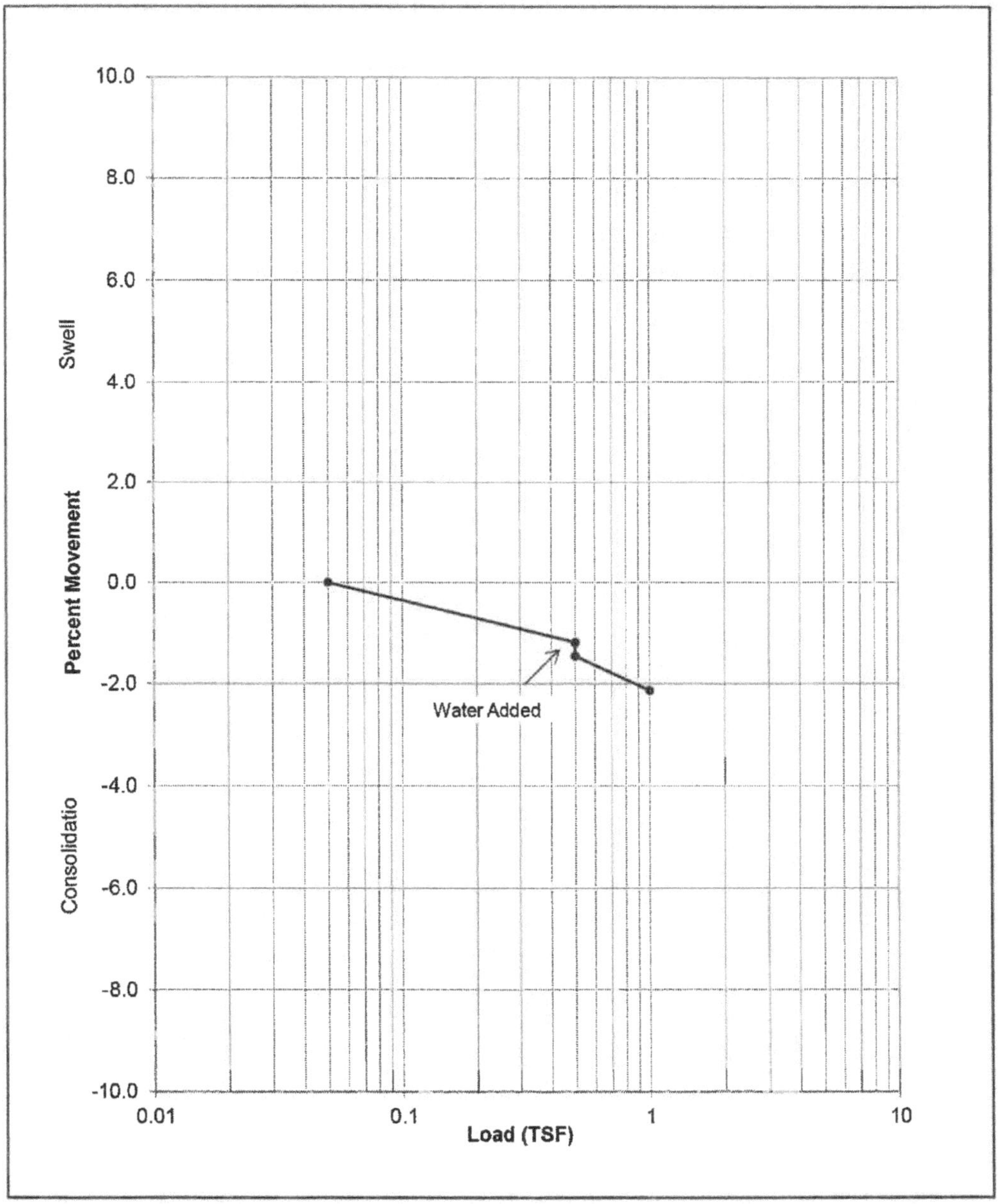

Project: Aims Community College - North Campus Refresh
Location: Greeley, Colorado
Project #: 1162022
Date: March 2016

SWELL / CONSOLIDATION TEST RESULTS

Material Description:	Brown Sandy Lean Clay to Clayey Sand (CL/SC)		
Sample Location:	Boring 21, Sample 1, Depth 4'		
Liquid Limit: 26	Plasticity Index: 11		% Passing #200: 37.8%
Beginning Moisture: 8.3%	Dry Density: 117.7 pcf		Ending Moisture: 15.0%
Swell Pressure: <500 psf		% Swell @ 500: None	

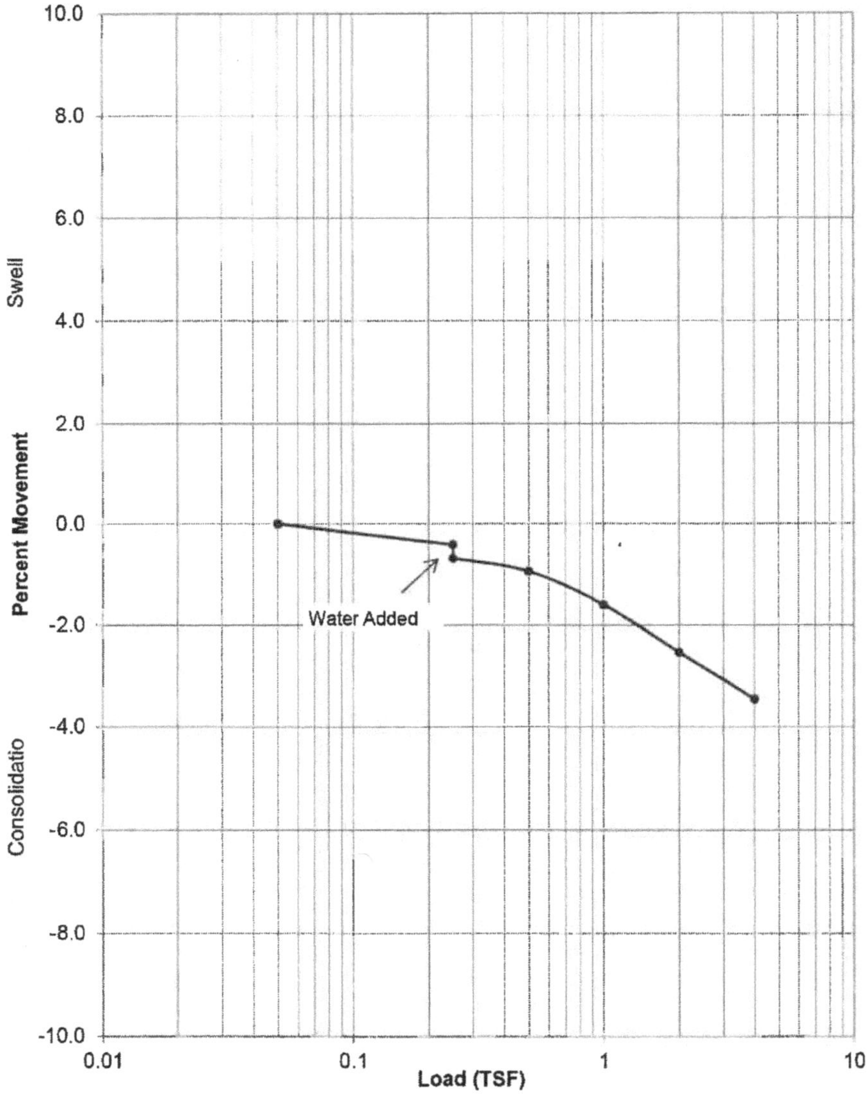

Project: Aims Community College - North Campus Refresh
Location: Greeley, Colorado
Project #: 1162022
Date: March 2016

EEC

SWELL / CONSOLIDATION TEST RESULTS

Material Description:	Brown / Tan Sandy Lean Clay to Clayey Sand (CL/SC)			
Sample Location:	Boring 22, Sample 1, Depth 4'			
Liquid Limit:		Plasticity Index:	% Passing #200:	
Beginning Moisture: 14.5%		Dry Density: 110.2 pcf	Ending Moisture: 15.8%	
Swell Pressure: <500 psf		% Swell @ 500:	None	

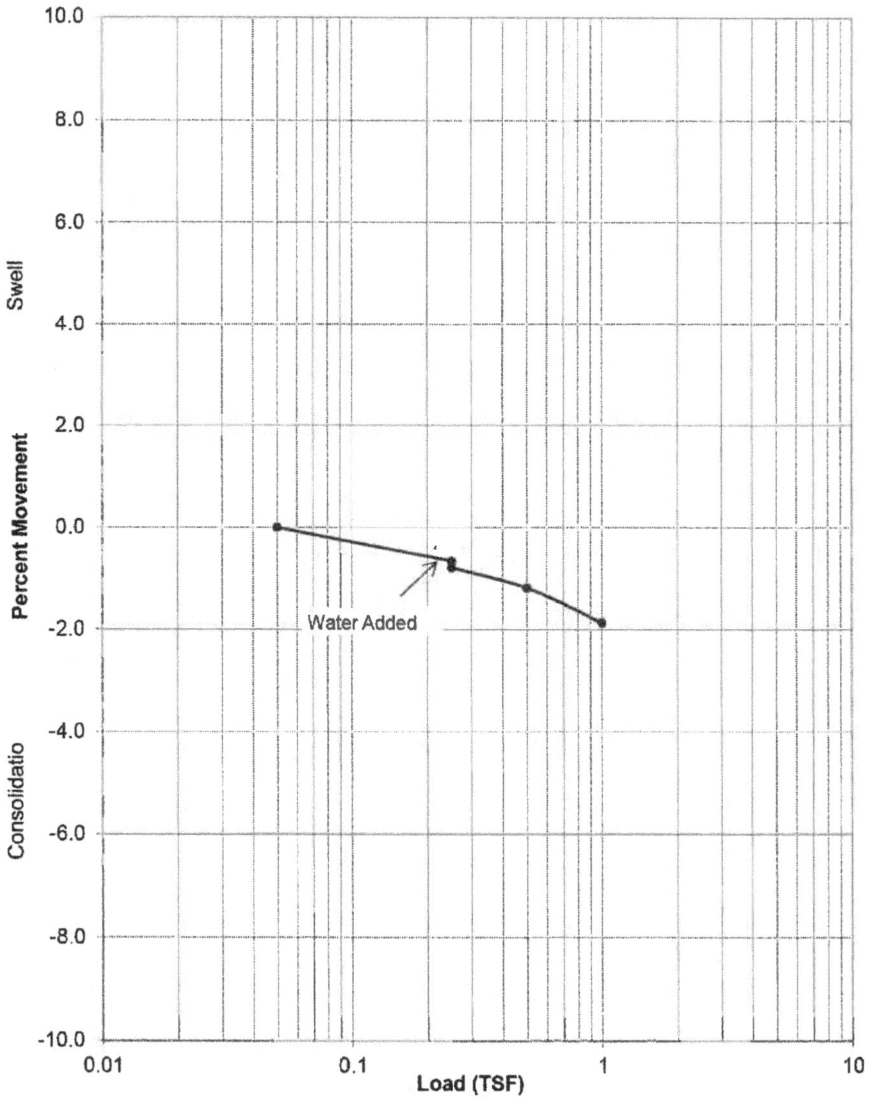

Project:	Aims Community College - North Campus Refresh
Location:	Greeley, Colorado
Project #:	1162022
Date:	March 2016

EEC

SWELL / CONSOLIDATION TEST RESULTS

Material Description: Sandstone		
Sample Location: Boring 22, Sample 5, Depth 24'		
Liquid Limit:	Plasticity Index:	% Passing #200:
Beginning Moisture: 25.0%	Dry Density: 96.5 pcf	Ending Moisture: 26.0%
Swell Pressure: <1000 psf	% Swell @ 1000: None	

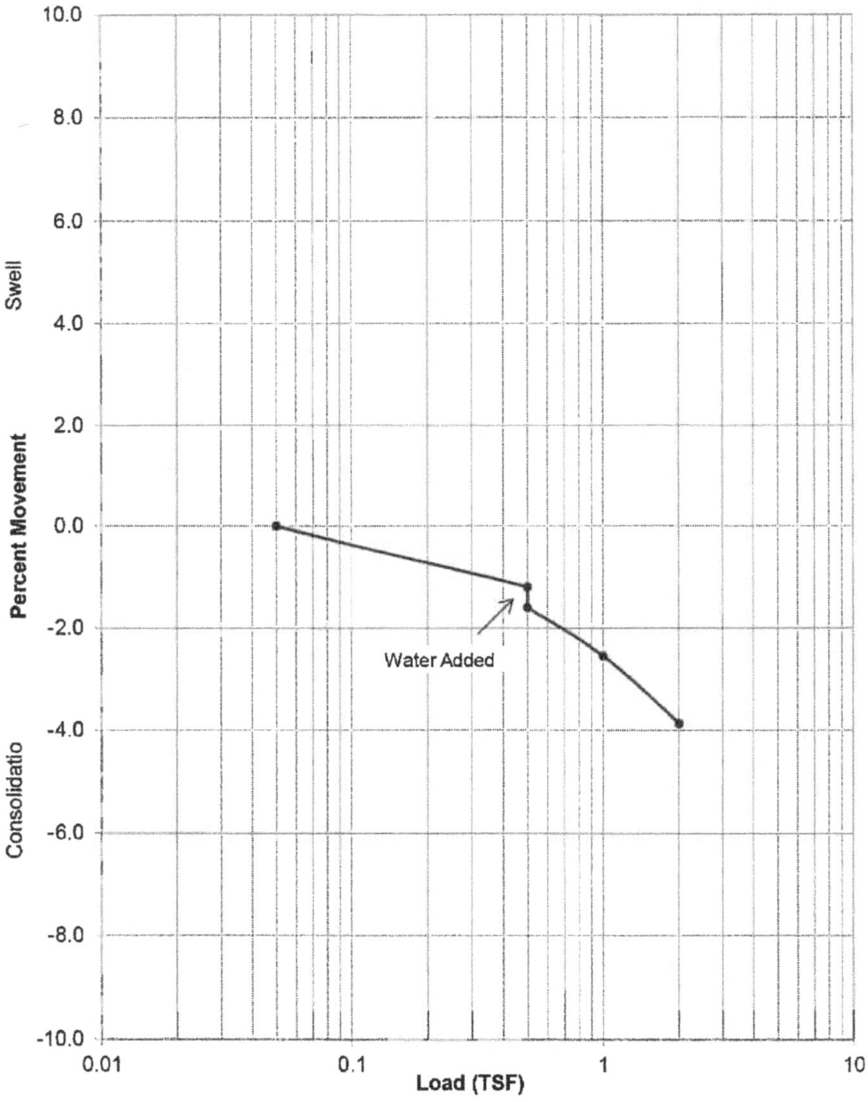

Project: Aims Community College - North Campus Refresh
Location: Greeley, Colorado
Project #: 1162022
Date: March 2016

EEC

SWELL / CONSOLIDATION TEST RESULTS

Material Description:	Brown / Tan Sandy Lean Clay (CL)		
Sample Location:	Boring 23, Sample 1, Depth 2'		
Liquid Limit:	Plasticity Index:	% Passing #200:	
Beginning Moisture: 17.1%	Dry Density: 113 pcf	Ending Moisture: 18.2%	
Swell Pressure: 200 psf		% Swell @ 150:	0.1%

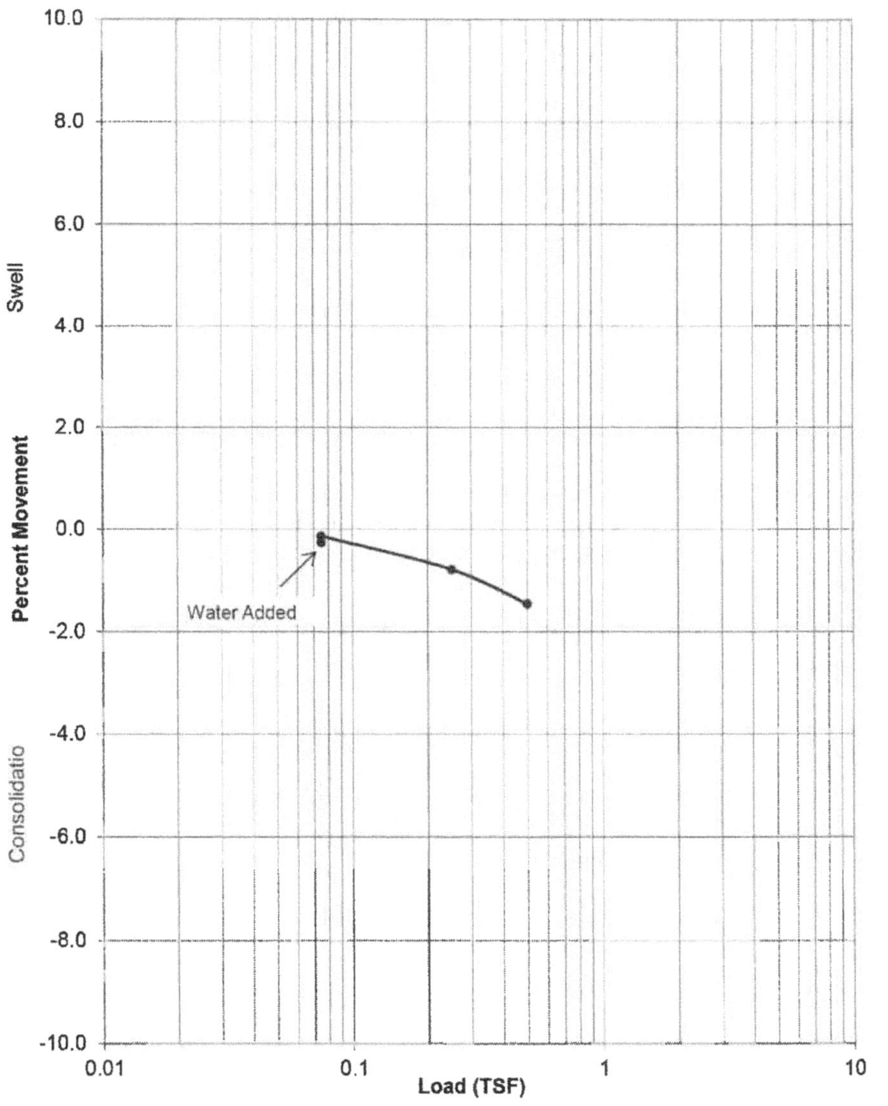

Project: Aims Community College - North Campus Refresh
Location: Greeley, Colorado
Project #: 1162022
Date: March 2016

EEC

SWELL / CONSOLIDATION TEST RESULTS

Material Description: Brown / Rust Sandy Lean Clay (CL)		
Sample Location: Boring 24, Sample 1, Depth 2'		
Liquid Limit: 33	Plasticity Index: 20	% Passing #200: 42.0%
Beginning Moisture: 20.3%	Dry Density: 106.6 pcf	Ending Moisture: 21.0%
Swell Pressure: <150 psf	% Swell @ 150: None	

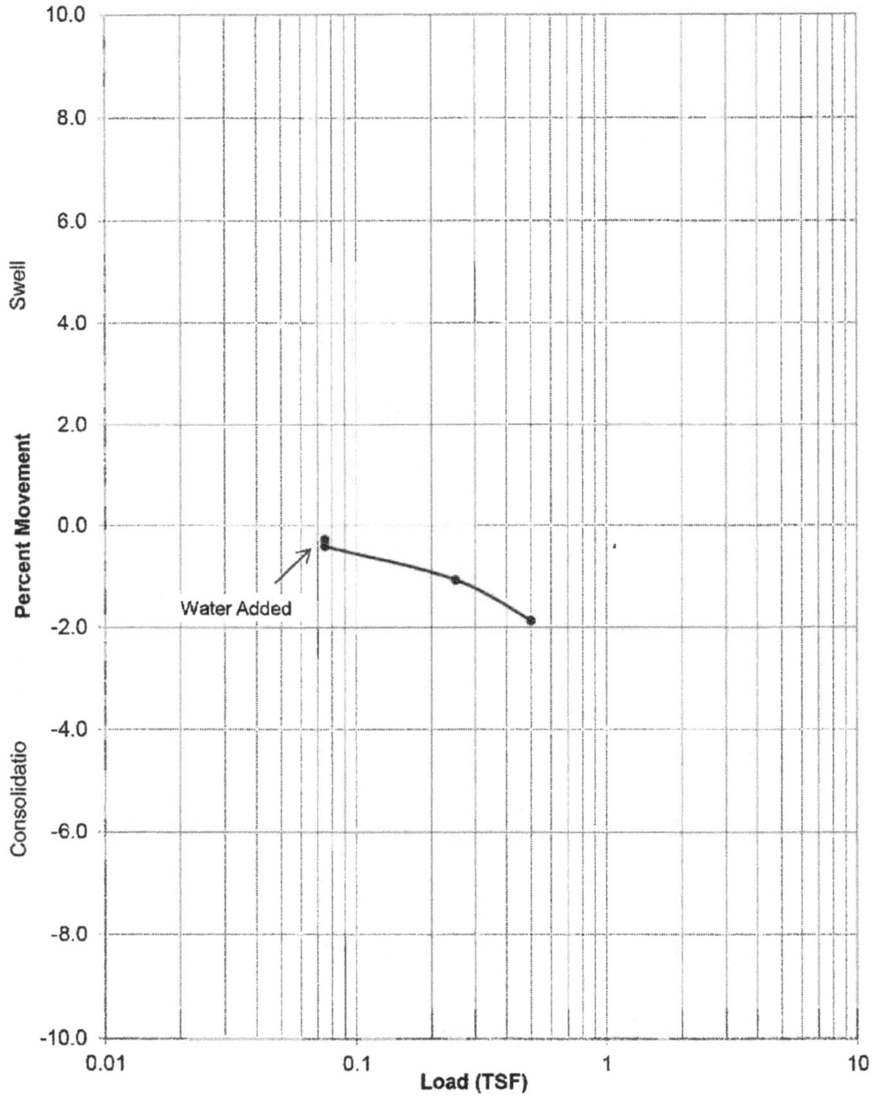

Project: Aims Community College - North Campus Refresh
Location: Greeley, Colorado
Project #: 1162022
Date: March 2016

EEC

EARTH ENGINEERING CONSULTANTS, LLC

SUMMARY OF LABORATORY TEST RESULTS

Sieve Analysis (AASHTO T 11 & T 27 / ASTM C 117 & C 136)		
Sieve Size		Percent Passing
3/4"	(19 mm)	100
1/2"	(12.5 mm)	100
3/8"	(9.5 mm)	95
No. 4	(4.75 mm)	80
No. 8	(2.36 mm)	62
No. 10	(2 mm)	57
No. 16	(1.18 mm)	42
No. 30	(0.6 mm)	23
No. 40	(0.425 mm)	16
No. 50	(0.3 mm)	12
No. 100	(0.15 mm)	8
No. 200	(0.075 mm)	5.3

Project: Aims Community College - North Campus Refresh
Location: Greeley, Colorado
Project No: 1162022
Sample ID: B1, S2, 9'
Sample Desc.: Sand & Gravel (SP/GP)
Date: March 2016

EEC

EARTH ENGINEERING CONSULTANTS, LLC

Summary of Washed Sieve Analysis Tests (ASTM C117 & C136)

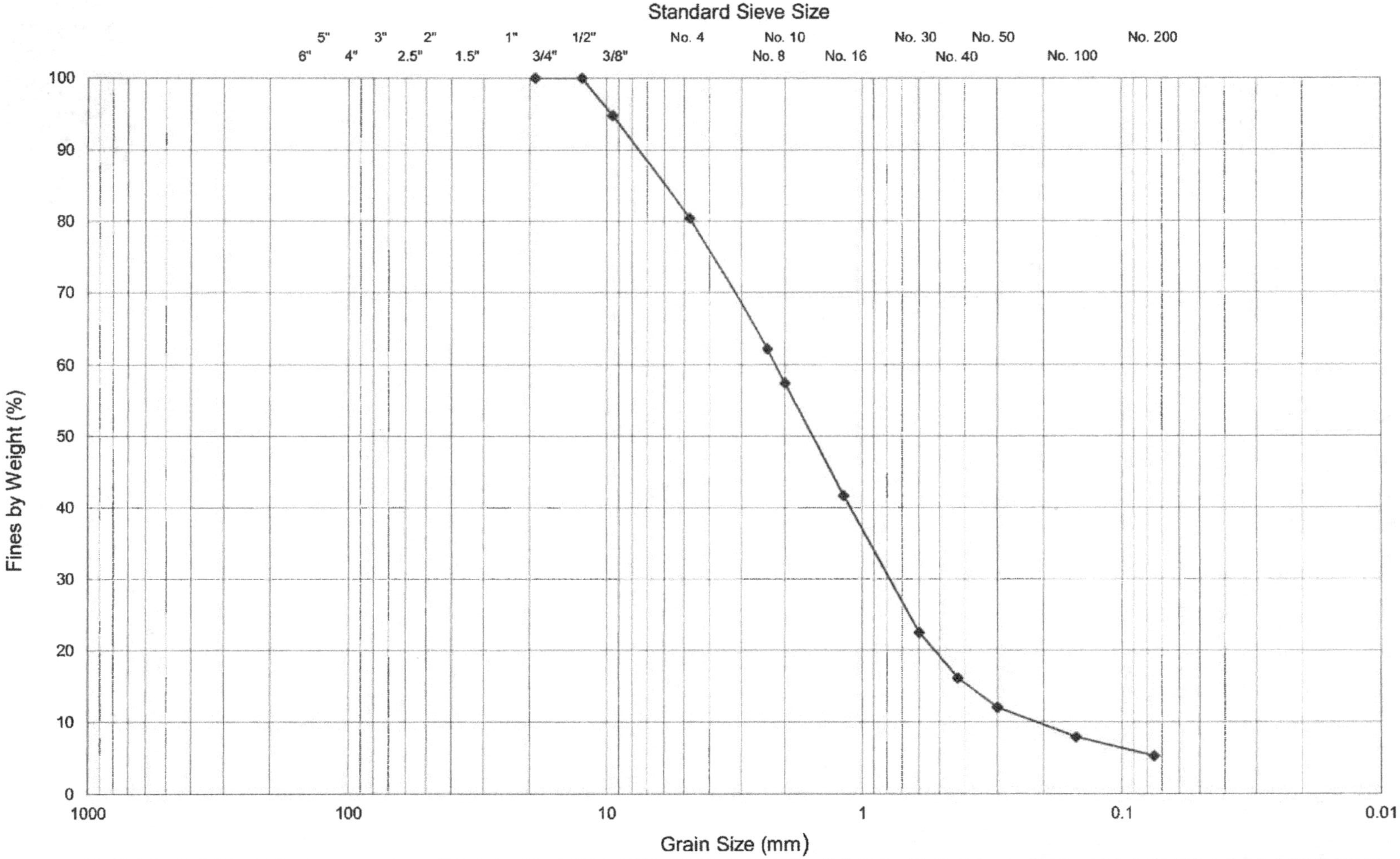

Cobble	Gravel		Sand			Silt or Clay
	Coarse	Fine	Coarse	Medium	Fine	

Project: Aims Community College - North Campus Refresh
Location: Greeley, Colorado
Project No: 1162022
Sample ID: B1, S2, 9'
Sample Desc.: Sand & Gravel (SP/GP)
Date: March 2016

EARTH ENGINEERING CONSULTANTS, LLC

SUMMARY OF LABORATORY TEST RESULTS

Sieve Analysis (AASHTO T 11 & T 27 / ASTM C 117 & C 136)

Sieve Size		Percent Passing
2 1/2"	(63 mm)	100
2"	(50 mm)	100
1 1/2"	(37.5 mm)	90
1"	(25 mm)	72
3/4"	(19 mm)	66
1/2"	(12.5 mm)	60
3/8"	(9.5 mm)	55
No. 4	(4.75 mm)	44
No. 8	(2.36 mm)	35
No. 10	(2 mm)	33
No. 16	(1.18 mm)	28
No. 30	(0.6 mm)	22
No. 40	(0.425 mm)	15
No. 50	(0.3 mm)	13
No. 100	(0.15 mm)	10
No. 200	(0.075 mm)	6.1

Project: Aims Community College - North Campus Refresh
Location: Greeley, Colorado
Project No: 1162022
Sample ID: B3, S3, 9'
Sample Desc.: Sand & Gravel (SP/GP)
Date: March 2016

EARTH ENGINEERING CONSULTANTS, LLC

Summary of Washed Sieve Analysis Tests (ASTM C117 & C136)

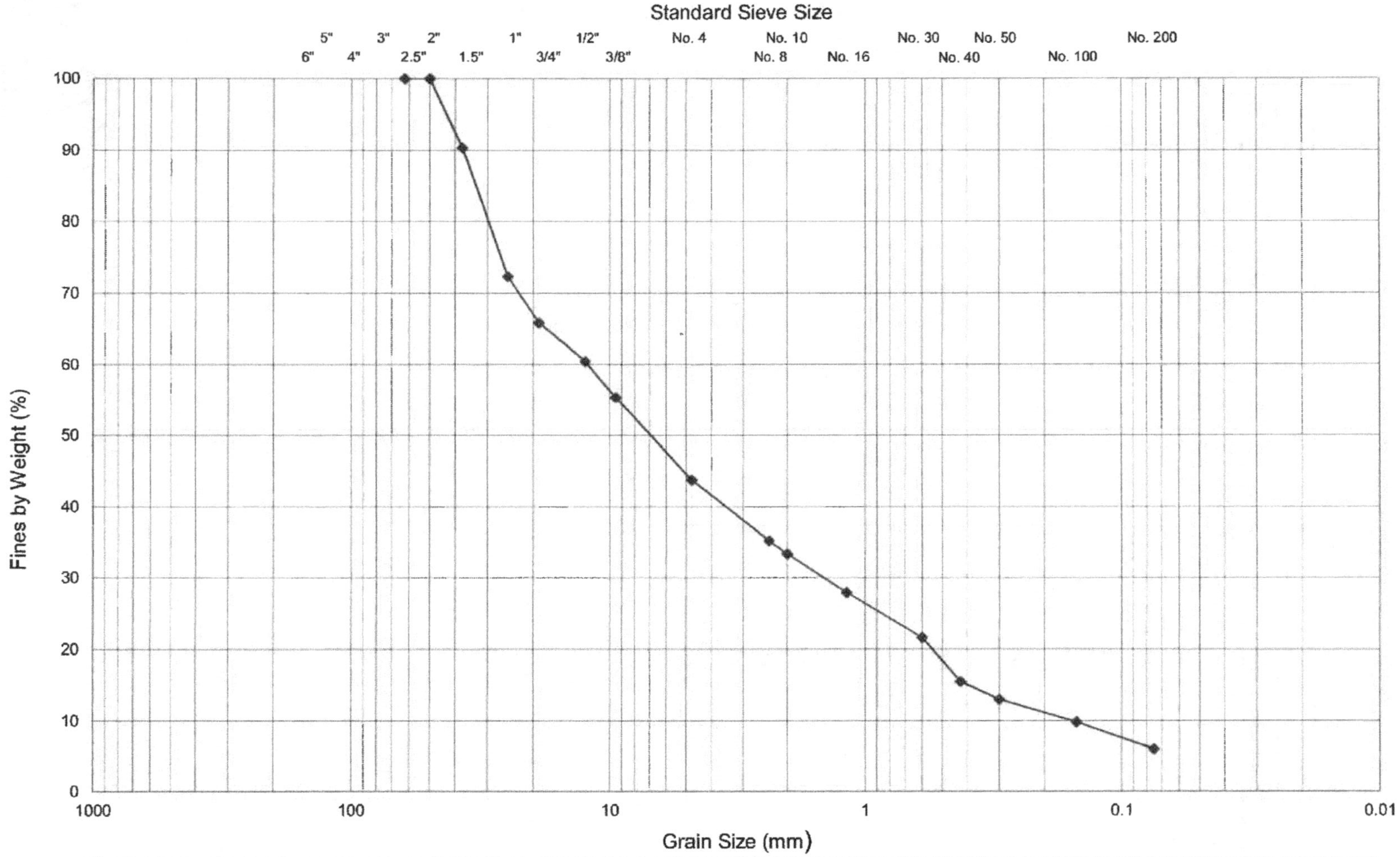

Cobble	Gravel		Sand			Silt or Clay
	Coarse	Fine	Coarse	Medium	Fine	

Project: Aims Community College - North Campus Refresh
Location: Greeley, Colorado
Project No: 1162022
Sample ID: B3, S3, 9'
Sample Desc.: Sand & Gravel (SP/GP)
Date: March 2016

EARTH ENGINEERING CONSULTANTS, LLC

SUMMARY OF LABORATORY TEST RESULTS

Sieve Analysis (AASHTO T 11 & T 27 / ASTM C 117 & C 136)		
Sieve Size		Percent Passing
1 1/2"	(37.5 mm)	100
1"	(25 mm)	100
3/4"	(19 mm)	86
1/2"	(12.5 mm)	78
3/8"	(9.5 mm)	70
No. 4	(4.75 mm)	59
No. 8	(2.36 mm)	51
No. 10	(2 mm)	49
No. 16	(1.18 mm)	43
No. 30	(0.6 mm)	31
No. 40	(0.425 mm)	27
No. 50	(0.3 mm)	23
No. 100	(0.15 mm)	17
No. 200	(0.075 mm)	11.5

Project: Aims Community College - North Campus Refresh
Location: Greeley, Colorado
Project No: 1162022
Sample ID: B4, S4, 14'
Sample Desc.: Poorly Graded Gravel (GP)
Date: March 2016

EEC

EARTH ENGINEERING CONSULTANTS, LLC

Summary of Washed Sieve Analysis Tests (ASTM C117 & C136)

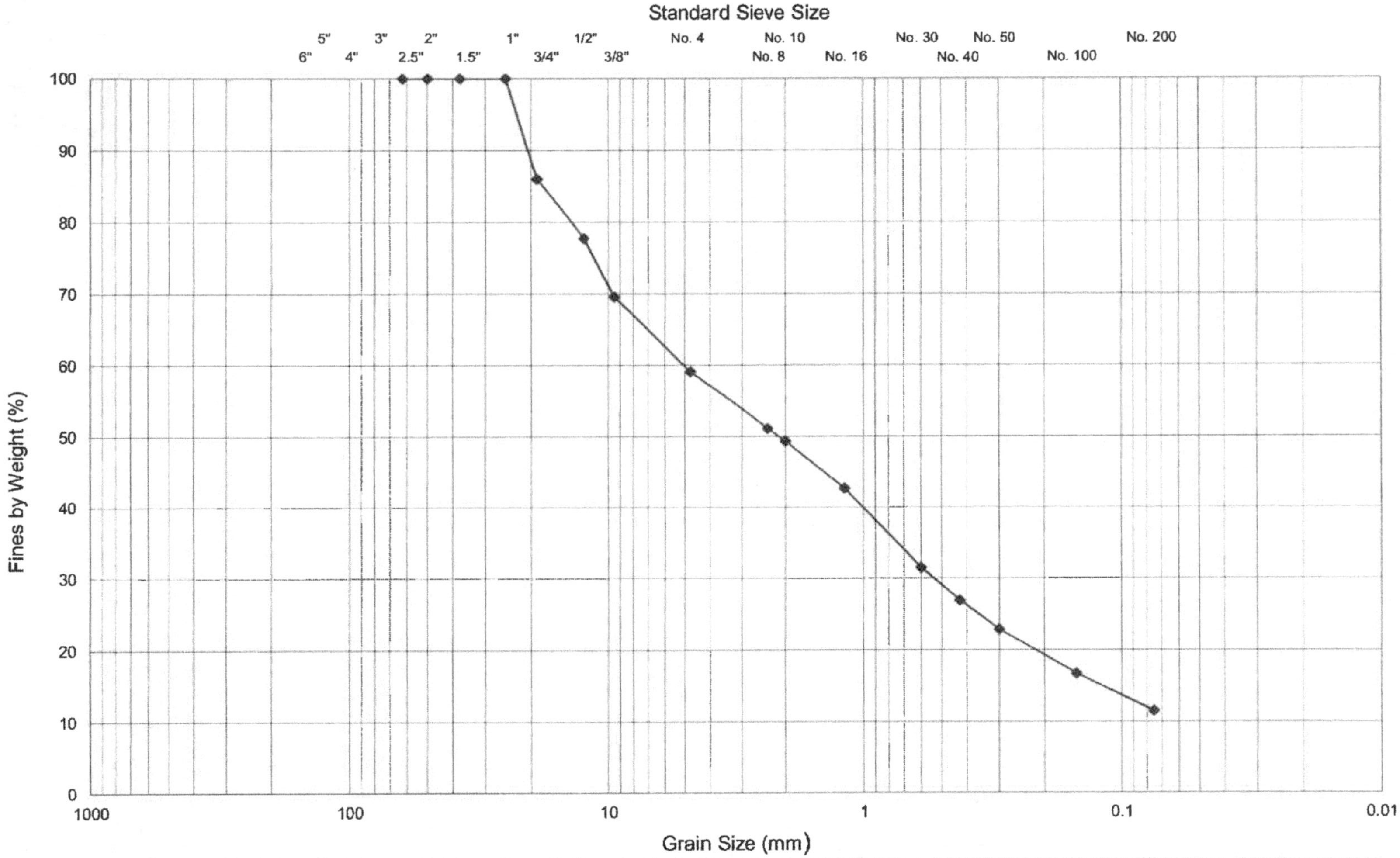

Cobble	Gravel		Sand			Silt or Clay
	Coarse	Fine	Coarse	Medium	Fine	

Project: Aims Community College - North Campus Refresh
Location: Greeley, Colorado
Project No: 1162022
Sample ID: B4, S4, 14'
Sample Desc.: Poorly Graded Gravel (GP)
Date: March 2016

EARTH ENGINEERING CONSULTANTS, LLC

SUMMARY OF LABORATORY TEST RESULTS

Sieve Analysis (AASHTO T 11 & T 27 / ASTM C 117 & C 136)		
Sieve Size		Percent Passing
No. 10	(2 mm)	100
No. 16	(1.18 mm)	100
No. 30	(0.6 mm)	99
No. 40	(0.425 mm)	98
No. 50	(0.3 mm)	96
No. 100	(0.15 mm)	84
No. 200	(0.075 mm)	60.6

Project: Aims Community College - North Campus Refresh
Location: Greeley, Colorado
Project No: 1162022
Sample ID: B5, S4, 19'
Sample Desc.: Grey / Rust Claystone
Date: March 2016

EARTH ENGINEERING CONSULTANTS, LLC

Summary of Washed Sieve Analysis Tests (ASTM C117 & C136)

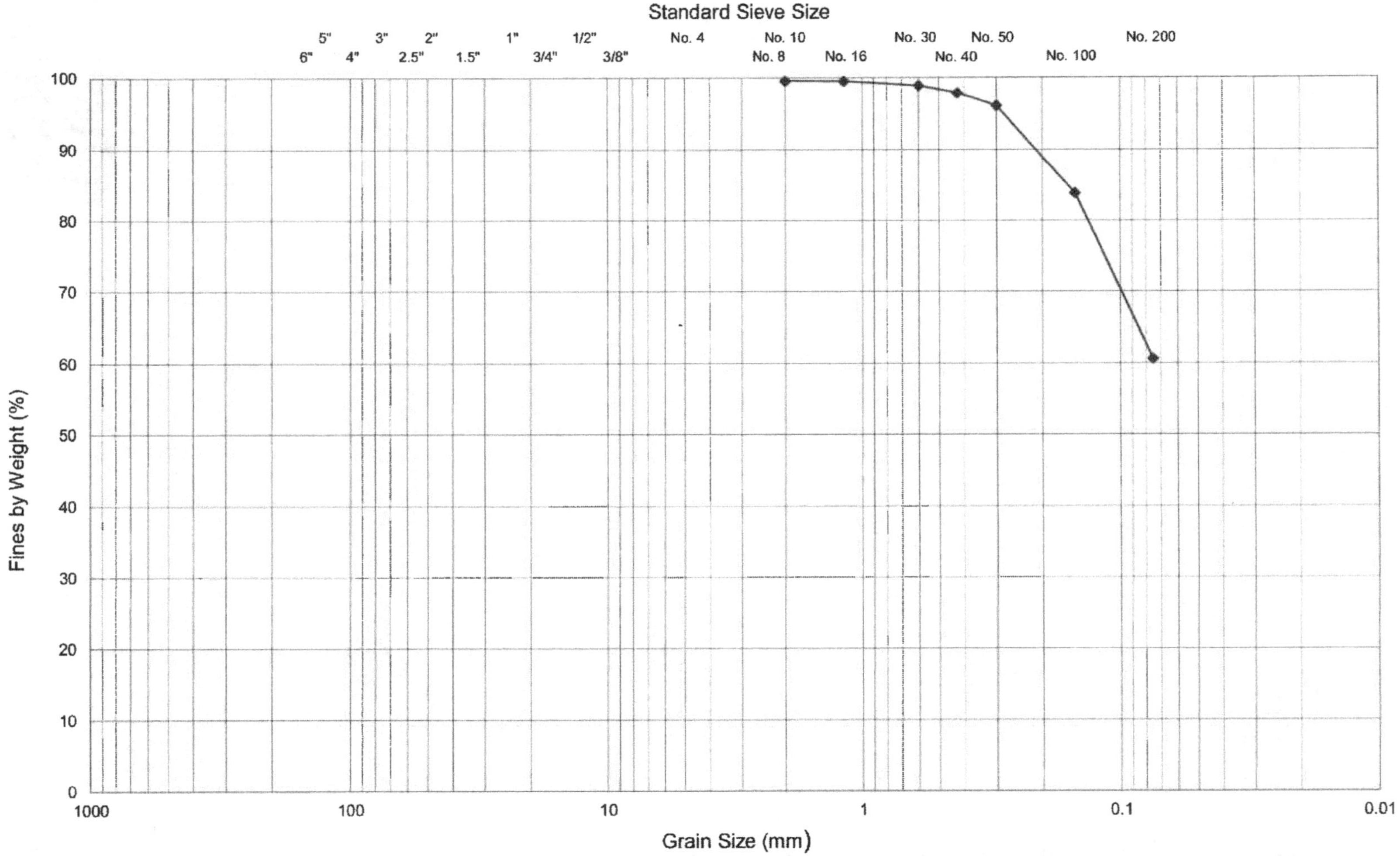

Cobble	Gravel		Sand			Silt or Clay
	Coarse	Fine	Coarse	Medium	Fine	

Project: Aims Community College - North Campus Refresh
Location: Greeley, Colorado
Project No: 1162022
Sample ID: B5, S4, 19'
Sample Desc.: Grey / Rust Claystone
Date: March 2016

EARTH ENGINEERING CONSULTANTS, LLC

SUMMARY OF LABORATORY TEST RESULTS

Sieve Analysis (AASHTO T 11 & T 27 / ASTM C 117 & C 136)		
Sieve Size		Percent Passing
1"	(25 mm)	100
3/4"	(19 mm)	100
1/2"	(12.5 mm)	94
3/8"	(9.5 mm)	89
No. 4	(4.75 mm)	78
No. 8	(2.36 mm)	65
No. 10	(2 mm)	63
No. 16	(1.18 mm)	57
No. 30	(0.6 mm)	46
No. 40	(0.425 mm)	41
No. 50	(0.3 mm)	35
No. 100	(0.15 mm)	27
No. 200	(0.075 mm)	20.6

Project: Aims Community College - North Campus Refresh
Location: Greeley, Colorado
Project No: 1162022
Sample ID: B14, S4, 14'
Sample Desc.: Clayey Sand & Gravel (SC/GP)
Date: March 2016

EARTH ENGINEERING CONSULTANTS, LLC

Summary of Washed Sieve Analysis Tests (ASTM C117 & C136)

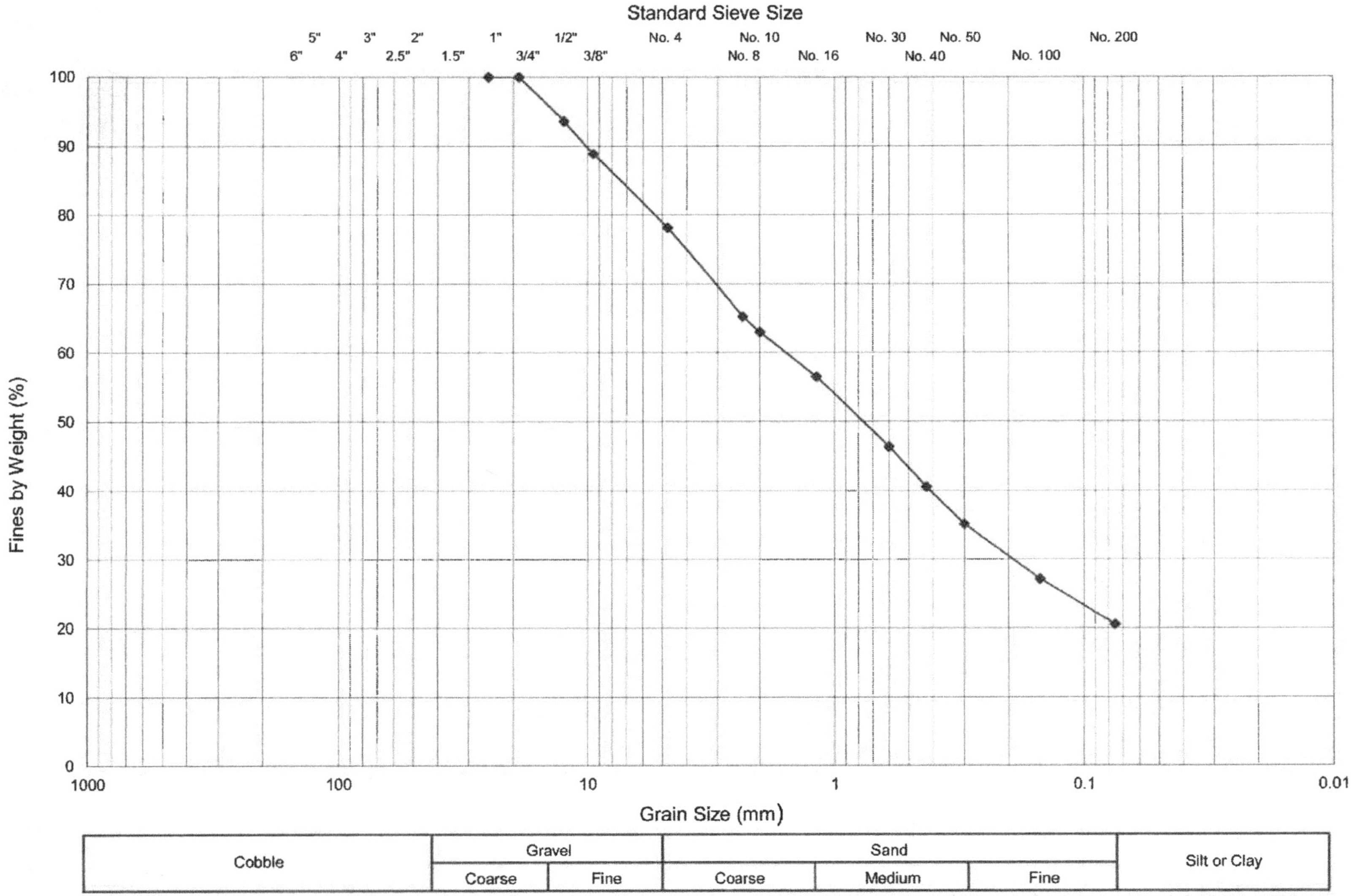

Project:	Aims Community College - North Campus Refresh
Location:	Greeley, Colorado
Project No:	1162022
Sample ID:	B14, S4, 14'
Sample Desc.:	Clayey Sand & Gravel (SC/GP)
Date:	March 2016

EARTH ENGINEERING CONSULTANTS, LLC

SUMMARY OF LABORATORY TEST RESULTS

Sieve Analysis (AASHTO T 11 & T 27 / ASTM C 117 & C 136)		
Sieve Size		Percent Passing
1"	(25 mm)	100
3/4"	(19 mm)	100
1/2"	(12.5 mm)	90
3/8"	(9.5 mm)	83
No. 4	(4.75 mm)	70
No. 8	(2.36 mm)	59
No. 10	(2 mm)	56
No. 16	(1.18 mm)	50
No. 30	(0.6 mm)	39
No. 40	(0.425 mm)	31
No. 50	(0.3 mm)	24
No. 100	(0.15 mm)	16
No. 200	(0.075 mm)	10.7

Project: Aims Community College - North Campus Refresh
Location: Greeley, Colorado
Project No: 1162022
Sample ID: B18, S3, 9'
Sample Desc.: Gravel (GP)
Date: March 2016

EARTH ENGINEERING CONSULTANTS, LLC

Summary of Washed Sieve Analysis Tests (ASTM C117 & C136)

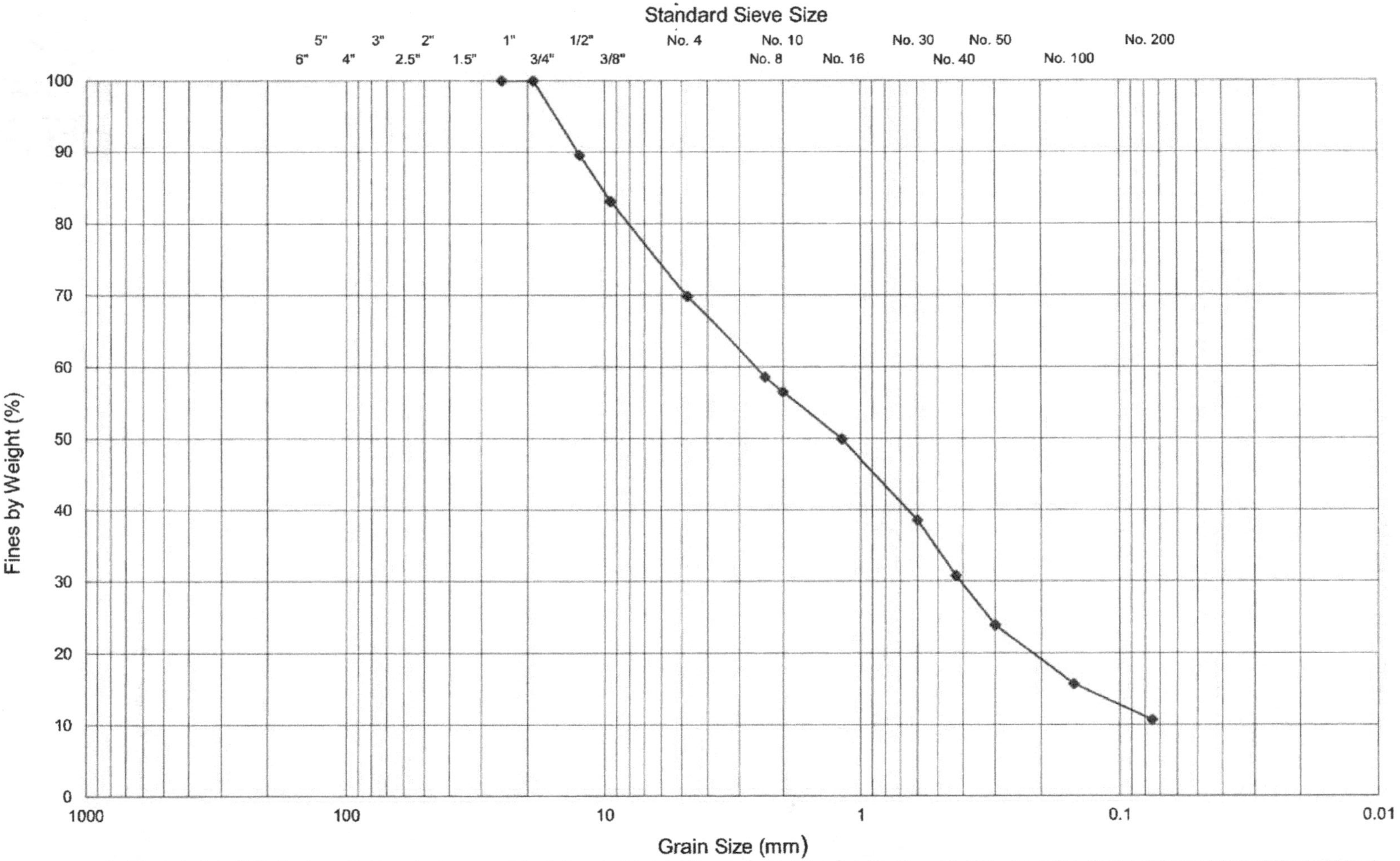

Cobble	Gravel		Sand			Silt or Clay
	Coarse	Fine	Coarse	Medium	Fine	

Project: Aims Community College - North Campus Refresh
Location: Greeley, Colorado
Project No: 1162022
Sample ID: B18, S3, 9'
Sample Desc.: Gravel (GP)
Date: March 2016

EARTH ENGINEERING CONSULTANTS, LLC

SUMMARY OF LABORATORY TEST RESULTS

Sieve Analysis (AASHTO T 11 & T 27 / ASTM C 117 & C 136)		
Sieve Size		Percent Passing
1 1/2"	(37.5 mm)	100
1"	(25 mm)	100
3/4"	(19 mm)	94
1/2"	(12.5 mm)	85
3/8"	(9.5 mm)	76
No. 4	(4.75 mm)	62
No. 8	(2.36 mm)	51
No. 10	(2 mm)	49
No. 16	(1.18 mm)	43
No. 30	(0.6 mm)	34
No. 40	(0.425 mm)	29
No. 50	(0.3 mm)	25
No. 100	(0.15 mm)	18
No. 200	(0.075 mm)	11.9

Project: Aims Community College - North Campus Refresh
Location: Greeley, Colorado
Project No: 1162022
Sample ID: B21, S3, 14'
Sample Desc.: Sand & Gravel (SP/GP)
Date: March 2016

EARTH ENGINEERING CONSULTANTS, LLC

Summary of Washed Sieve Analysis Tests (ASTM C117 & C136)

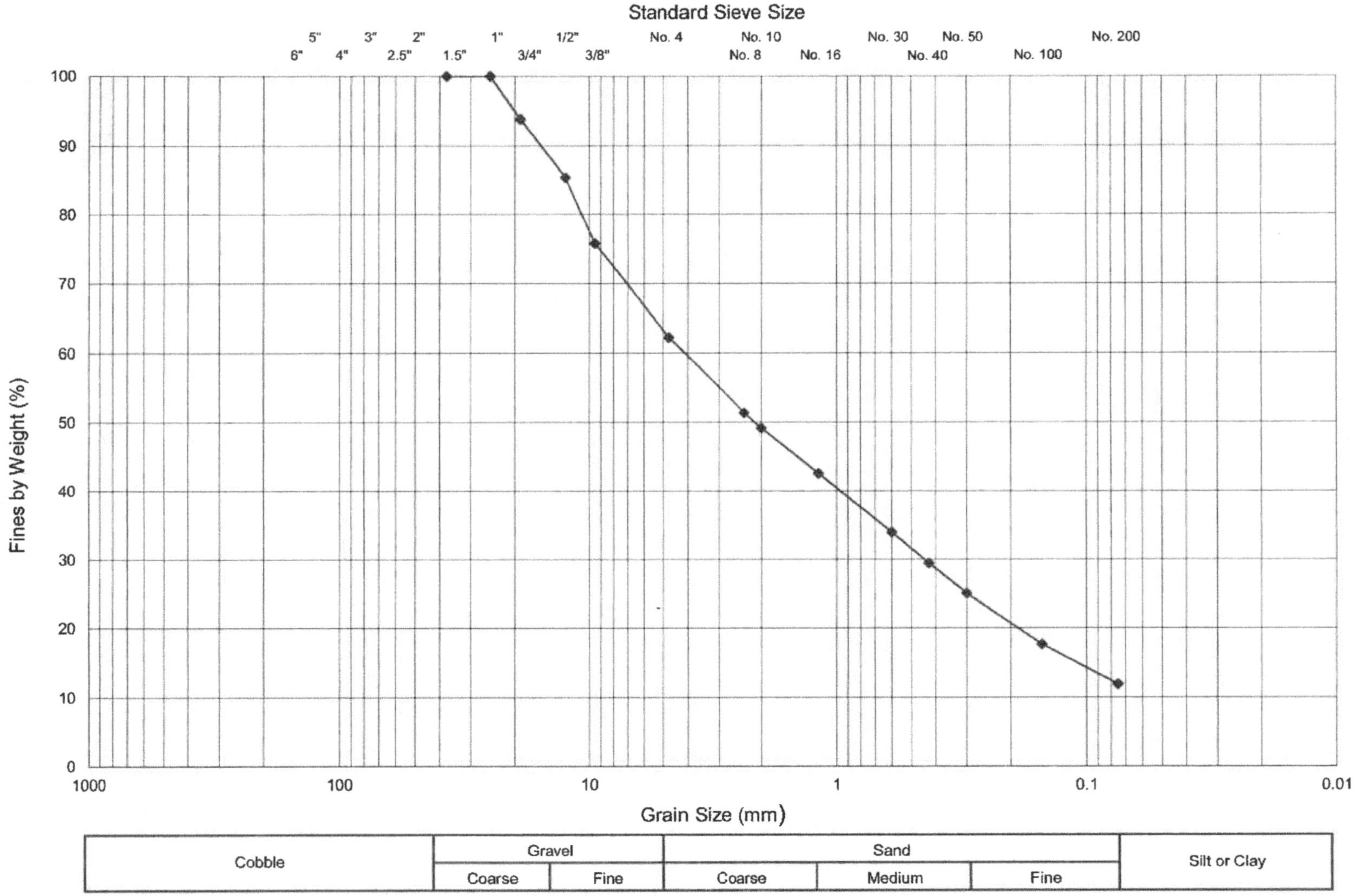

Project: Aims Community College - North Campus Refresh
Location: Greeley, Colorado
Project No: 1162022
Sample ID: B21, S3, 14'
Sample Desc.: Sand & Gravel (SP/GP)
Date: March 2016

Water Soluble Sulfate Ion and pH - Measurement

Project No: 1162022

Project Name: Aims Community College - N Exp

No. of Samples: 3

Test Standards: CP-L2103 / ASTM-C1580

Measurement Date: 3/28/2016

	Sample ID	Soluble Sulfate Content (SO_4)		pH
		(mg/l or ppm)	(% of Soil by Wt)	units
1	B-3 S-2 4'	70	0.01	
2	B-16 S-5 24'	160	0.02	
3	B-22 S-4 19'	130	0.01	

May 25, 2016

EARTH ENGINEERING
CONSULTANTS, LLC

Facilities and Operations Department
Aims Community College
Greeley, Colorado 80632

Attn: Mr. Michael Millsapps (michael.millsapps@aims.edu)

Re: Geotechnical Exploration Report – Addendum No. 1
Aims Community College
North Campus Refresh
Greeley, Colorado
EEC Project No. 1162022

Mr. Millsapps:

Our geotechnical exploration report for the referenced project was submitted to your attention April 14, 2016. Within that report, we recommended supporting individual column loads up to 260 kips on conventional footing foundations bearing on approved native soils or moisture-density conditioned soils with a maximum net allowable total load bearing pressure of 2,000 psf. At that time, we understood that maximum column loads in the three (3) story structure area could reach in the 260-kip range depending on final column configuration. Our report suggested using a deep foundation or overexcavation and backfill procedure to support loads in excess of 260 kips.

Based on discussions with the project structural engineer subsequent to completing that report, we understand that maximum column loads in the three (3) story portion of the proposed addition will be in the range of 320 kips. Based on those maximum column loads, we recommend overexcavating the proposed AET 3-story building area to a depth of at least 5 feet below top of subgrade and placing back the excavated material at 98% of standard Proctor maximum dry density with a moisture content of ±2% of standard Proctor optimum. Prior to the fill placement, the exposed subgrades below the overexcavation zone should be scarified and compacted to a similar requirement. The overexcavation zone should extend at least 3 feet laterally beyond the outside edges of the perimeter footings. For design of the footing foundations bearing on the subgrade materials placed and compacted as outlined, we recommend using a net allowable total load soil bearing pressure not to exceed 2,500 psf.

4396 GREENFIELD DRIVE
WINDSOR, COLORADO 80550
(970) 545-3908 FAX (970) 663-0282

As an alternative to overexcavating the entire building footprint, overexcavation could be completed solely under the interior column line with the higher loads with overexcavation extending at least 3 feet laterally in all directions beyond the edges of the footings. The column backfill materials should be placed and compacted as outlined above. In this instance, we recommend using the 2,000 psf allowable bearing pressure at these column locations as well as the remainder of the building area.

Additional recommendations concerning design and construction of the foundations and support of floor slabs and pavements were outlined in our April 14, 2016 report. We appreciate the opportunity to be of service to you on this project. If you have any questions concerning the enclosed report, or if we can be of further service to you in any other way, please do not hesitate to contact us.

Very truly yours,
Earth Engineering Consultants, LLC

Ethan P. Wiechert, P.E.
Senior Project Engineer

Reviewed by: Lester L. Litton, P.E.
Principal Engineer

EPW/LLL/dla

cc: DLR Group – Ryan D. Robley (rrobley@DLRGROUP.com)

Appendix H
Web Soil Survey - Sample

United States Department of Agriculture

NRCS

Natural Resources Conservation Service

A product of the National Cooperative Soil Survey, a joint effort of the United States Department of Agriculture and other Federal agencies, State agencies including the Agricultural Experiment Stations, and local participants

Custom Soil Resource Report for Weld County, Colorado, Southern Part

Clearview PUD - 4th Filing

October 29, 2012

Preface

Soil surveys contain information that affects land use planning in survey areas. They highlight soil limitations that affect various land uses and provide information about the properties of the soils in the survey areas. Soil surveys are designed for many different users, including farmers, ranchers, foresters, agronomists, urban planners, community officials, engineers, developers, builders, and home buyers. Also, conservationists, teachers, students, and specialists in recreation, waste disposal, and pollution control can use the surveys to help them understand, protect, or enhance the environment.

Various land use regulations of Federal, State, and local governments may impose special restrictions on land use or land treatment. Soil surveys identify soil properties that are used in making various land use or land treatment decisions. The information is intended to help the land users identify and reduce the effects of soil limitations on various land uses. The landowner or user is responsible for identifying and complying with existing laws and regulations.

Although soil survey information can be used for general farm, local, and wider area planning, onsite investigation is needed to supplement this information in some cases. Examples include soil quality assessments (http://soils.usda.gov/sqi/) and certain conservation and engineering applications. For more detailed information, contact your local USDA Service Center (http://offices.sc.egov.usda.gov/locator/app?agency=nrcs) or your NRCS State Soil Scientist (http://soils.usda.gov/contact/state_offices/).

Great differences in soil properties can occur within short distances. Some soils are seasonally wet or subject to flooding. Some are too unstable to be used as a foundation for buildings or roads. Clayey or wet soils are poorly suited to use as septic tank absorption fields. A high water table makes a soil poorly suited to basements or underground installations.

The National Cooperative Soil Survey is a joint effort of the United States Department of Agriculture and other Federal agencies, State agencies including the Agricultural Experiment Stations, and local agencies. The Natural Resources Conservation Service (NRCS) has leadership for the Federal part of the National Cooperative Soil Survey.

Information about soils is updated periodically. Updated information is available through the NRCS Soil Data Mart Web site or the NRCS Web Soil Survey. The Soil Data Mart is the data storage site for the official soil survey information.

The U.S. Department of Agriculture (USDA) prohibits discrimination in all its programs and activities on the basis of race, color, national origin, age, disability, and where applicable, sex, marital status, familial status, parental status, religion, sexual orientation, genetic information, political beliefs, reprisal, or because all or a part of an individual's income is derived from any public assistance program. (Not all prohibited bases apply to all programs.) Persons with disabilities who require alternative means for communication of program information (Braille, large print, audiotape, etc.) should contact USDA's TARGET Center at (202) 720-2600 (voice and TDD). To file a complaint of discrimination, write to USDA, Director, Office of Civil Rights, 1400 Independence Avenue, S.W., Washington, D.C. 20250-9410 or call (800) 795-3272 (voice) or (202) 720-6382 (TDD). USDA is an equal opportunity provider and employer.

Contents

How Soil Surveys Are Made

Soil surveys are made to provide information about the soils and miscellaneous areas in a specific area. They include a description of the soils and miscellaneous areas and their location on the landscape and tables that show soil properties and limitations affecting various uses. Soil scientists observed the steepness, length, and shape of the slopes; the general pattern of drainage; the kinds of crops and native plants; and the kinds of bedrock. They observed and described many soil profiles. A soil profile is the sequence of natural layers, or horizons, in a soil. The profile extends from the surface down into the unconsolidated material in which the soil formed or from the surface down to bedrock. The unconsolidated material is devoid of roots and other living organisms and has not been changed by other biological activity.

Currently, soils are mapped according to the boundaries of major land resource areas (MLRAs). MLRAs are geographically associated land resource units that share common characteristics related to physiography, geology, climate, water resources, soils, biological resources, and land uses (USDA, 2006). Soil survey areas typically consist of parts of one or more MLRA.

The soils and miscellaneous areas in a survey area occur in an orderly pattern that is related to the geology, landforms, relief, climate, and natural vegetation of the area. Each kind of soil and miscellaneous area is associated with a particular kind of landform or with a segment of the landform. By observing the soils and miscellaneous areas in the survey area and relating their position to specific segments of the landform, a soil scientist develops a concept, or model, of how they were formed. Thus, during mapping, this model enables the soil scientist to predict with a considerable degree of accuracy the kind of soil or miscellaneous area at a specific location on the landscape.

Commonly, individual soils on the landscape merge into one another as their characteristics gradually change. To construct an accurate soil map, however, soil scientists must determine the boundaries between the soils. They can observe only a limited number of soil profiles. Nevertheless, these observations, supplemented by an understanding of the soil-vegetation-landscape relationship, are sufficient to verify predictions of the kinds of soil in an area and to determine the boundaries.

Soil scientists recorded the characteristics of the soil profiles that they studied. They noted soil color, texture, size and shape of soil aggregates, kind and amount of rock fragments, distribution of plant roots, reaction, and other features that enable them to identify soils. After describing the soils in the survey area and determining their properties, the soil scientists assigned the soils to taxonomic classes (units). Taxonomic classes are concepts. Each taxonomic class has a set of soil characteristics with precisely defined limits. The classes are used as a basis for comparison to classify soils systematically. Soil taxonomy, the system of taxonomic classification used in the United States, is based mainly on the kind and character of soil properties and the arrangement of horizons within the profile. After the soil scientists classified and named the soils in the survey area, they compared the individual soils with similar soils in the same taxonomic class in other areas so that they could confirm data and assemble additional data based on experience and research.

The objective of soil mapping is not to delineate pure map unit components; the objective is to separate the landscape into landforms or landform segments that have similar use and management requirements. Each map unit is defined by a unique combination of soil components and/or miscellaneous areas in predictable proportions. Some components may be highly contrasting to the other components of the map unit. The presence of minor components in a map unit in no way diminishes the usefulness or accuracy of the data. The delineation of such landforms and landform segments on the map provides sufficient information for the development of resource plans. If intensive use of small areas is planned, onsite investigation is needed to define and locate the soils and miscellaneous areas.

Soil scientists make many field observations in the process of producing a soil map. The frequency of observation is dependent upon several factors, including scale of mapping, intensity of mapping, design of map units, complexity of the landscape, and experience of the soil scientist. Observations are made to test and refine the soil-landscape model and predictions and to verify the classification of the soils at specific locations. Once the soil-landscape model is refined, a significantly smaller number of measurements of individual soil properties are made and recorded. These measurements may include field measurements, such as those for color, depth to bedrock, and texture, and laboratory measurements, such as those for content of sand, silt, clay, salt, and other components. Properties of each soil typically vary from one point to another across the landscape.

Observations for map unit components are aggregated to develop ranges of characteristics for the components. The aggregated values are presented. Direct measurements do not exist for every property presented for every map unit component. Values for some properties are estimated from combinations of other properties.

While a soil survey is in progress, samples of some of the soils in the area generally are collected for laboratory analyses and for engineering tests. Soil scientists interpret the data from these analyses and tests as well as the field-observed characteristics and the soil properties to determine the expected behavior of the soils under different uses. Interpretations for all of the soils are field tested through observation of the soils in different uses and under different levels of management. Some interpretations are modified to fit local conditions, and some new interpretations are developed to meet local needs. Data are assembled from other sources, such as research information, production records, and field experience of specialists. For example, data on crop yields under defined levels of management are assembled from farm records and from field or plot experiments on the same kinds of soil.

Predictions about soil behavior are based not only on soil properties but also on such variables as climate and biological activity. Soil conditions are predictable over long periods of time, but they are not predictable from year to year. For example, soil scientists can predict with a fairly high degree of accuracy that a given soil will have a high water table within certain depths in most years, but they cannot predict that a high water table will always be at a specific level in the soil on a specific date.

After soil scientists located and identified the significant natural bodies of soil in the survey area, they drew the boundaries of these bodies on aerial photographs and identified each as a specific map unit. Aerial photographs show trees, buildings, fields, roads, and rivers, all of which help in locating boundaries accurately.

Soil Map

The soil map section includes the soil map for the defined area of interest, a list of soil map units on the map and extent of each map unit, and cartographic symbols displayed on the map. Also presented are various metadata about data used to produce the map, and a description of each soil map unit.

Custom Soil Resource Report
Soil Map

Map Scale: 1:4,480 if printed on A size (8.5" x 11") sheet.

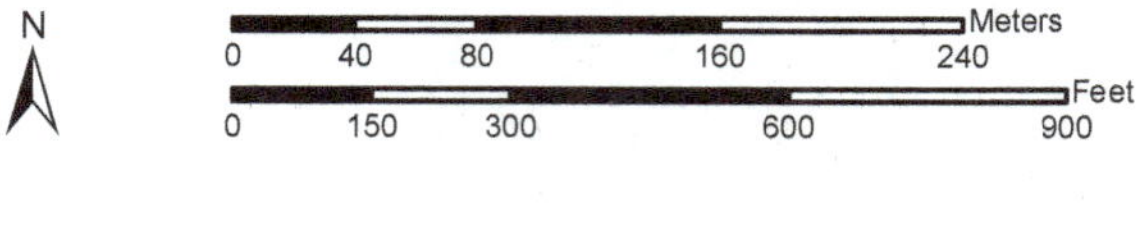

MAP LEGEND

Area of Interest (AOI)

- Area of Interest (AOI)

Soils

- Soil Map Units

Special Point Features

- Blowout
- Borrow Pit
- Clay Spot
- Closed Depression
- Gravel Pit
- Gravelly Spot
- Landfill
- Lava Flow
- Marsh or swamp
- Mine or Quarry
- Miscellaneous Water
- Perennial Water
- Rock Outcrop
- Saline Spot
- Sandy Spot
- Severely Eroded Spot
- Sinkhole
- Slide or Slip
- Sodic Spot
- Spoil Area
- Stony Spot
- Very Stony Spot
- Wet Spot
- Other

Special Line Features

- Gully
- Short Steep Slope
- Other

Political Features

- Cities
- PLSS Township and Range
- PLSS Section

Water Features

- Streams and Canals

Transportation

- Rails
- Interstate Highways
- US Routes
- Major Roads
- Local Roads

MAP INFORMATION

Map Scale: 1:4,480 if printed on A size (8.5" × 11") sheet.

The soil surveys that comprise your AOI were mapped at 1:24,000.

Warning: Soil Map may not be valid at this scale.

Enlargement of maps beyond the scale of mapping can cause misunderstanding of the detail of mapping and accuracy of soil line placement. The maps do not show the small areas of contrasting soils that could have been shown at a more detailed scale.

Please rely on the bar scale on each map sheet for accurate map measurements.

Source of Map: Natural Resources Conservation Service
Web Soil Survey URL: http://websoilsurvey.nrcs.usda.gov
Coordinate System: UTM Zone 13N NAD83

This product is generated from the USDA-NRCS certified data as of the version date(s) listed below.

Soil Survey Area: Weld County, Colorado, Southern Part
Survey Area Data: Version 11, Aug 27, 2009

Date(s) aerial images were photographed: 8/6/2005

The orthophoto or other base map on which the soil lines were compiled and digitized probably differs from the background imagery displayed on these maps. As a result, some minor shifting of map unit boundaries may be evident.

Map Unit Legend

Weld County, Colorado, Southern Part (CO618)			
Map Unit Symbol	**Map Unit Name**	**Acres in AOI**	**Percent of AOI**
16	Colby loam, 3 to 5 percent slopes	32.6	39.1%
42	Nunn clay loam, 1 to 3 percent slopes	22.7	27.2%
82	Wiley-Colby complex, 1 to 3 percent slopes	28.1	33.7%
Totals for Area of Interest		**83.4**	**100.0%**

Map Unit Descriptions

The map units delineated on the detailed soil maps in a soil survey represent the soils or miscellaneous areas in the survey area. The map unit descriptions, along with the maps, can be used to determine the composition and properties of a unit.

A map unit delineation on a soil map represents an area dominated by one or more major kinds of soil or miscellaneous areas. A map unit is identified and named according to the taxonomic classification of the dominant soils. Within a taxonomic class there are precisely defined limits for the properties of the soils. On the landscape, however, the soils are natural phenomena, and they have the characteristic variability of all natural phenomena. Thus, the range of some observed properties may extend beyond the limits defined for a taxonomic class. Areas of soils of a single taxonomic class rarely, if ever, can be mapped without including areas of other taxonomic classes. Consequently, every map unit is made up of the soils or miscellaneous areas for which it is named and some minor components that belong to taxonomic classes other than those of the major soils.

Most minor soils have properties similar to those of the dominant soil or soils in the map unit, and thus they do not affect use and management. These are called noncontrasting, or similar, components. They may or may not be mentioned in a particular map unit description. Other minor components, however, have properties and behavioral characteristics divergent enough to affect use or to require different management. These are called contrasting, or dissimilar, components. They generally are in small areas and could not be mapped separately because of the scale used. Some small areas of strongly contrasting soils or miscellaneous areas are identified by a special symbol on the maps. If included in the database for a given area, the contrasting minor components are identified in the map unit descriptions along with some characteristics of each. A few areas of minor components may not have been observed, and consequently they are not mentioned in the descriptions, especially where the pattern was so complex that it was impractical to make enough observations to identify all the soils and miscellaneous areas on the landscape.

The presence of minor components in a map unit in no way diminishes the usefulness or accuracy of the data. The objective of mapping is not to delineate pure taxonomic classes but rather to separate the landscape into landforms or landform segments that have similar use and management requirements. The delineation of such segments on the map provides sufficient information for the development of resource plans. If intensive use of small areas is planned, however, onsite investigation is needed to define and locate the soils and miscellaneous areas.

An identifying symbol precedes the map unit name in the map unit descriptions. Each description includes general facts about the unit and gives important soil properties and qualities.

Soils that have profiles that are almost alike make up a *soil series*. Except for differences in texture of the surface layer, all the soils of a series have major horizons that are similar in composition, thickness, and arrangement.

Soils of one series can differ in texture of the surface layer, slope, stoniness, salinity, degree of erosion, and other characteristics that affect their use. On the basis of such differences, a soil series is divided into *soil phases*. Most of the areas shown on the detailed soil maps are phases of soil series. The name of a soil phase commonly indicates a feature that affects use or management. For example, Alpha silt loam, 0 to 2 percent slopes, is a phase of the Alpha series.

Some map units are made up of two or more major soils or miscellaneous areas. These map units are complexes, associations, or undifferentiated groups.

A *complex* consists of two or more soils or miscellaneous areas in such an intricate pattern or in such small areas that they cannot be shown separately on the maps. The pattern and proportion of the soils or miscellaneous areas are somewhat similar in all areas. Alpha-Beta complex, 0 to 6 percent slopes, is an example.

An *association* is made up of two or more geographically associated soils or miscellaneous areas that are shown as one unit on the maps. Because of present or anticipated uses of the map units in the survey area, it was not considered practical or necessary to map the soils or miscellaneous areas separately. The pattern and relative proportion of the soils or miscellaneous areas are somewhat similar. Alpha-Beta association, 0 to 2 percent slopes, is an example.

An *undifferentiated group* is made up of two or more soils or miscellaneous areas that could be mapped individually but are mapped as one unit because similar interpretations can be made for use and management. The pattern and proportion of the soils or miscellaneous areas in a mapped area are not uniform. An area can be made up of only one of the major soils or miscellaneous areas, or it can be made up of all of them. Alpha and Beta soils, 0 to 2 percent slopes, is an example.

Some surveys include *miscellaneous areas*. Such areas have little or no soil material and support little or no vegetation. Rock outcrop is an example.

Weld County, Colorado, Southern Part

16—Colby loam, 3 to 5 percent slopes

Map Unit Setting

Elevation: 4,850 to 5,050 feet
Mean annual precipitation: 12 to 16 inches
Mean annual air temperature: 48 to 50 degrees F
Frost-free period: 135 to 155 days

Map Unit Composition

Colby and similar soils: 85 percent
Minor components: 15 percent

Description of Colby

Setting

Landform: Ridges, hills
Down-slope shape: Linear
Across-slope shape: Linear
Parent material: Calcareous eolian deposits

Properties and qualities

Slope: 3 to 5 percent
Depth to restrictive feature: More than 80 inches
Drainage class: Well drained
Capacity of the most limiting layer to transmit water (Ksat): Moderately high to high (0.57 to 2.00 in/hr)
Depth to water table: More than 80 inches
Frequency of flooding: None
Frequency of ponding: None
Calcium carbonate, maximum content: 15 percent
Available water capacity: High (about 10.6 inches)

Interpretive groups

Land capability classification (irrigated): 3e
Land capability (nonirrigated): 4e
Ecological site: Loamy Plains (R067BY002CO)

Typical profile

0 to 7 inches: Loam
7 to 60 inches: Silt loam

Minor Components

Wiley

Percent of map unit: 8 percent

Keith

Percent of map unit: 7 percent

42—Nunn clay loam, 1 to 3 percent slopes

Map Unit Setting

Elevation: 4,550 to 5,150 feet
Mean annual precipitation: 12 to 18 inches
Mean annual air temperature: 46 to 54 degrees F
Frost-free period: 115 to 180 days

Map Unit Composition

Nunn and similar soils: 85 percent
Minor components: 15 percent

Description of Nunn

Setting

Landform: Plains, terraces
Down-slope shape: Linear
Across-slope shape: Linear
Parent material: Mixed alluvium and/or eolian deposits

Properties and qualities

Slope: 1 to 3 percent
Depth to restrictive feature: More than 80 inches
Drainage class: Well drained
Capacity of the most limiting layer to transmit water (Ksat): Moderately low to moderately high (0.06 to 0.20 in/hr)
Depth to water table: More than 80 inches
Frequency of flooding: None
Frequency of ponding: None
Calcium carbonate, maximum content: 15 percent
Maximum salinity: Nonsaline (0.0 to 2.0 mmhos/cm)
Available water capacity: High (about 9.1 inches)

Interpretive groups

Land capability classification (irrigated): 2e
Land capability (nonirrigated): 4c
Ecological site: Clayey Plains (R067BY042CO)

Typical profile

0 to 9 inches: Clay loam
9 to 29 inches: Clay loam
29 to 60 inches: Sandy loam

Minor Components

Haverson

Percent of map unit: 9 percent

Heldt

Percent of map unit: 6 percent

82—Wiley-Colby complex, 1 to 3 percent slopes

Map Unit Setting

Elevation: 4,850 to 5,000 feet
Mean annual precipitation: 12 to 16 inches
Mean annual air temperature: 48 to 54 degrees F
Frost-free period: 135 to 170 days

Map Unit Composition

Wiley and similar soils: 60 percent
Colby and similar soils: 30 percent
Minor components: 10 percent

Description of Wiley

Setting

Landform: Plains
Down-slope shape: Linear
Across-slope shape: Linear
Parent material: Calcareous eolian deposits

Properties and qualities

Slope: 1 to 3 percent
Depth to restrictive feature: More than 80 inches
Drainage class: Well drained
Capacity of the most limiting layer to transmit water (Ksat): Moderately high to high (0.60 to 2.00 in/hr)
Depth to water table: More than 80 inches
Frequency of flooding: None
Frequency of ponding: None
Calcium carbonate, maximum content: 15 percent
Maximum salinity: Nonsaline (0.0 to 2.0 mmhos/cm)
Available water capacity: High (about 11.7 inches)

Interpretive groups

Land capability classification (irrigated): 2e
Land capability (nonirrigated): 4e
Ecological site: Loamy Plains (R067BY002CO)

Typical profile

0 to 11 inches: Silt loam
11 to 60 inches: Silty clay loam
60 to 64 inches: Silty clay loam

Description of Colby

Setting

Landform: Plains
Down-slope shape: Linear
Across-slope shape: Linear

Parent material: Calcareous eolian deposits

Properties and qualities

Slope: 1 to 3 percent
Depth to restrictive feature: More than 80 inches
Drainage class: Well drained
Capacity of the most limiting layer to transmit water (Ksat): Moderately high to high (0.57 to 2.00 in/hr)
Depth to water table: More than 80 inches
Frequency of flooding: None
Frequency of ponding: None
Calcium carbonate, maximum content: 15 percent
Available water capacity: High (about 10.6 inches)

Interpretive groups

Land capability classification (irrigated): 3e
Land capability (nonirrigated): 4e
Ecological site: Loamy Plains (R067BY002CO)

Typical profile

0 to 7 inches: Loam
7 to 60 inches: Silt loam

Minor Components

Heldt

Percent of map unit: 4 percent

Weld

Percent of map unit: 4 percent

Keith

Percent of map unit: 2 percent

Soil Information for All Uses

Suitabilities and Limitations for Use

The Suitabilities and Limitations for Use section includes various soil interpretations displayed as thematic maps with a summary table for the soil map units in the selected area of interest. A single value or rating for each map unit is generated by aggregating the interpretive ratings of individual map unit components. This aggregation process is defined for each interpretation.

Building Site Development

Building site development interpretations are designed to be used as tools for evaluating soil suitability and identifying soil limitations for various construction purposes. As part of the interpretation process, the rating applies to each soil in its described condition and does not consider present land use. Example interpretations can include corrosion of concrete and steel, shallow excavations, dwellings with and without basements, small commercial buildings, local roads and streets, and lawns and landscaping.

Corrosion of Concrete

"Risk of corrosion" pertains to potential soil-induced electrochemical or chemical action that corrodes or weakens concrete. The rate of corrosion of concrete is based mainly on the sulfate and sodium content, texture, moisture content, and acidity of the soil. Special site examination and design may be needed if the combination of factors results in a severe hazard of corrosion. The concrete in installations that intersect soil boundaries or soil layers is more susceptible to corrosion than the concrete in installations that are entirely within one kind of soil or within one soil layer.

The risk of corrosion is expressed as "low," "moderate," or "high."

Map—Corrosion of Concrete

Map Scale: 1:4,480 if printed on A size (8.5" x 11") sheet.

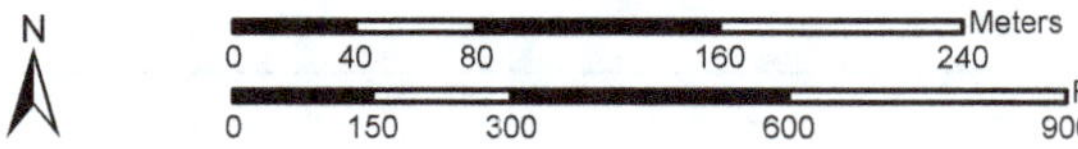

MAP LEGEND

Area of Interest (AOI)

- Area of Interest (AOI)

Soils

- Soil Map Units

Soil Ratings

- High
- Moderate
- Low
- Not rated or not available

Political Features

- Cities
- PLSS Township and Range
- PLSS Section

Water Features

- Streams and Canals

Transportation

- Rails
- Interstate Highways
- US Routes
- Major Roads
- Local Roads

MAP INFORMATION

Map Scale: 1:4,480 if printed on A size (8.5" × 11") sheet.

The soil surveys that comprise your AOI were mapped at 1:24,000.

Warning: Soil Map may not be valid at this scale.

Enlargement of maps beyond the scale of mapping can cause misunderstanding of the detail of mapping and accuracy of soil line placement. The maps do not show the small areas of contrasting soils that could have been shown at a more detailed scale.

Please rely on the bar scale on each map sheet for accurate map measurements.

Source of Map: Natural Resources Conservation Service
Web Soil Survey URL: http://websoilsurvey.nrcs.usda.gov
Coordinate System: UTM Zone 13N NAD83

This product is generated from the USDA-NRCS certified data as of the version date(s) listed below.

Soil Survey Area: Weld County, Colorado, Southern Part
Survey Area Data: Version 11, Aug 27, 2009

Date(s) aerial images were photographed: 8/6/2005

The orthophoto or other base map on which the soil lines were compiled and digitized probably differs from the background imagery displayed on these maps. As a result, some minor shifting of map unit boundaries may be evident.

Table—Corrosion of Concrete

Corrosion of Concrete— Summary by Map Unit — Weld County, Colorado, Southern Part (CO618)				
Map unit symbol	**Map unit name**	**Rating**	**Acres in AOI**	**Percent of AOI**
16	Colby loam, 3 to 5 percent slopes	Low	32.6	39.1%
42	Nunn clay loam, 1 to 3 percent slopes	Low	22.7	27.2%
82	Wiley-Colby complex, 1 to 3 percent slopes	Low	28.1	33.7%
Totals for Area of Interest			**83.4**	**100.0%**

Rating Options—Corrosion of Concrete

Aggregation Method: Dominant Condition

Component Percent Cutoff: None Specified

Tie-break Rule: Higher

Corrosion of Steel

"Risk of corrosion" pertains to potential soil-induced electrochemical or chemical action that corrodes or weakens uncoated steel. The rate of corrosion of uncoated steel is related to such factors as soil moisture, particle-size distribution, acidity, and electrical conductivity of the soil. Special site examination and design may be needed if the combination of factors results in a severe hazard of corrosion. The steel in installations that intersect soil boundaries or soil layers is more susceptible to corrosion than the steel in installations that are entirely within one kind of soil or within one soil layer.

The risk of corrosion is expressed as "low," "moderate," or "high."

Custom Soil Resource Report
Map—Corrosion of Steel

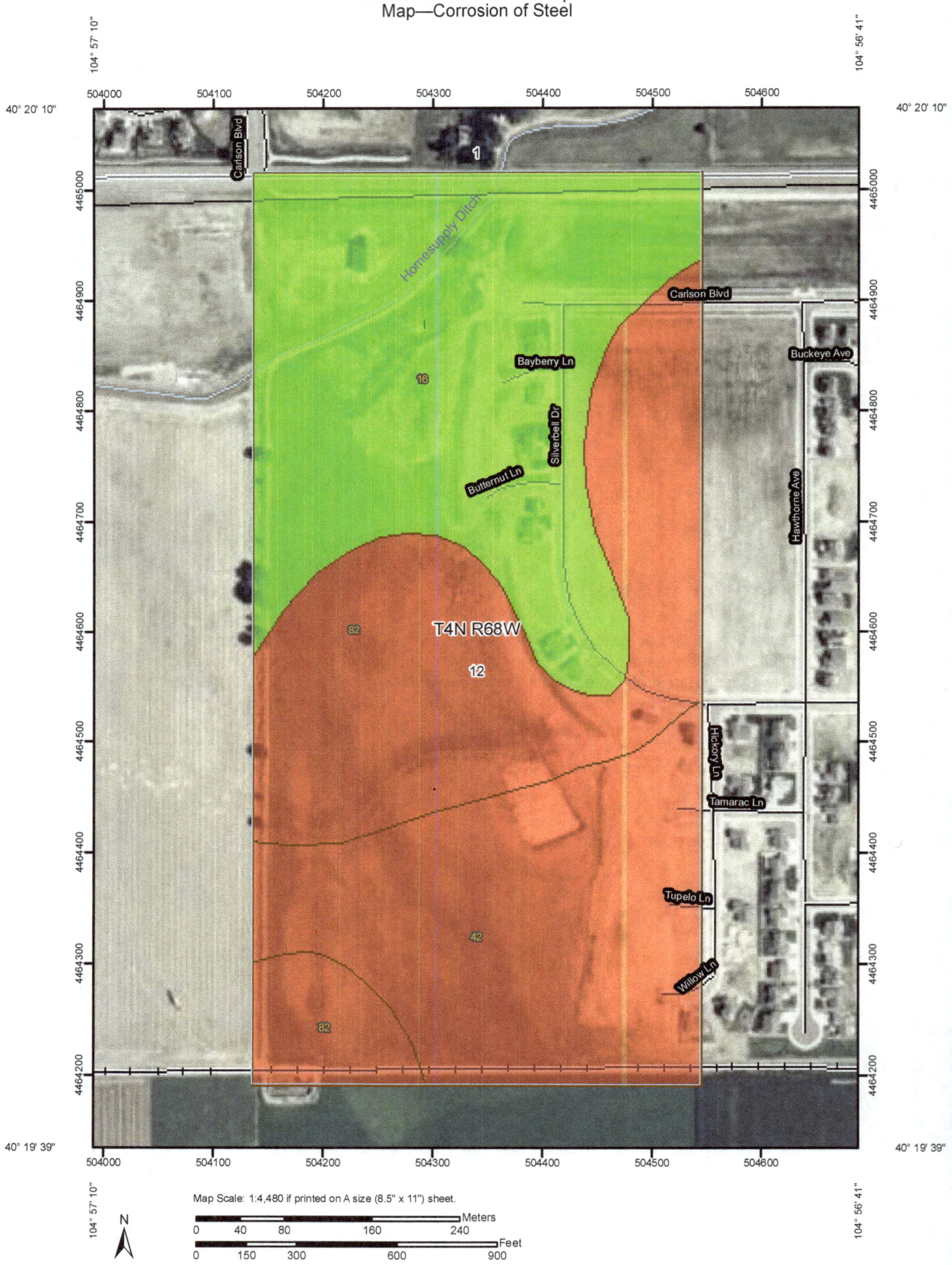

MAP LEGEND

Area of Interest (AOI)

Area of Interest (AOI)

Soils

Soil Map Units

Soil Ratings

High

Moderate

Low

Not rated or not available

Political Features

Cities

PLSS Township and Range

PLSS Section

Water Features

Streams and Canals

Transportation

Rails

Interstate Highways

US Routes

Major Roads

Local Roads

MAP INFORMATION

Map Scale: 1:4,480 if printed on A size (8.5" × 11") sheet.

The soil surveys that comprise your AOI were mapped at 1:24,000.

Warning: Soil Map may not be valid at this scale.

Enlargement of maps beyond the scale of mapping can cause misunderstanding of the detail of mapping and accuracy of soil line placement. The maps do not show the small areas of contrasting soils that could have been shown at a more detailed scale.

Please rely on the bar scale on each map sheet for accurate map measurements.

Source of Map: Natural Resources Conservation Service
Web Soil Survey URL: http://websoilsurvey.nrcs.usda.gov
Coordinate System: UTM Zone 13N NAD83

This product is generated from the USDA-NRCS certified data as of the version date(s) listed below.

Soil Survey Area: Weld County, Colorado, Southern Part
Survey Area Data: Version 11, Aug 27, 2009

Date(s) aerial images were photographed: 8/6/2005

The orthophoto or other base map on which the soil lines were compiled and digitized probably differs from the background imagery displayed on these maps. As a result, some minor shifting of map unit boundaries may be evident.

Table—Corrosion of Steel

Corrosion of Steel— Summary by Map Unit — Weld County, Colorado, Southern Part (CO618)				
Map unit symbol	Map unit name	Rating	Acres in AOI	Percent of AOI
16	Colby loam, 3 to 5 percent slopes	Low	32.6	39.1%
42	Nunn clay loam, 1 to 3 percent slopes	High	22.7	27.2%
82	Wiley-Colby complex, 1 to 3 percent slopes	High	28.1	33.7%
Totals for Area of Interest			**83.4**	**100.0%**

Rating Options—Corrosion of Steel

Aggregation Method: Dominant Condition

Component Percent Cutoff: None Specified

Tie-break Rule: Higher

Dwellings With Basements

Dwellings are single-family houses of three stories or less. For dwellings with basements, the foundation is assumed to consist of spread footings of reinforced concrete built on undisturbed soil at a depth of about 7 feet.

The ratings for dwellings are based on the soil properties that affect the capacity of the soil to support a load without movement and on the properties that affect excavation and construction costs. The properties that affect the load-supporting capacity include depth to a water table, ponding, flooding, subsidence, linear extensibility (shrink-swell potential), and compressibility. Compressibility is inferred from the Unified classification of the soil. The properties that affect the ease and amount of excavation include depth to a water table, ponding, flooding, slope, depth to bedrock or a cemented pan, hardness of bedrock or a cemented pan, and the amount and size of rock fragments.

The ratings are both verbal and numerical. Rating class terms indicate the extent to which the soils are limited by all of the soil features that affect the specified use. "Not limited" indicates that the soil has features that are very favorable for the specified use. Good performance and very low maintenance can be expected. "Somewhat limited" indicates that the soil has features that are moderately favorable for the specified use. The limitations can be overcome or minimized by special planning, design, or installation. Fair performance and moderate maintenance can be expected. "Very limited" indicates that the soil has one or more features that are unfavorable for the specified use. The limitations generally cannot be overcome without major soil reclamation, special design, or expensive installation procedures. Poor performance and high maintenance can be expected.

Numerical ratings indicate the severity of individual limitations. The ratings are shown as decimal fractions ranging from 0.01 to 1.00. They indicate gradations between the point at which a soil feature has the greatest negative impact on the use (1.00) and the point at which the soil feature is not a limitation (0.00).

The map unit components listed for each map unit in the accompanying Summary by Map Unit table in Web Soil Survey or the Aggregation Report in Soil Data Viewer are determined by the aggregation method chosen. An aggregated rating class is shown for each map unit. The components listed for each map unit are only those that have the same rating class as listed for the map unit. The percent composition of each component in a particular map unit is presented to help the user better understand the percentage of each map unit that has the rating presented.

Other components with different ratings may be present in each map unit. The ratings for all components, regardless of the map unit aggregated rating, can be viewed by generating the equivalent report from the Soil Reports tab in Web Soil Survey or from the Soil Data Mart site. Onsite investigation may be needed to validate these interpretations and to confirm the identity of the soil on a given site.

Custom Soil Resource Report
Map—Dwellings With Basements

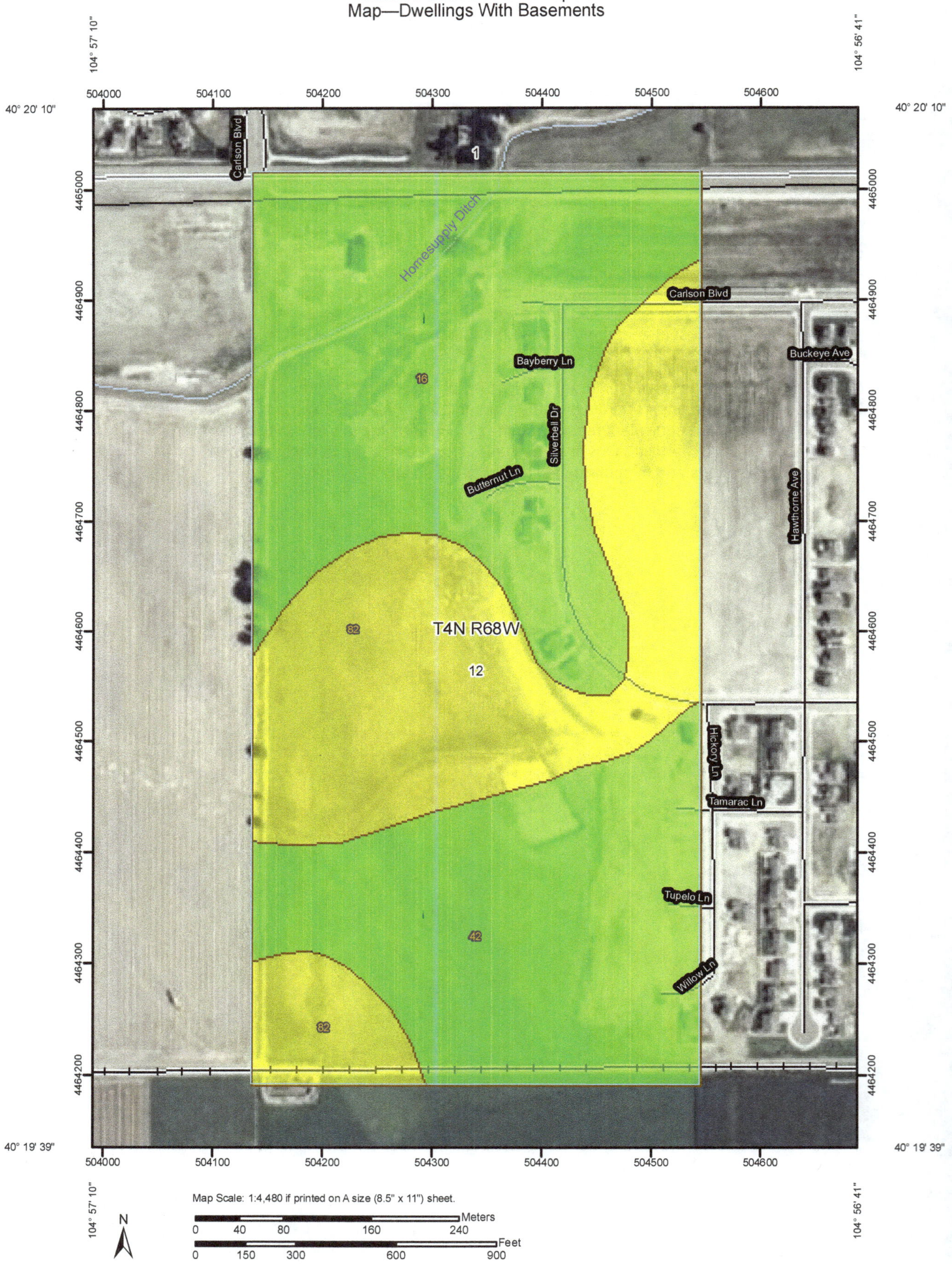

MAP LEGEND

Area of Interest (AOI)
- Area of Interest (AOI)

Soils
- Soil Map Units

Soil Ratings
- Very limited
- Somewhat limited
- Not limited
- Not rated or not available

Political Features
- Cities
- PLSS Township and Range
- PLSS Section

Water Features
- Streams and Canals

Transportation
- Rails
- Interstate Highways
- US Routes
- Major Roads
- Local Roads

MAP INFORMATION

Map Scale: 1:4,480 if printed on A size (8.5" × 11") sheet.

The soil surveys that comprise your AOI were mapped at 1:24,000.

Warning: Soil Map may not be valid at this scale.

Enlargement of maps beyond the scale of mapping can cause misunderstanding of the detail of mapping and accuracy of soil line placement. The maps do not show the small areas of contrasting soils that could have been shown at a more detailed scale.

Please rely on the bar scale on each map sheet for accurate map measurements.

Source of Map: Natural Resources Conservation Service
Web Soil Survey URL: http://websoilsurvey.nrcs.usda.gov
Coordinate System: UTM Zone 13N NAD83

This product is generated from the USDA-NRCS certified data as of the version date(s) listed below.

Soil Survey Area: Weld County, Colorado, Southern Part
Survey Area Data: Version 11, Aug 27, 2009

Date(s) aerial images were photographed: 8/6/2005

The orthophoto or other base map on which the soil lines were compiled and digitized probably differs from the background imagery displayed on these maps. As a result, some minor shifting of map unit boundaries may be evident.

Tables—Dwellings With Basements

Dwellings With Basements— Summary by Map Unit — Weld County, Colorado, Southern Part (CO618)						
Map unit symbol	**Map unit name**	**Rating**	**Component name (percent)**	**Rating reasons (numeric values)**	**Acres in AOI**	**Percent of AOI**
16	Colby loam, 3 to 5 percent slopes	Not limited	Colby (85%)		32.6	39.1%
42	Nunn clay loam, 1 to 3 percent slopes	Not limited	Nunn (85%)		22.7	27.2%
82	Wiley-Colby complex, 1 to 3 percent slopes	Somewhat limited	Wiley (60%)	Shrink-swell (0.50)	28.1	33.7%
Totals for Area of Interest					**83.4**	**100.0%**

Dwellings With Basements— Summary by Rating Value		
Rating	**Acres in AOI**	**Percent of AOI**
Not limited	55.3	66.3%
Somewhat limited	28.1	33.7%
Totals for Area of Interest	**83.4**	**100.0%**

Rating Options—Dwellings With Basements

Aggregation Method: Dominant Condition

Component Percent Cutoff: None Specified

Tie-break Rule: Higher

Dwellings Without Basements

Dwellings are single-family houses of three stories or less. For dwellings without basements, the foundation is assumed to consist of spread footings of reinforced concrete built on undisturbed soil at a depth of 2 feet or at the depth of maximum frost penetration, whichever is deeper.

The ratings for dwellings are based on the soil properties that affect the capacity of the soil to support a load without movement and on the properties that affect excavation and construction costs. The properties that affect the load-supporting capacity include depth to a water table, ponding, flooding, subsidence, linear extensibility (shrink-swell potential), and compressibility. Compressibility is inferred from the Unified classification of the soil. The properties that affect the ease and amount of excavation include depth to a water table, ponding, flooding, slope, depth to bedrock or a cemented pan, hardness of bedrock or a cemented pan, and the amount and size of rock fragments.

The ratings are both verbal and numerical. Rating class terms indicate the extent to which the soils are limited by all of the soil features that affect the specified use. "Not limited" indicates that the soil has features that are very favorable for the specified use. Good performance and very low maintenance can be expected. "Somewhat limited" indicates that the soil has features that are moderately favorable for the specified use. The limitations can be overcome or minimized by special planning, design, or installation. Fair performance and moderate maintenance can be expected. "Very limited" indicates that the soil has one or more features that are unfavorable for the specified use. The limitations generally cannot be overcome without major soil reclamation, special design, or expensive installation procedures. Poor performance and high maintenance can be expected.

Numerical ratings indicate the severity of individual limitations. The ratings are shown as decimal fractions ranging from 0.01 to 1.00. They indicate gradations between the point at which a soil feature has the greatest negative impact on the use (1.00) and the point at which the soil feature is not a limitation (0.00).

The map unit components listed for each map unit in the accompanying Summary by Map Unit table in Web Soil Survey or the Aggregation Report in Soil Data Viewer are determined by the aggregation method chosen. An aggregated rating class is shown for each map unit. The components listed for each map unit are only those that have the same rating class as listed for the map unit. The percent composition of each component in a particular map unit is presented to help the user better understand the percentage of each map unit that has the rating presented.

Other components with different ratings may be present in each map unit. The ratings for all components, regardless of the map unit aggregated rating, can be viewed by generating the equivalent report from the Soil Reports tab in Web Soil Survey or from the Soil Data Mart site. Onsite investigation may be needed to validate these interpretations and to confirm the identity of the soil on a given site.

Custom Soil Resource Report
Map—Dwellings Without Basements

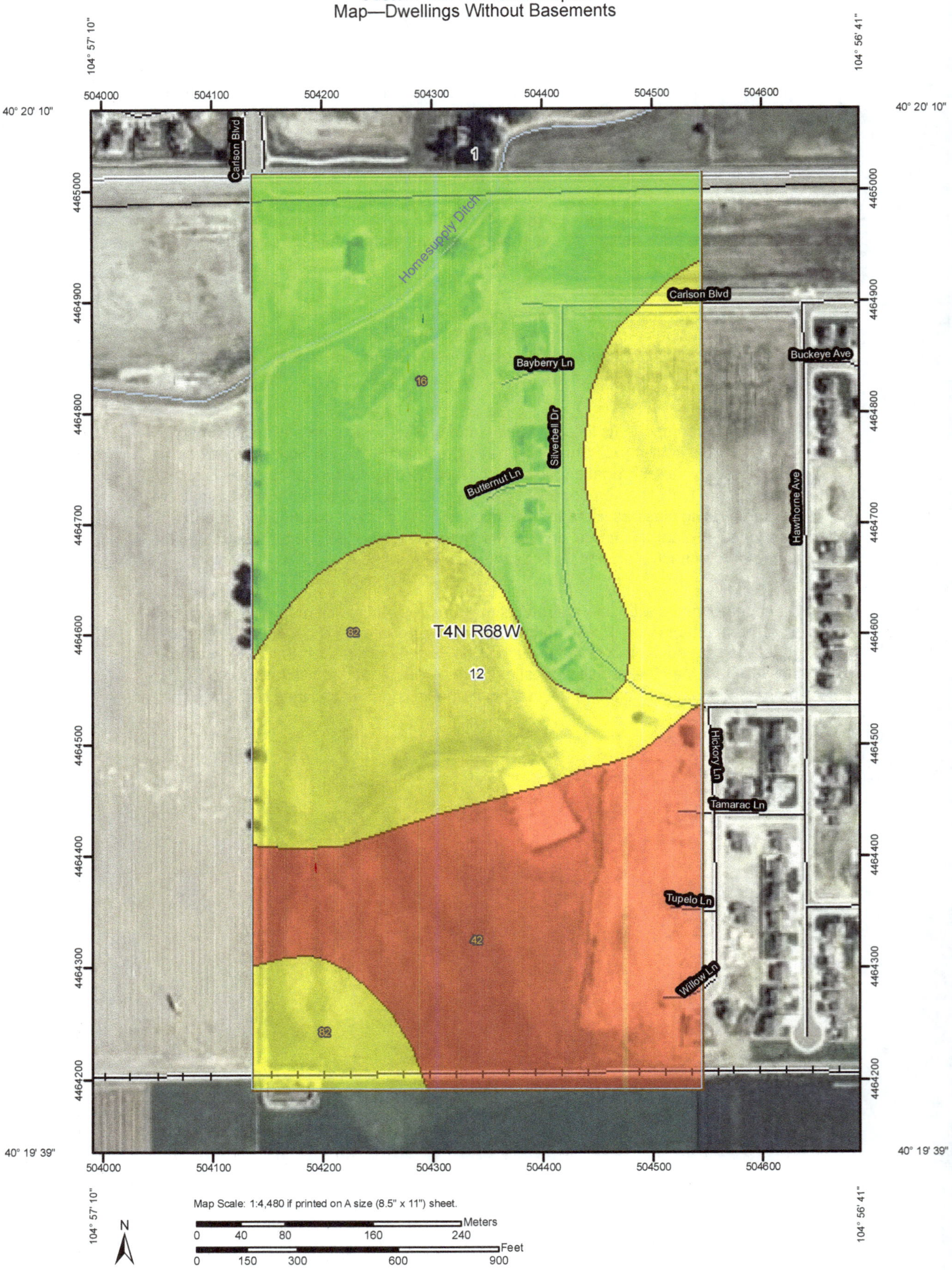

MAP LEGEND

Area of Interest (AOI)

- Area of Interest (AOI)

Soils

- Soil Map Units

Soil Ratings

- Very limited
- Somewhat limited
- Not limited
- Not rated or not available

Political Features

- Cities
- PLSS Township and Range
- PLSS Section

Water Features

- Streams and Canals

Transportation

- Rails
- Interstate Highways
- US Routes
- Major Roads
- Local Roads

MAP INFORMATION

Map Scale: 1:4,480 if printed on A size (8.5" × 11") sheet.

The soil surveys that comprise your AOI were mapped at 1:24,000.

Warning: Soil Map may not be valid at this scale.

Enlargement of maps beyond the scale of mapping can cause misunderstanding of the detail of mapping and accuracy of soil line placement. The maps do not show the small areas of contrasting soils that could have been shown at a more detailed scale.

Please rely on the bar scale on each map sheet for accurate map measurements.

Source of Map: Natural Resources Conservation Service
Web Soil Survey URL: http://websoilsurvey.nrcs.usda.gov
Coordinate System: UTM Zone 13N NAD83

This product is generated from the USDA-NRCS certified data as of the version date(s) listed below.

Soil Survey Area: Weld County, Colorado, Southern Part
Survey Area Data: Version 11, Aug 27, 2009

Date(s) aerial images were photographed: 8/6/2005

The orthophoto or other base map on which the soil lines were compiled and digitized probably differs from the background imagery displayed on these maps. As a result, some minor shifting of map unit boundaries may be evident.

Tables—Dwellings Without Basements

Dwellings Without Basements— Summary by Map Unit — Weld County, Colorado, Southern Part (CO618)						
Map unit symbol	Map unit name	Rating	Component name (percent)	Rating reasons (numeric values)	Acres in AOI	Percent of AOI
16	Colby loam, 3 to 5 percent slopes	Not limited	Colby (85%)		32.6	39.1%
42	Nunn clay loam, 1 to 3 percent slopes	Very limited	Nunn (85%)	Shrink-swell (1.00)	22.7	27.2%
82	Wiley-Colby complex, 1 to 3 percent slopes	Somewhat limited	Wiley (60%)	Shrink-swell (0.50)	28.1	33.7%
Totals for Area of Interest					**83.4**	**100.0%**

Dwellings Without Basements— Summary by Rating Value		
Rating	Acres in AOI	Percent of AOI
Not limited	32.6	39.1%
Somewhat limited	28.1	33.7%
Very limited	22.7	27.2%
Totals for Area of Interest	**83.4**	**100.0%**

Rating Options—Dwellings Without Basements

Aggregation Method: Dominant Condition

Component Percent Cutoff: None Specified

Tie-break Rule: Higher

Soil Properties and Qualities

The Soil Properties and Qualities section includes various soil properties and qualities displayed as thematic maps with a summary table for the soil map units in the selected area of interest. A single value or rating for each map unit is generated by aggregating the interpretive ratings of individual map unit components. This aggregation process is defined for each property or quality.

Soil Chemical Properties

Soil Chemical Properties are measured or inferred from direct observations in the field or laboratory. Examples of soil chemical properties include pH, cation exchange capacity, calcium carbonate, gypsum, and electrical conductivity.

pH (1 to 1 Water)

Soil reaction is a measure of acidity or alkalinity. It is important in selecting crops and other plants, in evaluating soil amendments for fertility and stabilization, and in determining the risk of corrosion. In general, soils that are either highly alkaline or highly acid are likely to be very corrosive to steel. The most common soil laboratory measurement of pH is the 1:1 water method. A crushed soil sample is mixed with an equal amount of water, and a measurement is made of the suspension.

For each soil layer, this attribute is actually recorded as three separate values in the database. A low value and a high value indicate the range of this attribute for the soil component. A "representative" value indicates the expected value of this attribute for the component. For this soil property, only the representative value is used.

Custom Soil Resource Report
Map—pH (1 to 1 Water)

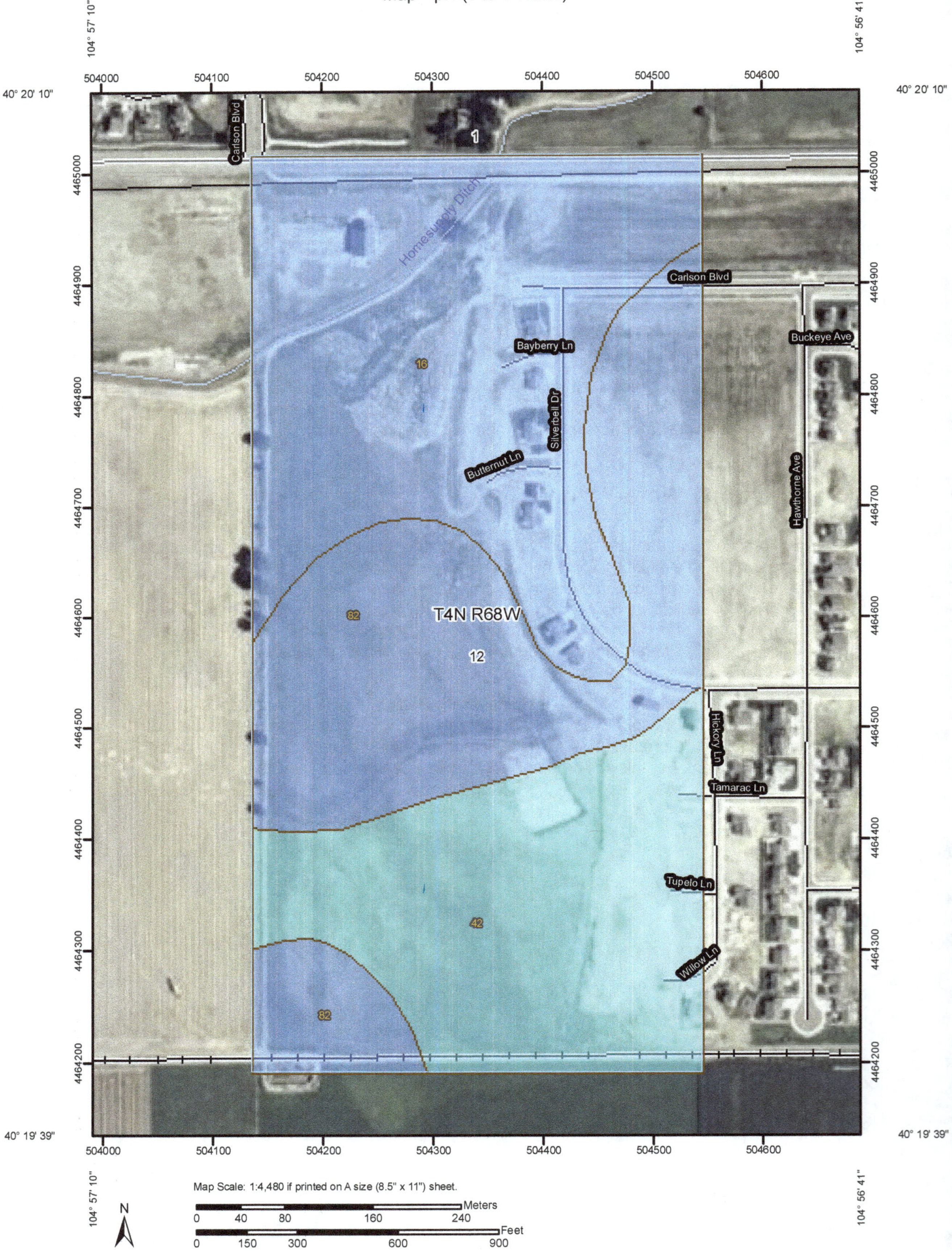

MAP LEGEND

Area of Interest (AOI)

- Area of Interest (AOI)

Soils

- Soil Map Units

Soil Ratings

- Ultra acid (ph < 3.5)
- Extremely acid (pH 3.5 - 4.4)
- Very strongly acid (pH 4.5 - 5.0)
- Strongly acid (pH 5.1 - 5.5)
- Moderately acid (pH 5.6 - 6.0)
- Slightly acid (pH 6.1 - 6.5)
- Neutral (pH 6.6 - 7.3)
- Slightly alkaline (pH 7.4 - 7.8)
- Moderately alkaline (pH 7.9 - 8.4)
- Strongly alkaline (pH 8.5 - 9.0)
- Very strongly alkaline (pH > 9.0)
- Not rated or not available

Political Features

- Cities
- PLSS Township and Range
- PLSS Section

Water Features

- Streams and Canals

Transportation

- Rails
- Interstate Highways
- US Routes
- Major Roads
- Local Roads

MAP INFORMATION

Map Scale: 1:4,480 if printed on A size (8.5" × 11") sheet.

The soil surveys that comprise your AOI were mapped at 1:24,000.

Warning: Soil Map may not be valid at this scale.

Enlargement of maps beyond the scale of mapping can cause misunderstanding of the detail of mapping and accuracy of soil line placement. The maps do not show the small areas of contrasting soils that could have been shown at a more detailed scale.

Please rely on the bar scale on each map sheet for accurate map measurements.

Source of Map: Natural Resources Conservation Service
Web Soil Survey URL: http://websoilsurvey.nrcs.usda.gov
Coordinate System: UTM Zone 13N NAD83

This product is generated from the USDA-NRCS certified data as of the version date(s) listed below.

Soil Survey Area: Weld County, Colorado, Southern Part
Survey Area Data: Version 11, Aug 27, 2009

Date(s) aerial images were photographed: 8/6/2005

The orthophoto or other base map on which the soil lines were compiled and digitized probably differs from the background imagery displayed on these maps. As a result, some minor shifting of map unit boundaries may be evident.

Table—pH (1 to 1 Water)

pH (1 to 1 Water)— Summary by Map Unit — Weld County, Colorado, Southern Part (CO618)				
Map unit symbol	**Map unit name**	**Rating**	**Acres in AOI**	**Percent of AOI**
16	Colby loam, 3 to 5 percent slopes	7.9	32.6	39.1%
42	Nunn clay loam, 1 to 3 percent slopes	7.6	22.7	27.2%
82	Wiley-Colby complex, 1 to 3 percent slopes	8.2	28.1	33.7%
Totals for Area of Interest			**83.4**	**100.0%**

Rating Options—pH (1 to 1 Water)

Aggregation Method: Dominant Component

Component Percent Cutoff: None Specified

Tie-break Rule: Higher

Interpret Nulls as Zero: No

Layer Options: Depth Range

Top Depth: 0

Bottom Depth: 84

Units of Measure: Inches

Soil Qualities and Features

Soil qualities are behavior and performance attributes that are not directly measured, but are inferred from observations of dynamic conditions and from soil properties. Example soil qualities include natural drainage, and frost action. Soil features are attributes that are not directly part of the soil. Example soil features include slope and depth to restrictive layer. These features can greatly impact the use and management of the soil.

Hydrologic Soil Group

Hydrologic soil groups are based on estimates of runoff potential. Soils are assigned to one of four groups according to the rate of water infiltration when the soils are not protected by vegetation, are thoroughly wet, and receive precipitation from long-duration storms.

The soils in the United States are assigned to four groups (A, B, C, and D) and three dual classes (A/D, B/D, and C/D). The groups are defined as follows:

Group A. Soils having a high infiltration rate (low runoff potential) when thoroughly wet. These consist mainly of deep, well drained to excessively drained sands or gravelly sands. These soils have a high rate of water transmission.

Group B. Soils having a moderate infiltration rate when thoroughly wet. These consist chiefly of moderately deep or deep, moderately well drained or well drained soils that have moderately fine texture to moderately coarse texture. These soils have a moderate rate of water transmission.

Group C. Soils having a slow infiltration rate when thoroughly wet. These consist chiefly of soils having a layer that impedes the downward movement of water or soils of moderately fine texture or fine texture. These soils have a slow rate of water transmission.

Group D. Soils having a very slow infiltration rate (high runoff potential) when thoroughly wet. These consist chiefly of clays that have a high shrink-swell potential, soils that have a high water table, soils that have a claypan or clay layer at or near the surface, and soils that are shallow over nearly impervious material. These soils have a very slow rate of water transmission.

If a soil is assigned to a dual hydrologic group (A/D, B/D, or C/D), the first letter is for drained areas and the second is for undrained areas. Only the soils that in their natural condition are in group D are assigned to dual classes.

Custom Soil Resource Report
Map—Hydrologic Soil Group

MAP LEGEND

Area of Interest (AOI)

- Area of Interest (AOI)

Soils

- Soil Map Units

Soil Ratings

- A
- A/D
- B
- B/D
- C
- C/D
- D
- Not rated or not available

Political Features

- Cities
- PLSS Township and Range
- PLSS Section

Water Features

- Streams and Canals

Transportation

- Rails
- Interstate Highways
- US Routes
- Major Roads
- Local Roads

MAP INFORMATION

Map Scale: 1:4,480 if printed on A size (8.5" × 11") sheet.

The soil surveys that comprise your AOI were mapped at 1:24,000.

Warning: Soil Map may not be valid at this scale.

Enlargement of maps beyond the scale of mapping can cause misunderstanding of the detail of mapping and accuracy of soil line placement. The maps do not show the small areas of contrasting soils that could have been shown at a more detailed scale.

Please rely on the bar scale on each map sheet for accurate map measurements.

Source of Map: Natural Resources Conservation Service
Web Soil Survey URL: http://websoilsurvey.nrcs.usda.gov
Coordinate System: UTM Zone 13N NAD83

This product is generated from the USDA-NRCS certified data as of the version date(s) listed below.

Soil Survey Area: Weld County, Colorado, Southern Part
Survey Area Data: Version 11, Aug 27, 2009

Date(s) aerial images were photographed: 8/6/2005

The orthophoto or other base map on which the soil lines were compiled and digitized probably differs from the background imagery displayed on these maps. As a result, some minor shifting of map unit boundaries may be evident.

Table—Hydrologic Soil Group

Hydrologic Soil Group— Summary by Map Unit — Weld County, Colorado, Southern Part (CO618)				
Map unit symbol	**Map unit name**	**Rating**	**Acres in AOI**	**Percent of AOI**
16	Colby loam, 3 to 5 percent slopes	B	32.6	39.1%
42	Nunn clay loam, 1 to 3 percent slopes	C	22.7	27.2%
82	Wiley-Colby complex, 1 to 3 percent slopes	B	28.1	33.7%
Totals for Area of Interest			**83.4**	**100.0%**

Rating Options—Hydrologic Soil Group

Aggregation Method: Dominant Condition

Component Percent Cutoff: None Specified

Tie-break Rule: Higher

References

American Association of State Highway and Transportation Officials (AASHTO). 2004. Standard specifications for transportation materials and methods of sampling and testing. 24th edition.

American Society for Testing and Materials (ASTM). 2005. Standard classification of soils for engineering purposes. ASTM Standard D2487-00.

Cowardin, L.M., V. Carter, F.C. Golet, and E.T. LaRoe. 1979. Classification of wetlands and deep-water habitats of the United States. U.S. Fish and Wildlife Service FWS/OBS-79/31.

Federal Register. July 13, 1994. Changes in hydric soils of the United States.

Federal Register. September 18, 2002. Hydric soils of the United States.

Hurt, G.W., and L.M. Vasilas, editors. Version 6.0, 2006. Field indicators of hydric soils in the United States.

National Research Council. 1995. Wetlands: Characteristics and boundaries.

Soil Survey Division Staff. 1993. Soil survey manual. Soil Conservation Service. U.S. Department of Agriculture Handbook 18. http://soils.usda.gov/

Soil Survey Staff. 1999. Soil taxonomy: A basic system of soil classification for making and interpreting soil surveys. 2nd edition. Natural Resources Conservation Service, U.S. Department of Agriculture Handbook 436. http://soils.usda.gov/

Soil Survey Staff. 2006. Keys to soil taxonomy. 10th edition. U.S. Department of Agriculture, Natural Resources Conservation Service. http://soils.usda.gov/

Tiner, R.W., Jr. 1985. Wetlands of Delaware. U.S. Fish and Wildlife Service and Delaware Department of Natural Resources and Environmental Control, Wetlands Section.

United States Army Corps of Engineers, Environmental Laboratory. 1987. Corps of Engineers wetlands delineation manual. Waterways Experiment Station Technical Report Y-87-1.

United States Department of Agriculture, Natural Resources Conservation Service. National forestry manual. http://soils.usda.gov/

United States Department of Agriculture, Natural Resources Conservation Service. National range and pasture handbook. http://www.glti.nrcs.usda.gov/

United States Department of Agriculture, Natural Resources Conservation Service. National soil survey handbook, title 430-VI. http://soils.usda.gov/

United States Department of Agriculture, Natural Resources Conservation Service. 2006. Land resource regions and major land resource areas of the United States, the Caribbean, and the Pacific Basin. U.S. Department of Agriculture Handbook 296. http://soils.usda.gov/

United States Department of Agriculture, Soil Conservation Service. 1961. Land capability classification. U.S. Department of Agriculture Handbook 210.

APPENDIX I—WORKSHEETS

Introduction

Worksheet 1	Chapter 1

Soil

Worksheet 2	Grain Size Analysis – Mechanical
Worksheet 3	Grain Size Analysis – Hydrometer
Worksheet 4	Grain Size Distribution
Worksheet 5	Atterberg Limits
Worksheet 6	USC Classification
Worksheet 7	AASHTO Classification
Worksheet 8	Proctor
Worksheet 9	Specifications
Worksheet 10	Nuclear Density Sheet
Worksheet 11	Subsurface Soil Investigation

Concrete

Worksheet 12	Mixed Concrete
Worksheet 13	Ingredients
Worksheet 14	Concrete Procedures
Worksheet 15	Testing and Sampling

WORKSHEET #1

Student Name:______________________________

1. Using the Glossary in Appendix A define the following:

    a. Consultant

    b. Contractor

    c. Geotechnical Engineer

    d. Owner

    e. Soil

2. Define the following symbols:

    a. F

    b. V

    c. ω

3. Convert the following:

    e. 16 in. into ft.

    f. 16 gal. into ft^3.

GRAIN SIZE ANALYSIS - MECHANICAL

Data Sheet 2

Project: Worksheet #2 **Project No.:** ____________

Location: ____________

Soil Description: ____________

Tested By: ____________ **Date of Test:** ____________

Total Soil		***Washed Soil***	
Wt. of dry sample + container (W_1)	766.8	Wt. of washed dry sample + container (W_2)	610.0
Container No.	AA	Container No.	AA
Wt. of container (W_c)	112.2	Wt. of container (W_c)	112.2
Wt. of dry sample (W_s) $W_s = W_1 - W_c$	654.6	Wt. of washed dry sample (W_{wds}) $W_{wds} = W_2 - W_c$	497.8

Sieve No.	Diameter (mm)	Weight Retained	Cumlative Weight Retained	Cumulative % Retained	Cumulative % Passing
3/4"		0			
3/8"		35.2			
#4		30.5			
#10		21.6			
#20		49.5			
#40		102.6			
#60		89.1			
#100		95.6			
#200		60.4			
Pan		11.2			

Cumlative % Retained = 100 x Cumlative Wt. Retained/Ws

Cumlative % Passing = 100 - Cumlative % Retained

GRAIN SIZE ANALYSIS - HYDROMETER

Data Sheet 3

Project: Worksheet #3 **Project No.:**

Location: **Bore No.** **Sample No.:**

Soil Description: **Sample Depth:**

Tested By: **Date of Test:**

Hydrometer No.: 152 H **Gs of solids:** 2.65 assumed **α =**

Dispersing agent Sodium Hexametaphosphate **Amount:** 5.0 g **Wt. of soil (Ws):**

Soil washed on sieve? Yes / No **If yes, Sieve No.** 10 / 40 / 200 **% Passing on Sieve** 24.3

Total Weight of Material	
Wt. of dry material + container (W_1)	162.9 g
Container No.	K7
Wt. of container (W_c)	108.4
Wt. of dry material (W_3) $W_3 = W_1 - W_c$	

NOTES: Assume G_s = 2.65 predominately granular soils.
Assume G_s = 2.70 predominately clayey soils.
Use adjusted % Finer only for washed sieve sample.

Wt. of dry soil (W_s) = W_3 - Dispersing agent =

Above equation is for a 152H hydrometer

Date	Time of Reading	Elapsed time (t) (min)	Temp (T) (°C)	Actual Hydrom. Reading (R_a)	Comp. Corr. Hydrom. (R_c)	Correct. Hydrom. (R)	% Finer (F)	Adjusted % F	Effective Depth (L) (cm)	Coor. Factor (K)	Particle Dia. (D) (mm)
		0.5	20	29							
		1	20	28							
		2	20	27							
		5	20	23							
		15	20	21							
		30	20	20							
		60	20	14							
		250	19	14							
		1440	19	12							

$R_c = 13 - 0.4T$ **$R = R_a - R_c$** **$F = 100R(\alpha)/W_s$** **$\alpha = 1.0 + 0.2(2.65 - G_s)$** **$D = K(L/t)^{1/2}$**

Adjusted % Finer = F x % Passing Sieve

$K = [13 + 0.15(24°C - T) + 4(2.65 - G_s)]/1000$ **$L = 16.3(1 - R_a/100)$**

Worksheet #4

Combine the results from worksheets #2 and #3.

GRAIN SIZE DISTRIBUTION **Data Sheet 4**

Project: ______________________ **Project No.:** __________

Location: ______________________

Soil Description: ______________________

Tested By: ______________________ **Date of Test:** __________

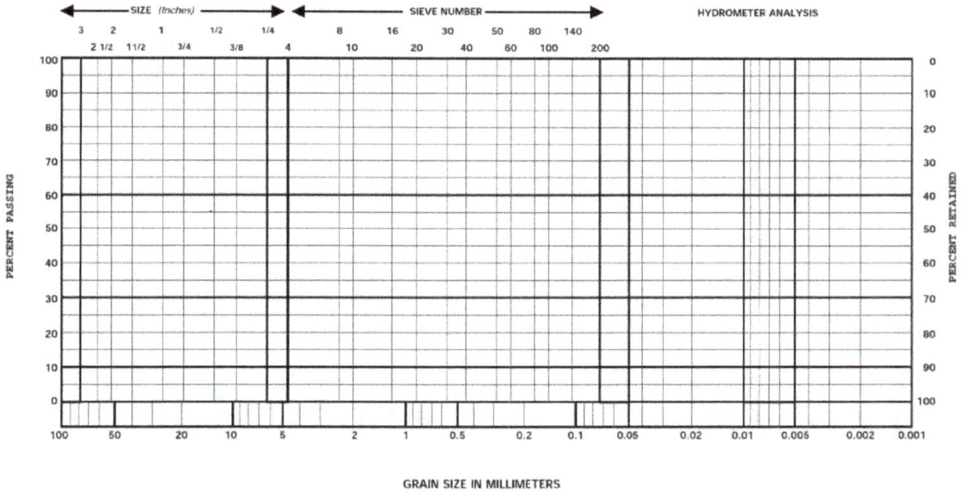

ATTERBERG LIMITS

Data Sheet 5

Project: Worksheet #5 **Project No.:**

Location: Wilson High School Addition **Bore No.** 3 **Sample No.:**

Soil Description: **Sample Depth:** 4'

Tested By: **Date of Test:**

Liquid Limit

Container No. (cup)	A6	A3	A4		
Wt. of cup + wet soil (W_1)	38.28	40.47	40.2		
Wt. of cup + dry soil (W_2)	33.01	35.40	35.88		
Wt. of cup (W_c)	13.58	12.98	13.02		
Wt. of dry soil (W_s) $W_s = W_2 - W_c$					
Wt. of water (W_w) $W_w = W_1 - W_2$					
Water content, ω% $\omega = 100W_w/W_s$					
No. of blows, N	17	23	28		

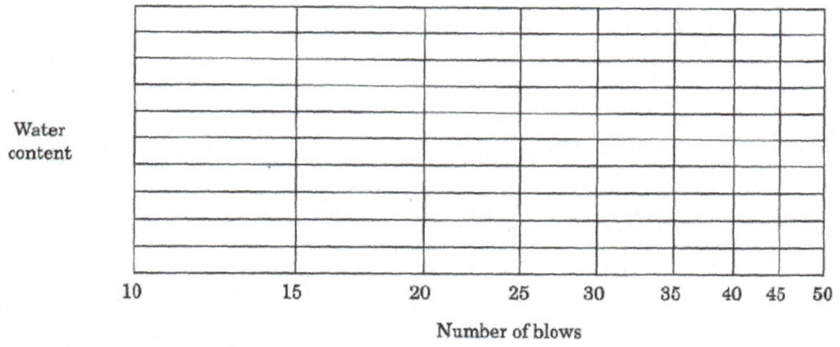

Plastic Limit

Container No. (cup)	G3	D6			
Wt. of cup + wet soil (W_1)	29.26	29.33			
Wt. of cup + dry soil (W_2)	24.54	24.48			
Wt. of cup (W_c)	13.45	13.05			
Wt. of dry soil (W_s) $W_s = W_2 - W_c$					
Wt. of water (W_w) $W_w = W_1 - W_2$					
Water content, ω% $\omega = 100W_w/W_s$					

Liquid Limit (LL) =	Plastic Limit (PL) =	Plasticity Index (PI) =

WORKSHEET #6

Student Name:____________________________

1. Classify the following soils using the USC method.

    a. SB-39 @ 4.0 – 8.0’

    b. SB-42 @ 4.0 – 8.0’

    c. SB-44 @ 4.0 – 8.0’

    d. SB-45 @ 4.0 – 8.0’

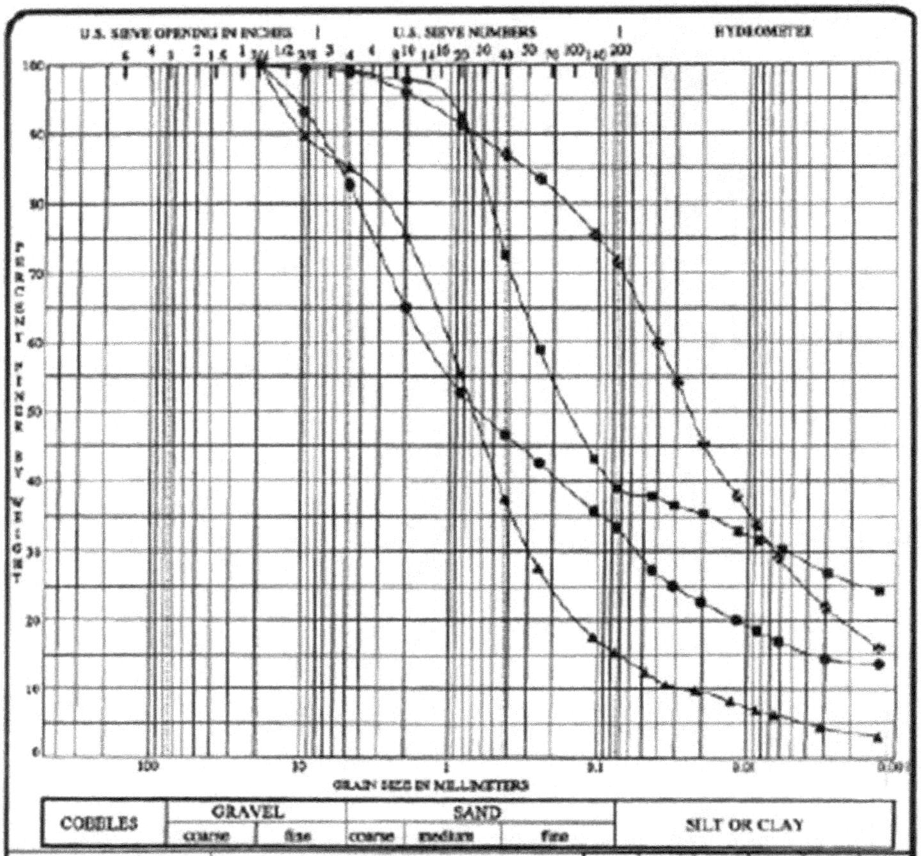

	Specimen Identification	Classification	MC%	LL	PL	PI	Cc	Cu
■	SB-39 @ 4.0-8.0'			58	17	41		
✲	SB-42 @ 4.0-8.0'			33	20	13		
▲	SB-44 @ 4.0-8.0'			NP	NP	NP		
●	SB-45 @ 4.0-8.0'			30	14	15		

Specimen Identification	D100(mm)	D60(mm)	D30(mm)	D10(mm)	%Gravel	%Sand	%Silt	%Clay
SB-39@ 4.0-8.0'	19.00	0.26	0.006		0.9	60.1	13.3	25.7
SB-42@ 4.0-8.0'	19.00	0.04	0.007		1.2	27.3	52.5	19.0
SB-44@ 4.0-8.0'	19.00	1.03	0.284	0.0254	15.0	69.7	11.6	3.7
SB-45@ 4.0-8.0'	19.00	1.41	0.055		17.5	49.1	19.3	14.1

WORKSHEET #7

Student Name:____________________________

1. Classify the following soils using the AASHTO method.

   a. SB-39 @ 4.0 – 8.0’

   b. SB-42 @ 4.0 – 8.0’

   c. SB-44 @ 4.0 – 8.0’

   d. SB-45 @ 4.0 – 8.0’

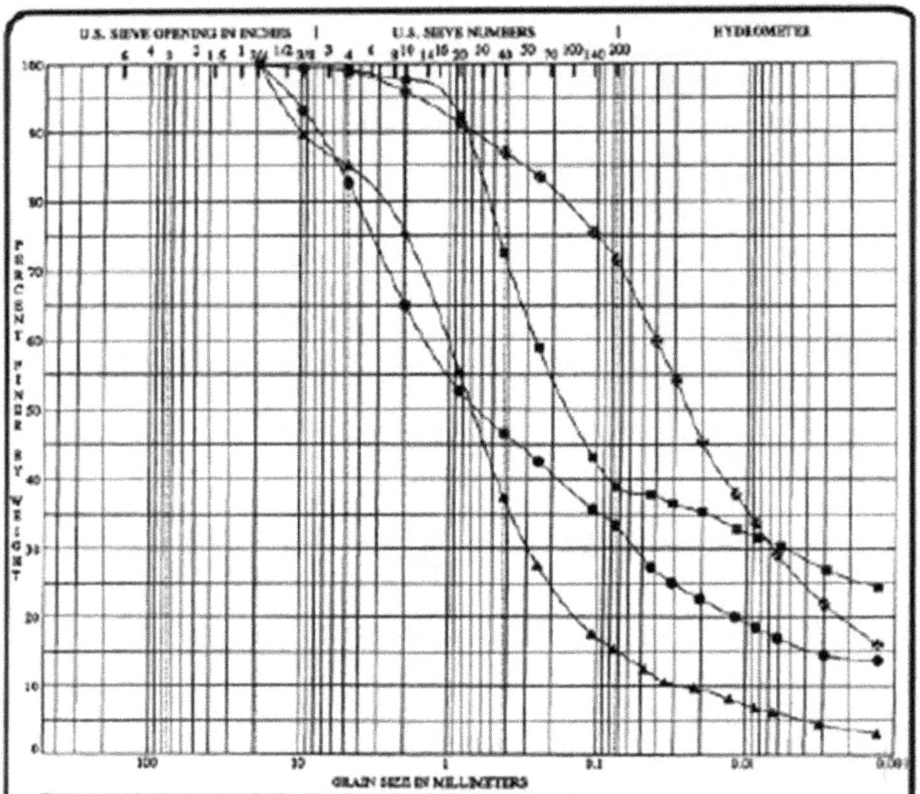

COBBLES	GRAVEL		SAND			SILT OR CLAY
	coarse	fine	coarse	medium	fine	

	Specimen Identification	Classification	MCN	LL	PL	PI	Cc	Cu
■	SB-39 @ 4.0-8.0'			58	17	41		
✲	SB-42 @ 4.0-8.0'			33	20	13		
▲	SB-44 @ 4.0-8.0'			NP	NP	NP		
●	SB-45 @ 4.0-8.0'			30	14	15		

Specimen Identification	D100(mm)	D60(mm)	D30(mm)	D10(mm)	%Gravel	%Sand	%Silt	%Clay
SB-39@ 4.0-8.0'	19.00	0.26	0.006		0.9	60.1	13.3	25.7
SB-42@ 4.0-8.0'	19.00	0.04	0.007		1.2	27.3	52.5	19.0
SB-44@ 4.0-8.0'	19.00	1.03	0.284	0.0254	15.0	69.7	11.6	3.7
SB-45@ 4.0-8.0'	19.00	1.41	0.055		17.5	49.1	19.3	14.1

WORKSHEET #8

Student Name:____________________________

Complete the proctor calculations for water content and dry density. Plot the points and determine the optimum moisture and maximum dry density.

PROCTOR TEST

Data Sheet 6

Project: Worksheet #8 __________ Project No.: __________

Location: __________ Bore No. ______ Sample No.: __________

Soil Description: __________ Sample Depth: __________

Tested By: __________ Date of Test: __________

Test Method: [Standard] [Modified] Procedure: [A] [B] [C]

Mold Diameter: [4"] [6"] Mold Volume: (V_m) [1/30 CF] [1/13 CF]

Water Content

Circle choice of boxes used above

Sample No.	1	2	3	4	5
Container No. (cup)	B1	B6	B2	A5	
Wt. of cup + wet soil (W_1)	624.8	618.7	632.5	621.3	
Wt. of cup + dry soil (W_2)	585.3	570.9	575.8	557.0	
Wt. of cup (W_c)	103.4	102.8	103.3	104.3	
Wt. of dry soil (W_s) $W_s = W_2 - W_c$					
Wt. of water (W_w) $W_w = W_1 - W_2$					
Water content, ω% $\omega\% = 100W_w/W_s$					

Density

Wt. of soil + mold (W_s+W_m)	8.73	8.88	8.94	8.85	
Wt. of mold (W_m)	4.40	4.40	4.40	4.40	
Wt. of soil in mold (W_s)					
Wet Density (PCF) $\gamma_{wet} = W_s/V_m$					
Dry Density (PCF) $\gamma_d = \gamma_{wet}/(1+\omega)$					

Dry Density γ_d PCF

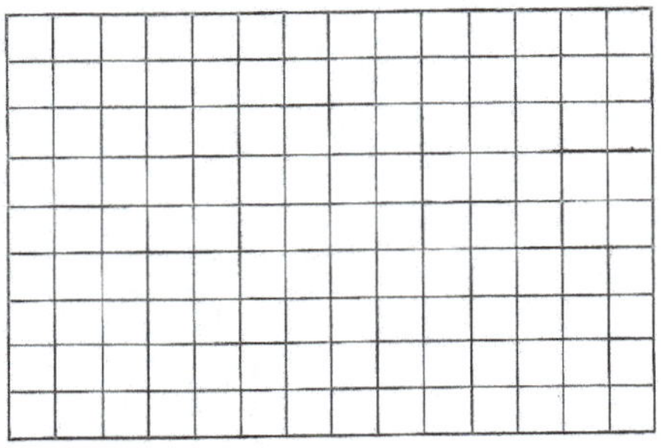

Water Content ω%

Optimum Moisture (%) =	Maximum Dry Density (PCF) =

WORKSHEET #9

Student Name:______________________________

1. Answer the following questions from Table 2.13.

   a. What type of quality assurance tests are required?

   b. What is the minimum sampling/testing frequency required?

2. Answer the following questions from Appendix D, Soil Specifications.

   a. What is the project name for these specifications?

   b. What is the definition of compacted backfill?

   c. What is the moisture content for backfilling? List the soil type and moisture content requirements.

   d. What is the density requirement for backfill? List all that apply.

WORKSHEET #10

Student Name:______________________________

1. Identify which curve (proctor) should be used on the 5 tests obtained.
2. Calculate the percent compaction.
3. Identify if the compaction test passed compaction and moisture specifications.

MOISTURE/DENSITY TEST RESULTS

TWO RIVERS TESTING, INC.

3310 State Street, Evans, CO 80620 Phone: (970) 339-4092 Fax: (970) 330-1252

Project Name: Worksheet #10 **Project No.:**
Client:
Contractor:

	Curve No. 1	Curve No. 2	Curve No. 3	Curve No.
ASTM Designation:	D-698	D-698	D-698	
Method:	A	C	A	
Maximum Density:	109.5	136.8	110.0	
Optimum Moisture:	22.7	7.0	17.2	

Test No.	Date	Location	Elevation Depth	Dry Density (PCF)	Moisture (%)	Compaction (%)	Compaction Req.	Compaction P/F	Moisture Req.	Moisture P/F	Curve No.
1	6/12/2022	Water Main - Sta 12+00	-2.0	106.8	15.9		95%		±2%		
2	6/12/2022	Water Main - Sta 14+00	-2.0	103.8	17.2						
3	6/12/2022	Water Service - Lot 2, Block 3	-0.5	131.3	4.9						
4	6/13/2022	Sewer Main - Sta 33+50	-8.5	110.1	21.4						
5	6/13/2022	Sewer Main - Sta 35+25	-7.0	109.5	25.1						

WORKSHEET #11

Student Name:____________________________

1. From the Subsurface Soil Investigation in Appendix G answer the following questions.

   a. What type of foundation system was recommended?

   b. What was the recommended maximum net allowable bearing capacity for the above foundation system?

   c. How many bores were drilled?

   d. Answer the following questions for Bore #11

      i. What was the total depth of the bore?

      ii. Was groundwater encountered? If yes, at what depth?

      iii. What depth was bedrock first logged?

      iv. What type of bedrock was first logged?

WORKSHEET #12

Student Name:___________________________

1. Concrete is the world's most used construction material. True or False

2. How does concrete help in the construction industry?
   a. Strength.
   b. Weight.
   c. Durability.
   d. Low permeability.
   e. All of the above.

3. What does an initial set mean in terms of concrete?
   a. The first songs played at a concert.
   b. When a ship first starts to sail.
   c. When the material starts, the phase changes from plastic to hardened.
   d. The first time you place silverware on the table.
   e. All of the above.

4. Name at least three items that play an important role in mixing concrete.

5. How does consolidation affect concrete?

6. What is the greatest attribute of concrete?

7. What is required for the hydration of concrete to continue to occur? List all that apply.
   a. Un-hydrated cement is present.
   b. Sufficient water in the mix to react to the cement.
   c. Temperature below freezing.
   d. Relative humidity above 80%.
   e. All of the above.

8. Concrete hardens by drying. True or False

WORKSHEET #13

Student Name:____________________________

1. In what manner is the load applied in compressive strength? List all that apply.
   a. Bending
   b. Pulling
   c. Twisting
   d. Crushing
   e. Sliding
   f. All of the above

2. How many days are required to achieve design compressive strength?
   a. 1
   b. 3
   c. 28
   d. 365

3. Causes of cracks in concrete are stress due to applied loads and stress due to drying shrinkage or temperature changes in restrained conditions. True or False?

4. What are the three main ingredients to make concrete?

5. What is the difference between hydraulic and non-hydraulic cement?

6. Aggregates generally make up what range of percentage of the concrete weight?

7. What properties may be altered by adding admixtures to the concrete mix? List all that are applicable.
   a. Set control
   b. color
   c. Air entrainment
   d. Damp proofing
   e. Water reduction
   f. All of the above.

8. List three qualities a concrete mixture should possess.

9. List the following ideas that are needed to design the best water-cement ratio. Circle all that apply.
   a. You must use the correct mixture of aggregate.
   b. There must be sufficient paste to surround the aggregate.
   c. There must be sufficient water to hydrate the cement fully.
   d. The curing environment must allow full hydration to occur.
   e. Time is required to allow the hardening to accept the load sufficiently.
   f. All of the above

10. How many pounds of cement is in a six-sack mix?

WORKSHEET #14

Student Name:______________________________

1. Concrete that has stiffened in the mixer drum can have water added as long as which conditions are met?
   a. Maximum allowable water-cement ratio is not exceeded
   b. Maximum allowable slump is not exceeded
   c. Maximum allowable mixing and agitating time or drum revolutions are not exceeded
   d. Concrete is remixed for at least half the minimum required mixing time or number of revolutions.
   e. All of the above.

2. Excessive movement of the concrete may cause what problem?

3. List at least four methods to finish concrete.

4. What is the main concern with hot weather concrete pours? List at least two items that may cause this situation.

5. In accordance with ACI 306, what defines cold weather concrete placement?
   a. Temperature drops below 40°F.
   b. Two successive days below 40°F.
   c. Three successive days below 40°F.
   d. Five successive days below 40°F.

WORKSHEET #15

Student Name:______________________

1. Who generally conducts Quality Control?

2. List the Quality Assurance tests and frequency for the following concrete test from Table No. 2.13.

    a. Curb and Gutter

    b. Sidewalks

3. Slump measures water content in the concrete. True or False?

4. When is sampling from the end of the pumper discharge hose acceptable? List all that apply.
    a. When additional water is added.
    b. The slump exceeds the specifications.
    c. The project specifications require it.
    d. The air temperature is below 40°F.
    e. None of the above.

5. Which comments are true while checking the concrete temperature? List all that apply.
    a. Submerge the thermometer at least 3 inches into the sample.
    b. Try to minimize the influence of the ambient temperature.
    c. Leave the thermometer in the concrete for at least 1 minute and not more than 5 minutes.
    d. Read the temperature to the nearest 1°F.

6. How many rod strokes per lift are required for a slump test?

7. How much time is allowed for lifting the slump cone?
   a. 2 seconds to 5 seconds
   b. 5 seconds ±3 seconds
   c. 5 seconds ±2 seconds
   d. 3 seconds to 10 seconds
   e. None of the above

8. The unit weight of the concrete should be reported to the nearest __________?

APPENDIX J—DATA SHEETS

The following blank data sheets have been provided for individuals using this book.

SOILS

Data Sheet 1	Water Content Determination
Data Sheet 2	Grain Size Analysis – Mechanical
Data Sheet 3	Grain Size Analysis – Hydrometer
Data Sheet 4	Grain Size Distribution
Data Sheet 5	Atterberg Limits
Data Sheet 6	Proctor Test

WATER CONTENT

Data Sheet 1

Project: ______________________ **Project No.:** __________

Location: ______________________

Soil Description: ______________________

Tested By: ______________________ **Date of Test:** __________

Sample/Boring No.					
Sample depth					
Soil description					
Container No.					
Wt. of container (W_c)					
Wt. of container + wet soil (W_{wc})					
Wt. of container + dry soil (W_{dc})					
Wt. of dry soil (W_s) $W_s = W_{dc} - W_c$					
Wt. of water (W_w) $W_w = W_{wc} - W_{dc}$					
Water content % (ω) $\omega = W_w/W_s$					

Sample/Boring No.					
Sample depth					
Soil description					
Container No.					
Wt. of container (W_c)					
Wt. of container + wet soil (W_{wc})					
Wt. of container + dry soil (W_{dc})					
Wt. of dry soil (W_s) $W_s = W_{dc} - W_c$					
Wt. of water (W_w) $W_w = W_{wc} - W_{dc}$					
Water content % (ω) $\omega = W_w/W_s$					

GRAIN SIZE ANALYSIS - MECHANICAL

Data Sheet 2

Project: ______________________ **Project No.:** __________

Location: ______________________

Soil Description: ______________________

Tested By: ______________________ **Date of Test:** __________

Total Soil		*Washed Soil*	
Wt. of dry sample + container (W_1)		Wt. of washed dry sample + container (W_2)	
Container No.		Container No.	
Wt. of container (W_c)		Wt. of container (W_c)	
Wt. of dry sample (W_s) $W_s = W_1 - W_c$		Wt. of washed dry sample (W_{wds}) $W_{wds} = W_2 - W_c$	

Sieve No.	Diameter (mm)	Weight Retained	Cumlative Weight Retained	Cumulative % Retained	Cumulative % Passing

Cumlative % Retained = 100 x Cumlative Wt. Retained/Ws

Cumlative % Passing = 100 - Cumlative % Retained

GRAIN SIZE ANALYSIS - HYDROMETER

Data Sheet 3

Project: ____________________ **Project No.:** __________

Location: ____________________ **Bore No.** ______ **Sample No.:** __________

Soil Description: ____________________ **Sample Depth:** __________

Tested By: ____________________ **Date of Test:** __________

Hydrometer No.: __________ **Gs of solids:** __________ **α =** __________

Dispersing agent __________ **Amount:** ______ **Wt. of soil (Ws):** __________

Soil washed on sieve? | Yes | No | **If yes, Sieve No.** | 10 | 40 | 200 | **% Passing on Sieve** __________

Total Weight of Material	
Wt. of dry material + container (W_1)	
Container No.	
Wt. of container (W_c)	
Wt. of dry material (W_3) $W_3 = W_1 - W_c$	

NOTES: Assume $G_s = 2.65$ predominately granular soils.
Assume $G_s = 2.70$ predominately clayey soils.
Use adjusted % Finer only for washed sieve sample.

Wt. of dry soil (W_s) = W_3 - Dispersing agent =	

Above equation is for a 152H hydrometer

Date	Time of Reading	Elapsed time (t) (min)	Temp (T) (°C)	Actual Hydrom. Reading (R_a)	Comp. Corr. Hydrom. (R_c)	Correct. Hydrom. (R)	% Finer (F)	Adjusted % F	Effective Depth (L) (cm)	Coor. Factor (K)	Particle Dia. (D) (mm)

$R_c = 13 - 0.4T$ $R = R_a - R_c$ $F = 100R(\alpha)/W_s$ $\alpha = 1.0 + 0.2(2.65 - G_s)$ $D = K(L/t)^{1/2}$

$K = [13 + 0.15(24°C - T) + 4(2.65 - G_s)]/1000$ $L = 16.3(1 - R_a/100)$

GRAIN SIZE DISTRIBUTION

Data Sheet 4

Project: Worksheet 4 **Project No.:** AEC 2530

Location:

Soil Description:

Tested By: **Date of Test:**

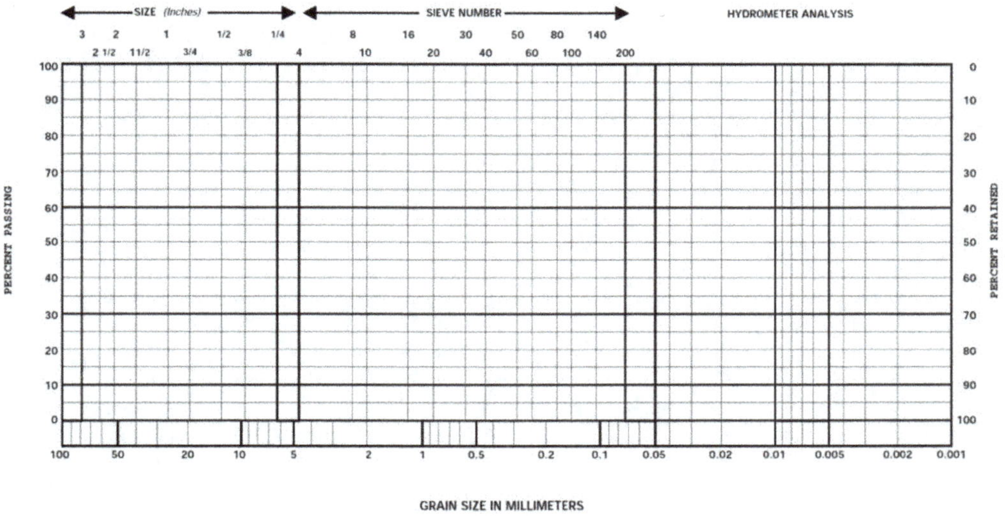

ATTERBERG LIMITS

Data Sheet 5

Project: ______________________ Project No.: __________

Location: ______________________ Bore No. ______ Sample No.: __________

Soil Description: ______________________ Sample Depth: __________

Tested By: ______________________ Date of Test: __________

Liquid Limit

Container No. (cup)					
Wt. of cup + wet soil (W_1)					
Wt. of cup + dry soil (W_2)					
Wt. of cup (W_c)					
Wt. of dry soil (W_s) $W_s = W_2 - W_c$					
Wt. of water (W_w) $W_w = W_1 - W_2$					
Water content, ω% $\omega = 100W_w/W_s$					
No. of blows, N					

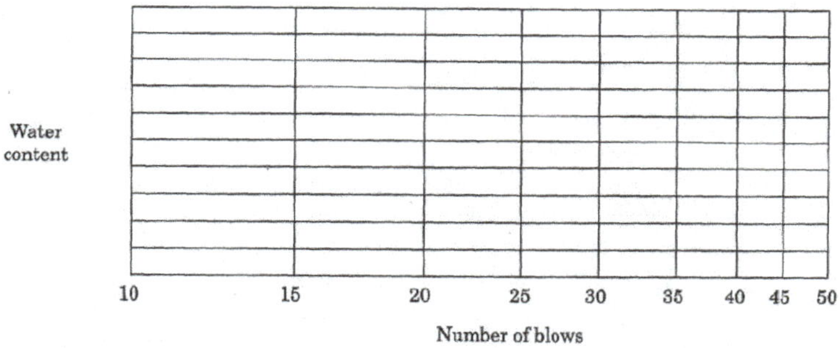

Plastic Limit

Container No. (cup)					
Wt. of cup + wet soil (W_1)					
Wt. of cup + dry soil (W_2)					
Wt. of cup (W_c)					
Wt. of dry soil (W_s) $W_s = W_2 - W_c$					
Wt. of water (W_w) $W_w = W_1 - W_2$					
Water content, ω% $\omega = 100W_w/W_s$					

Liquid Limit (LL) =	Plastic Limit (PL) =	Plasticity Index (PI) =

PROCTOR TEST

Data Sheet 6

Project: ______________________ Project No.: ____________

Location: ______________________ Bore No. ________ Sample No.: ____________

Soil Description: ______________________ Sample Depth: ____________

Tested By: ______________________ Date of Test: ____________

Test Method: Standard | Modified   Procedure: A | B | C

Mold Diameter: 4" | 6"   Mold Volume: (V_m) 1/30 CF | 1/13 CF

Water Content

Circle choice of boxes used above

Sample No.	1	2	3	4	5
Container No. (cup)					
Wt. of cup + wet soil (W_1)					
Wt. of cup + dry soil (W_2)					
Wt. of cup (W_c)					
Wt. of dry soil (W_s) $W_s = W_2 - W_c$					
Wt. of water (W_w) $W_w = W_1 - W_2$					
Water content, ω% $\omega\% = 100W_w/W_s$					

Density

Wt. of soil + mold (W_s+W_m)					
Wt. of mold (W_m)					
Wt. of soil in mold (W_s)					
Wet Density (PCF) $\gamma_{wet} = W_s/V_m$					
Dry Density (PCF) $\gamma_d = \gamma_{wet}/(1+\omega)$					

Dry Density γ_d PCF

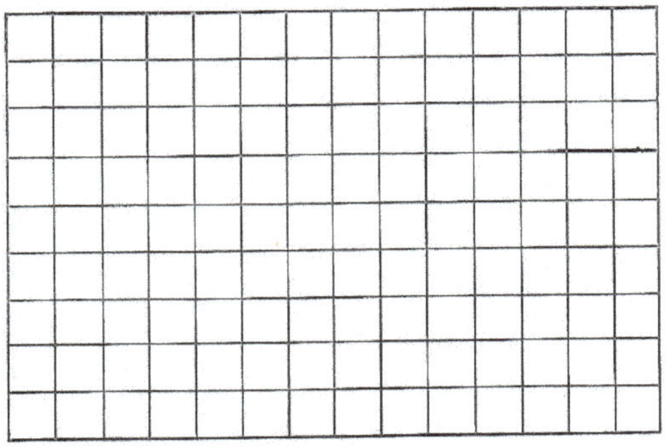

Water Content ω%

Optimum Moisture (%) =	Maximum Dry Density (PCF) =